The Biochemistry
of
Inorganic Polyphosphates

The Biochemistry
of
Inorganic Polyphosphates

I. S. Kulaev
Moscow State University
Institute of Biochemistry and Physiology of Microorganisms,
Academy of Sciences of the USSR

Translation:
R. F. Brookes

A Wiley–Interscience Publication

JOHN WILEY & SONS

Chichester · New York · Brisbane · Toronto

Library of Congress Cataloging in Publication Data:

Kulaev, Igor' Stepanovich.
    The biochemistry of inorganic polyphosphates.

    Translation of Biokhimia vysokomolekuliarnykh polifosfatov.
        'A Wiley–Interscience Publication.'
        Includes index.
        1. Polyphosphates—Metabolism.   2. Polyphosphates—
        Physiological effect.
    3. Biological chemistry.   I. Title.
    QP535.P1K8413   574.1'921   78-31627

    ISBN 0 471 27574 3

Photoset and printed in Malta by
Interprint Limited

Contents

Foreword

The presence of high molecular polyphosphates in many microorganisms such as yeast, fungi, and bacteria, has been known for a long time, but studies on the biochemical functions of these substances are of much more recent origin and still in a rudimentary state. Professor Igor S. Kulaev, one of the most eminent pupils of the late Professor A. N. Belosersky, who was an internationally known authority on nucleic acids, has dedicated in his laboratory at the University of Moscow, in collaboration with a large team of collaborators, intensive studies over many years to the somewhat neglected subject of the biochemical functions of polyphosphates. His group has studied the enzymes involved in the synthesis and breakdown of these compounds. There is no doubt that in some cases they can take over the phosphorylation functions of ATP, as the phosphate residues are linked together to form energy-rich phosphate bonds.

Professor Kulaev has taken the not inconsiderable trouble of collecting and critically reviewing the large literature on the subject in one monograph, at present the only one in existence on this important subject.

With this onerous and time consuming task he has rendered a signal service to the international biochemical community, which owes him a debt of gratitude for this work. Professor Kulaev has shown that the study of the biochemical functions of the high molecular polyphosphates is still a very active field of research offering a great challenge to the enterprising young biochemist in which many discoveries of general importance can still be made.

Professor Emeritus Ernst Chain, FRS
Imperial College of Science & Technology,
London. SW7.

The author dedicates this book to the respected memory of his teacher, **Andrei Nikolaevich Belozersky**, *an outstanding scientist, teacher, and man.*

Preface to the Russian Edition

The investigation of the role of phosphorus compounds in the metabolism of the organisms in which they occur is a major task in present-day biochemistry.

Inorganic polyphosphates play an important part in the metabolism of many of the lower organisms. They were first observed in living organisms at an early date,[453] but their presence in the cells of many microorganisms was not accepted by biochemists until recently, following the classical work of Wiame[802–810] and Ebel.[127–130] This protracted delay in the biochemical investigation of inorganic polyphosphates was due mainly to ignorance of the precise chemical structures of these compounds, and the unavailability to biochemists of rigorous methods for their identification in biological material.

Not until the early 1950s was it shown[129,736,788] that these compounds are polymers composed of orthophosphate molecules linked by phosphoric anhydride groups to form the unbranched, linear structures shown in Fig. 1. It thus became clear that the compounds are, in fact, condensed inorganic phosphates.

Although the chemical structures of the inorganic polyphosphates are now firmly established, their functions in the metabolism of the organisms in which they occur have been far from adequately explained, and the fact that

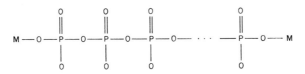

Figure 1. The structure of an inorganic polyphosphate

they are high-energy compounds (the hydrolysis of each phosphoric anhydride bond releases as much energy as the removal of the terminal phosphate group from ATP[514,832]) has attracted the attention of biochemists. It has been found that these phosphorus compounds are present in significant amounts in many organisms, especially bacteria, moulds, yeasts, and algae, and they appear to play an important role in the metabolism of these organisms.

However, mutants of some microorganisms have recently been obtained[221] which have lost the ability to synthesize polyphosphates within their cells by means of enzyme systems which are now well known, but which have not lost their ability to grow and develop normally. Further, there are a number of lower organisms in which these compounds are either not found at all, or are found only under very specific conditions which are unfavourable for the growth of the organisms. These observations create the impression that polyphosphates are, so to speak, not very essential, or at least not always essential to the normal vital activities of the organisms which contain them.

A dilemma now arises—either polyphosphates are not, in fact, essential to the vital processes of contemporary living organisms, and their formation in many organisms is a unique 'biochemical atavism' carried over from the remote ancestors of these organisms, or alternatively the actual presence of these compounds as such is not essential for the functioning of these organisms, but rather the ability to generate them at any time within their cells, thereby regulating their phosphorus and energy metabolism.

Consequently, a thorough investigation of the biochemistry of inorganic polyphosphates may throw light on such important questions as the evolution of phosphorus metabolism, and the mechanism by which it is regulated in contemporary living organisms. However, despite the considerable amount of work which has been carried out recently on the biochemistry of poly-phosphates (for reviews, see refs. 102, 145, 169, 220, 243, 363, 379, 380, 389, 443 and 714), the physiological role of these compounds in contemporary organisms remains obscure.

This book summarizes the more important literature on the chemistry and biochemistry of inorganic polyphosphates, together with the experimental data obtained by the author and his co-workers during 25 years' work in this area. The author's work on the biochemistry of the inorganic polyphosphates was begun in 1953 in the laboratory of the late Academician A. N. Belozersky, whose fruitful scientific activities have provided the basis for the development of wide-ranging investigations on this topic, both in Russia and abroad. The scientific interests of the author were formed under the guidance of A. N. Belozersky, and this book has been written under the direct influence of the ideas which he developed.

The author warmly thanks the following colleagues, whose self-denying efforts have been responsible in no small measure for experimental results of the highest significance in the biochemistry of polyphosphates, and for their creative interpretation: T. P. Afanas'eva, T. A. Belozerskaya, M. A. Bobyk, B. Brommer (German Federal Republic), V. M. Vagabov, M. N. Valikhanov,

S. N. Egorov, S. A. Ermakova, V. M. Kadomtseva, G. I. Konoshenko, I. A. Krasheninnikov, M. S. Kritskii, L. P. Lichko, S. É. Mansurova, V. I. Mel'gunov, M. A. Nesmeyanova, N. N. Nikolaev (Bulgaria), L. A. Okorokov, D. N. Ostrovskii, V. V. Rozhanets, P. M. Rubtsov, K. G. Skryabin, I. Tobek (Czechoslovakia), A. M. Umnov, H. Urbanek (Poland), S. O. Uryson, V. P. Kholodenko, A. B. Tsiomenko, E. K. Chernysheva, Yu. A. Shabalin, A. Shadi (Egypt), Yu. A. Shakhov, and O. V. Szymona (Poland).

It is the author's agreeable duty to thank V. V. Ryzhenkova and A. M. Mudrik for their great help in preparing the manuscript of this book.

Preface to the English Edition

This book seeks to present an account of contemporary ideas concerning the chemical structure and metabolism of condensed inorganic phosphates (polyphosphates) in the cells of organisms of different types. The author also examines contemporary views regarding the physiological functions of these energy-rich phosphorus compounds, and develops his own concept of their possible role in the evolution of phosphorus metabolism.

Inorganic polyphosphates for long remained in obscurity, outside the main areas of interest of biochemists. Until recently they were considered to be merely reserve materials, metabolically rather inert, which were specific to microorganisms. Some workers have regarded them as 'microbial phosphagens', able to be synthesized and utilized in the metabolism of bacteria and other microorganisms only via the ATP–ADP system. These remarks refer primarily to high-molecular-weight polyphosphates. The lowest homologue of this series of phosphorus compounds, pyrophosphate, is currently regarded by most biochemists as a by-product of a variety of anabolic processes, i.e. as a compound which has no independent function in the metabolism of living creatures.

In recent years, views on the possible physiological functions of inorganic polyphosphates (including pyrophosphate) have undergone profound changes. For instance, it has been shown that inorganic polyphosphates and pyrophosphate are accumulated in and utilized by the cells of a wide variety of organisms in the reactions in which it was earlier thought that ATP alone was accumulated and utilized.

By virtue of their role in the metabolism of the cell, and in particular in its bioenergetics, these inorganic phosphorus compounds fall into the same category of energy-rich compounds as nucleoside polyphosphates, acyl phosphates, phosphoamides, and phosphoenol pyruvate. No doubt now remains that one of the most important functions of inorganic polyphosphate in the cell is to provide bioenergetic metabolic pathways which are alternatives to the synthesis and utilization of ATP. That is, they appear to make a contribution towards ensuring the reliability and lability of the fundamental metabolic pathways in contemporary organisms. In lower organisms, they undoubtedly function in addition as valuable reserves of activated phosphate.

Since the efforts of biochemists are currently directed towards investigating the regulation of metabolic processes, it must also be said that, on the basis of current evidence, one of the most essential functions of inorganic polyphosphates in living organisms is their direct involvement in the regulation and coordination of a wide variety of cellular metabolic pathways. This function has been established most clearly in microorganisms, in which metabolic regulation assumes great importance in view of the absence of hormonal and nervous control mechanisms.

About 3 years have elapsed since the appearance of this book in Russian. During this time, a number of publications devoted to new work on the biochemistry of inorganic polyphosphates and pyrophosphate have appeared. These are largely concerned with the metabolism of inorganic polyphosphates in phototrophic organisms. All the publications of which the author has become aware during the last 2 years have been included in the bibliography. For the better guidance of the reader, the bibliography in this edition, unlike the Russian edition, includes the titles of the articles cited.

The greatest reward which the author of this book could receive, in recompense for the effort that he has expended in writing it, would be to arouse the interest of its readers in this hitherto neglected area of biochemistry. It is to be hoped that new scientific groups will take up the investigation of these interesting phosphorus compounds, which in turn would undoubtedly lead to rapid advances in the understanding of their physiological roles.

In conclusion, may I say that the appearance of my book in English gives me great pleasure. For this event, which is of great importance to me, I am wholly indebted to Sir E. B. Chain, without whose efforts and great good will I am sure this edition could not have appeared.

I also express my thanks to the Publisher, John Wiley and Sons Ltd., and above all to Dr. Howard Jones, for making it possible for this book to be published in English, and thus to become accessible to the wide circle of research workers acquainted with this language. It also gives me great pleasure to thank the translator, Mr. Robert F. Brookes, for his excellent translation from the Russian.

Chapter 1

The Chemical Structure and Properties of Condensed Inorganic Phosphates

For a proper understanding of the processes which take place in living organisms, a precise knowledge of the chemical structures of the compounds which participate in these processes is required. It is therefore deemed essential to present, even if only in brief, an account of present-day ideas of the chemical structures of the condensed phosphates, hitherto often known by the long-obsolete terms 'metaphosphates' and 'hexametaphosphates'.

THE STRUCTURE OF CONDENSED PHOSPHATES

The first mention of condensed inorganic phosphates dates back to 1816, when Berzelius showed that the vitreous product formed by the ignition of ortho-phosphoric acid was able to precipitate proteins.[790] In 1833, Graham[190] described a vitreous phosphate which he obtained by fusion of NaH_2PO_4. Believing that he had isolated a pure compound with the formula $NaPO_3$, Graham named it a 'metaphosphate'. Shortly afterwards, however, Fleitmann and Hennenberg,[171] working in Liebig's laboratory, demonstrated that the 'metaphosphates' having the general formula MPO_3 (where M is hydrogen or a monovalent metal) were mixtures of closely related compounds which differed mainly in their degree of polymerization. The numerous investigations which were carried out over the next 100 years (for reviews, see refs. 129, 295, 728, 751, and 790), although they provided a wealth of new data which shed much light on the structures and properties of this group of compounds, threw into perhaps even greater confusion both the chemical basis of the nomenclature of these compounds, and the names of the compounds themselves. This is perhaps hardly surprizing, since these investigations were carried out with

2

compounds of inadequate purity, using crude investigational methods. It was thanks to the work of Thilo,[735–744] Van Wazer,[788–790] Samuelson,[627] Ebel,[129–135] and Boulle[67] that the chemical structures and properties of this group of compounds were finally established, thus making it possible to bring order into their classification.[748,791]

According to the current classification, condensed phosphates are divided into metaphosphates, polyphosphates, and branched inorganic phosphates, or ultraphosphates.

Metaphosphates

The true metaphosphates have the composition which since the time of Graham has been incorrectly assigned to the whole group of condensed phosphates, i.e. MPO_3. These compounds are built up from cyclic anions. Only two representatives of this group have so far been investigated in detail— the trimetaphosphates, $M_3P_3O_9$, and the tetrametaphosphates, $M_4P_4O_{12}$, shown in Fig. 2.

The existence of mono- and dimetaphosphates has not been demonstrated in practice, and is theoretically unlikely.[129,740,790] The possible presence of pentametaphosphates and hexametaphosphates in a mixture of condensed sodium phosphates was shown by Van Wazer and Karl-Kroupa,[793] followed by Thilo and Schülke.[746] An octametaphosphate was obtained in a crystalline state by Schülke.[649] In addition, more highly polymerized cyclic phosphates containing as many as 10–15 orthophosphoric acid resideues have been observed in some samples of condensed phosphates prepared by Van Wazer.[790] In disucssing the occurrence of cyclic hexametaphosphates, it should be pointed out that the term 'hexametaphosphate', which is frequently encountered in the literature, refers in fact to the compound known as Graham's salt, which according to current ideas is a mixture of condensed sodium phosphates which contains cyclic metaphosphates (including hexametaphosphate), but which is mainly composed of highly polymerized linear polyphosphates.[748,791]

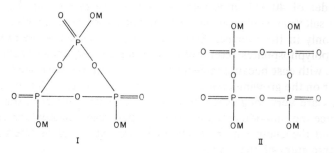

Figure 2. Structure of trimetaphosphate (I) and tetrametaphosphate (II)

Polyphosphates

Polyphosphates have the general formula $M_{(n+2)}P_nO_{(3n+1)}$. Their anions are composed of chains in which each phosphorus atom is linked to its neighbours through two oxygen atoms, thus forming a linear, unbranched structure which, according to Thilo,[740] may be represented schematically as shown in Fig. 1. The degree of polymerization, n, may possess values from 2 to 10^6, and as the value of n increases, the composition of the polyphosphates, i.e. the cation to P ratio, approximates to that of the metaphosphates, which explains the belief which prevailed until recently that 'polyphosphate' and 'metaphosphate' were equivalent terms. Polyphosphates in which $n = 2$–5 can be obtained in the pure, crystalline state,[790] but the members of this series in which n has higher values have been obtained in appreciable amounts only in admixture with each other.

In contrast to the metaphosphates, they are designated as 'tripolyphosphates', 'tetrapolyphosphates', etc., although the mono- and dimeric compounds are still called by their old names of 'orthophosphate' and 'pyrophosphate', respectively. In addition, the highly polymeric, water-insoluble potassium polyphosphate ($\bar{n} = 2 \times 10^4$), which has a fibrous structure of the asbestos type, is still called Kurrol's salt. We may mention in passing that the facile preparation of Kurrol's salt (by fusion of KH_2PO_4 at 260°C), and the ease with which it is converted into the water-soluble sodium form by means of cation-exchange materials, has led to its frequent preparation and use in chemical and biochemical work as an inorganic polyphosphate.

Even better known is Graham's salt, the vitreous sodium polyphosphate ($\bar{n} = 10^2$) obtained by fusion of NaH_2PO_4 at 700–800 °C for several hours followed by rapid cooling. Graham's salt is a mixture of linear polyphosphates with different chain lengths. Fractional precipitation from aqueous solution by means of acetone[790] affords less heterogeneous fractions with different molecular weights. For example, a sample of Graham's salt in which the chains on average have 193 P atoms (i.e. $\bar{n} = 193$) can be separated by this method, as shown in Fig. 3.

As can be seen from Fig. 3, this sample contains molecules of different sizes. The fraction of highest molecular weight has $\bar{n} = 500$, i.e. its molecular weight is of the order of 40 000. It is interesting that the reason for the failure of Graham's salt to crystallize is that it consists of a mixture of homologous chains differing only in their length. Since all the components of the homologous series of polyphosphates resemble each other closely, crystallization cannot take place with ease because molecules of different dimensions seek to displace each other on the growing crystal, thereby bringing its growth to a stop. When the chains are very long (such as is the case in Kurrol's salt), this does not occur, since the individual chains pass through many elementary cells of the crystal, and the chain length is not an important factor in determining the lattice parameters of the crystal.[790]

A second factor which determines the maximum chain length of the poly-

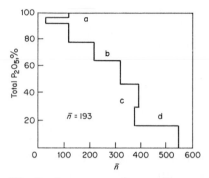

Figure 3. Distribution curve for sodium polyphosphate molecules (Graham's salt, $\bar{n} = 193$) by size on fractional precipitation, after Van Wazer:[790] (a) cyclic metaphosphates; (b), (c), and (d) linear polyphosphates

phosphates which are able to crystallize is the increase in polarity of the molecule which takes place as the degree of polymerization increases.

Two factors thus appear to be responsible for the failure so far to obtain linear polyphosphates containing 6–200 phosphorus atoms in a crystalline state: (1) the difficulty of crystallization from a mixture of similar compounds, and (2) the effect of polar groups on the molecule.

In addition to linear polyphosphates, Graham's salt usually contains very small amounts of cyclic metaphosphates (Fig. 3). For example, a sample of Graham's salt with $\bar{n} = 100$–125 was shown by Van Wazer[790] to contain 4% of trimetaphosphate, 2.5% of tetrametaphosphate, 0.8% of pentametaphosphate, 0.5% of hexametaphosphate, and fractional percentages of higher polymeric metaphosphates. The compositions of two samples of Graham's salt obtained by Dirheimer[106] are shown in Table 1.

The high-molecular-weight polyphosphates have a secondary structure. Thus, X-ray structural data show that the polyphosphate chains in crystals of the macromolecular sodium salt (Maddrell's salt) form helices with a period of 7.3 Å,[736,742] and in macromolecular potassium phosphate (Kurrol's salt) the helices have a period of 6.1 Å.[282,742] The structures of the anions of the

Table 1. Composition of synthetic samples of Graham's salt.[106] P content of the poly- and metaphosphates expressed as a percentage of the total P content of the compound

Polyphosphates and metaphosphates	Sample 1	Sample 2
High-molecular-weight polyphosphate	68.1	75.1
Polyphosphates ($n = 5$–10)	17.3	13.6
Tetrapolyphosphate + trimetaphosphate	7.8	7.0
Tripolyphosphate	4.5	2.8
Pyrophosphate	2.3	1.5

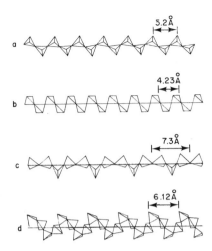

Figure 4. Structure of anions of crystalline polyphosphates: (a) $(LiPO_3)_x$; (b) $(PbPO_3)_x$; (c) $(NaPO_3)_x$, Maddrell's salt; (d) $(KPO_3)_x$, Kurrol's salt[743]

crystalline polyphosphates are shown in Fig. 4. It has been suggested that Graham's salt also possesses a secondary structure.

Branched inorganic phosphates, or ultraphosphates

High-molecular-weight condensed phosphates which, unlike the linear polyphosphates, contain 'branching points', i.e. phosphorus atoms which are linked to three rather than two neighbouring phosphorus atoms, are known as branched phosphates or ultraphosphates. These branched phosphates have a branched structure, a fragment of which is shown in Fig. 5. In this type of structure, the individual polyphosphate chains are linked to form a 'network', which is the reason for the name given to this type of condensed phosphates.

The existence of this group of phosphorus compounds was observed only recently in some samples of Kurrol's salt and Graham's salt which were obtained by chemical methods.[698–701,737,739,740,790,792] Thus, Strauss and co-workers[698–701] showed that, in samples of Graham's salt with very long chains (of the order of several hundred phosphorus atoms), approximately one in

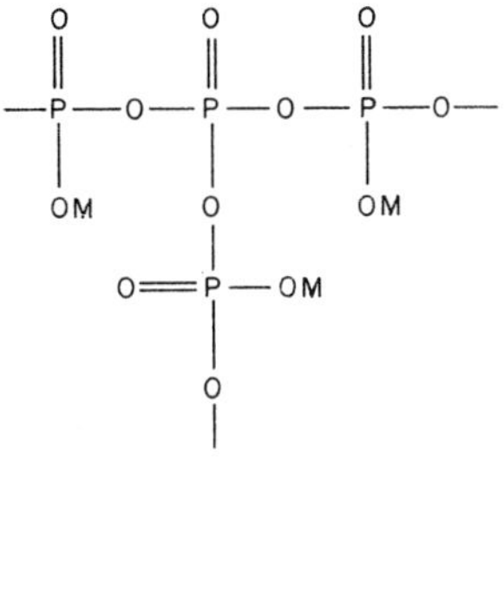

Figure 5. Structure of a reticular polyphosphate

every thousand phosphorus atoms is a branching point. The presence of branching in polyphosphate chains, or in other words the presence of a reticular structure, can be detected by the decrease in the viscosity of aqueous solutions which occurs following solution of the compounds in water (owing to rapid hydrolysis of the lateral bonds, which are very unstable).

Figure 6 shows how the proportion of lateral bonds in Graham's salt increases as the chain length is increased.

Although branched phosphates have not yet been found in living organisms (perhaps as a consequence of their unusually rapid hydrolysis in aqueous solution, irrespective of pH, even at room temperature), it is believed that their presence in biological materials cannot be excluded.

Information on the chemical composition of the condensed inorganic phosphates, together with descriptions of their chemical and physicochemical properties, can be found in several reviews and monographs.[67,129,195,196,566,738,739,742–744,788–790,792] We shall dwell here very briefly on those properties of condensed phosphates which are useful for their identification and chemical determination in living organisms.

SOME CHEMICAL PROPERTIES OF CONDENSED INORGANIC PHOSPHATES

1. Polyphosphates are salts of acids which, in solution, contain two types of hydroxyl groups that differ in their tendency to dissociate. The terminal hydroxy groups (two per molecule of polyphosphoric acid) are weakly acidic, whereas the intermediate hydroxyl groups, of which there are a number equal to the number of phosphorus atoms in the molecule, are strongly acidic.[790] Cyclic metaphosphates do not contain terminal hydroxyl groups and, for this reason, the corresponding metaphosphoric acids possess only strongly acidic groups which in solution are dissociated to approximately the same extent. Thus, titration of weakly and strongly acidic groups is a convenient means of determining whether a given condensed phosphate is a meta- or a polyphosphate. Moreover, this method provides a means of determining the average chain length of linear polyphosphates.[88,129,442,444,627,788] It is interesting that this was the method

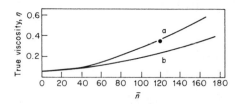

Figure 6. Changes in the viscosity of solutions of polyphosphates of differing chain lengths on keeping for 12 h. The abscissa represents the mean chain length as determined by end-group titration: (a) immediately after solution; (b) after keeping for 12 h[701]

used by Samuelson[627] in showing for the first time that Graham's salt was not a cyclic metaphosphate, as had been believed for almost 100 years, but a mixture of linear polyphosphates.

2. All alkali metal salts of condensed polyphosphoric acids are soluble in water. Potassium pyrophosphate is especially soluble, 100 g of water dissolving 187.4 g of $K_4P_2O_7$ at 25°C, 207 g at 50°C, and 240 g at 75°C. Exceptions to this rule are the water-insoluble Kurrol's salt (a macromoleculear crystalline potassium polyphosphate), and the compounds known as Maddrell's salts (crystalline sodium polyphosphates of very high molecular weight). Kurrol's salt is readily soluble in dilute solutions of salts containing cations of univalent metals (but not K^+), for example 0.2 M NaCl. It is worth mentioning that Graham's salt dissolves in water only when it is stirred rapidly. Without stirring, the compound forms a glue-like mass in water.

Polyphosphates of divalent metals such as Ba^{2+}, Pb^{2+}, and Mg^{2+} are either completely insoluble or dissolve to only a very limited extent in aqueous solutions. The polyphosphates of certain organic bases such as guanidine are also sparingly soluble in water.[665] Other solvents (liquid ammonia, anhydrous formic acid, and organic solvents such as ethanol and acetone) dissolve only trace amounts of sodium and ammonium polyphosphates. Low-molecular-weight polyphosphates dissolve readily in very dilute aqueous–alcoholic solutions, but addition of alcohol to these solutions rapidly reduces their solubility. Figure 7 shows that an ethanol–water mixture containing 40% of ehtanol is a very poor solvent for potassium pyrophosphate and tripolyphosphate (1.5 g per 100 g of solution).

3. Condensed phosphates other than branched phosphates are stable in neutral aqueous solution at room temperature.

It is interesting that hydrolysis of the P–O–P bond in linear polyphosphates such as Graham's salt liberates energy equivalent to approximately 10 kcal/mol,[832,790] i.e. the same amount of energy as is liberated in the hydrolysis of the terminal phosphoric anhydride bonds in the ATP molecule. Hydrolysis of the cyclic trimetaphosphate[514] also liberates this amount of energy.

The branching points in branched phosphates, in which one phosphorus

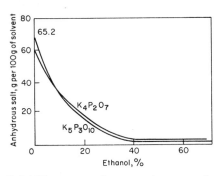

Figure 7. Solubility curves for potassium pyrophosphate and tripolyphosphate in ethanol–water mixtures at 25 °C[790]

atom is bonded through oxygen to three other phosphorus atoms, are extremely labile. The rate of hydrolysis of the branching points in the reticular phosphates in aqueous solution at 25 °C, resulting in the formation of linear polyphosphates, is about 1000 times greater than that of the P–O–P bonds in the linear polyphosphates. Hydrolysis of the branching points, according to Van Wazer,[790] liberates 28 kcal/mol, which is much more than that liberated in the hydrolysis of the 'central' phosphoric anhydride bonds.

The linear polyphosphates and cyclic metaphosphates are hydrolysed extremely slowly at neutral pH and room temperature in comparison with other polyacids such as polyarsenates and polyvanadates, and they are unique in this respect. The half-hydrolysis time for the P–O–P bonds in linear polyphosphates at pH 7 and 25 °C is several years.[790] The rate of hydrolysis of these bonds is increased by raising the temperature, reducing the pH, and by the presence in the solution of colloidal gels and complex cations. The hydrolysis of these bonds is dependent on the ionic strength of the solutions.[790]

When neutral solutions of polyphosphates are heated at 60–70°C for 1 h, they are broken down quantitatively to cyclic trimetaphosphate and orthophosphate. It has been shown that this hydrolysis does not occur randomly, but rather from the end of the polyphosphate chain.[742,749]

Thilo[742] related the formation of cyclic trimetaphosphates during the hydrolysis of linear polyphosphates in neutral solution (and even in non-aqueous solution) to the presence of a particular type of spiral secondary structure which makes it sterically possible for a rearrangement of the bonds to occur within the molecule with the formation of small closed chains (see fig. 8).

In alkaline solutions, cyclic metaphosphates undergo ring fission even on gentle warming to form linear polyphosphates with corresponding chain lengths.[129] Linear polyphosphates also undergo hydrolysis under alkaline conditions,[551,552] but more particularly under acidic conditions (pH 3.5–4). Under these conditions, significant hydrolysis of the P–O–P bonds takes place even at room temperature, and here breakdown occurs along the length of the chains rather than from the ends of the chains, to form polymers with increasingly lower molecular weights down to orthophosphate. The results of

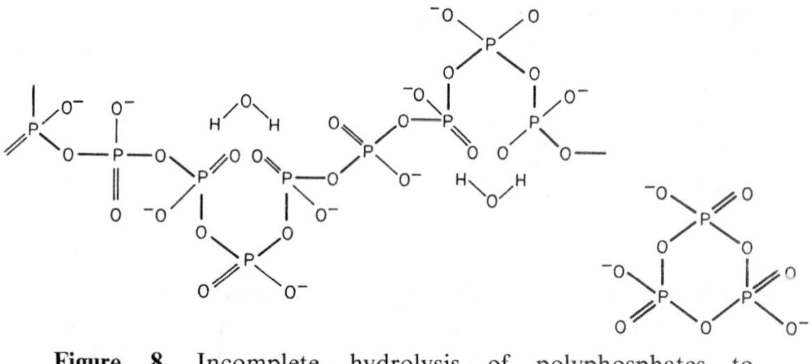

Figure 8. Incomplete hydrolysis of polyphosphates to trimetaphosphate and orthophosphate[739]

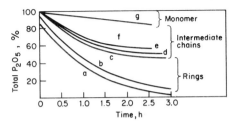

Figure 9. Results of a chromatographic examination of the hydrolysis products of Graham's salt at pH 4 and 90 °C: (a) high-molecular-weight polyphosphates; (b) cyclic metaphosphates containing 4–6 atoms of phosphorus; (c) trimetaphosphate; (d) pyrophosphate; (e) tripolyphosphate; (f) linear polyphosphates containing 4–15 P atoms; (g) orthophosphate[790]

an investigation of the hydrolysis products of Graham's salt at pH 4.0 and 90 °C are shown in Fig. 9.

It can be seen from Fig. 9 that the proportions of the hydrolysis products (linear polyphosphates, cyclic metaphosphates, and orthophosphate) are very dependent on the duration of hydrolysis. When the reaction time is increased to 3 h, the higher polymeric polyphosphates disappear altogether, the mixture consisting entirely of low-molecular-weight poly- and metaphosphates and orthophosphate. When the pH of the solution is reduced to 1 and below, the extent of hydrolysis of polyphosphates to orthophosphate increases rapidly. Linear polyphosphates such as Graham's salt are completely hydrolysed after 7–15 min at 100 °C in 1 N HCl.[740]

PHYSICOCHEMICAL PROPERTIES OF CONDENSED INORGANIC PHOSPHATES

Apart from the low-molecular-weight polyphosphates and metaphosphates, condensed inorganic phosphates are macromolecular compounds, and this affects their properties and behaviour in solution.

1. Aqueous solutions of polyphosphates of low ionic strength and pH values near neutral are very viscous, the viscosity increasing with increasing mean chain length.[267,480,788] The presence of branched phosphates in any given sample of condensed phosphates results, as we have seen, in a very high initial viscosity which decreases rapidly following solution in water, even at room temperature (see Fig. 6).

2. Polyphosphates in aqueous solutions of low ionic strength are capable of forming complexes with other polymers, especially proteins,[300] basic polypeptides,[665] and nucleic acids.[78,368] This ability increases as the chain length of the polyphosphate molecule increases. In acidic solution, these complexes separate as precipitates. The ability of condensed phosphates to precipitate proteins from acidic solutions has been known from a very early date.[129,591,790,809] It has been shown that this property of polyphosphates is due to a simple

interaction between cations and anions. At pH values below the isolectric point, proteins behave like macromolecular cationic electrolytes, having a total charge which is dependent on the pH. It was shown by Katchman and Van Wazer[300] that the higher the molecular weight of a water-soluble protein, the less polyphosphate is required. In other words, the higher the molecular weight of the phosphate or the protein, the lower is the concentration at which the polyphosphate-protein reaction takes place.

3. A similar polycation–polyanion interaction is found in the metachromatic reaction, in which high-molecular-weight polyphosphates cause a shift in the absorption maxima of cationic dyes, such as toluidine blue, towards shorter wavelengths.[65,129] This reaction essentially involves polymerization of the dye on the macromolecular anion.[804,810] In the case of toluidine blue, addition of polyphosphate to the solution results in a change in colour from blue to violet–red, and a shift in the position of the absorption maximum from 630 nm (which is characteristic of solutions of the monomeric form of toluidine blue) to 530 nm (typical of the complex of polyphosphate and the polymerised dye).[11,94,99,732] However, only comparatively high-molecular-weight polyphosphates are capable of undergoing the metachromatic reaction, either in solution or on paper.[94,152,732] Tripoly- and trimetaphosphates, for example, do not react with toluidine blue.[330]

4. Linear polyphosphates possess properties very similar to those of cross-linked, solid ion-exchange agents.[736] The behaviour of polyphosphates as 'dissolved ion-exchange agents' is yet further evidence of their ability to form complexes with counter ions. Polyphosphates are known to be very good complexing agents for many metal ions.[790] This property is widely exploited in the fractionation of polyphosphates, and for other analytical purposes.

The information given above concerning the chemical and physicochemical properties of the inorganic polyphosphates will assist in the better understanding and prediction of the behaviour of these compounds during their extraction from cells and their subsequent fractionation. A knowledge of these properties will facilitate the development and use of efficient and reliable biochemical procedures for the isolation, purification, identification, and determination of polyphosphates.

Chapter 2

Methods for the Detection and Extraction of Condensed Inorganic Phosphates in Biological Materials

CYTOCHEMICAL METHODS OF DETECTION

The oldest and most extensively used methods for the determination of condensed phosphates in biological materials, although of course the least accurate, are based on the staining of cells and tissues by certain basic dyes, such as toluidine blue, neutral red, and methylene blue. The presence of condensed phosphates in the organisms is judged from the appearance in the cells of metachromatically stained granules, or, as they are also known, volutin granules.[114,122–124,130,152,276,606,732,780,802–805,807,812] In some investigations, attempts were even made to assess quantitatively the metachromatic granules.[529,626] Further, Ebel and co-workers[148–150] and Rosenberg[619] used, in addition, other cytochemical methods for the detection of polyphosphates. It must be said, however, that, despite the fact that in most cases the cytochemical detection of 'polyphosphate' granules is associated with the actual presence of condensed phosphates in the organism, the use of such methods must nevertheless be attended with great caution.[114,320,495,619,814] This is due primarily to the fact that the basic dyes used to identify the 'polyphosphate' granules are also capable of metachromatically staining other polymeric compounds which are encountered in biological material,[495,814] in addition to high-molecular-weight polyphosphates.

CHEMICAL METHODS OF DETECTION

Chemical methods for the detection of condensed phosphates are more precise and reliable. The most widely used of these methods involves precipitation of these compounds from cell extracts in the form of the sparingly soluble Ba^{2+} salts, or the salts of other divalent metals (Pb^{2+}, Mg^{2+}, etc.), followed by hydrolysis with 1M HCl for 7–10 min at 100 °C. The amount of orthophosphate formed is a measure of the amount of condensed phosphates present.[50,106,127,129,130,220,335,483,484,504,807-809]

By precipitating the barium salts at different pH values (for example, 2.5, 4.5, and 7), it is possible to distinguish between condensed phosphates which differ in their degree of polymerization, and which are differently bound in the cell.[129,130,442,444] It is also possible to use organic bases such as guanidine to precipitate polyphosphates selectively from their aqueous solutions.[139,146-148,665] However, the detection condensed phosphates according to the labile phosphorus of the barium and other water-insoluble salts is likewise not fully reliable.[220] Further, not all condensed phosphates are precipitated by barium salts. Oligomeric poly- and metaphosphates, in particular tripolyphosphate and trimetaphosphate, are not precipitated by barium at any pH.[330,745] Again, precipitation by barium salts may result in some degradation of polyphosphates.[220,444,742] A method for the determination of the intracellular concentration of pyrophosphate has only recently been developed by Heinonen.[231] Chromatography and electrophoresis appear to offer the only completely reliable methods for the detection of condensed phosphates in cell extracts from living organisms. However, the paper chromatographic separation of condensed phosphates only permits the separation and determination of comparatively low-molecular-weight poly and metaphosphates ($n = 2$–9).[63,66,96,106,129,131-135,154,200,229,258,296,389,391,405,469,566,594,746,793,800] This method can be used for the determinatioj of various polyphosphates,[106,129,135] and even for the determination of the structures of condensed phosphates.[106,129,135] Using Ebel's method of two-dimensional paper chromatography,[129] it is possible to separate the oligomeric poly- and metaphosphates. An example of such a separation, carried out by Dirheimer,[106] is shown in Fig. 10.

In addition to paper chromatography, chromatography on ion-exchange resins has recently been employed with success.[78,129,139,187,330,332,439,460,496,497,566,592,813] Using this method, it is possible to separate polyphosphates with values of n from 2 to 12[497] (Fig. 11).

Paper electrophoresis has also been employed to separate oligomeric poly- and metaphosphates,[628,687,784] together with thin-layer chromatography.[422,425,725] A convenient method for the simultaneous determination of pyrophosphate and ATP has been described by Guillory and Fisher.[205]

However, these methods are only capable of separating polyphosphates of fairly low molecular weight. Two methods are currently available for the separation of high-molecular-weight polyphosphates. The first of these is Van Wazer's method for the fractional precipitation of polyphosphates of different chain lengths from aqueous solutions with acetone.[790] This method affords a large number of fractions which differ from each other in the lengths of the polyphosphate chains (Fig. 3). The second method, also developed in the laboratory of Van Wazer,[567] enables similar results to be obtained by the use of paper chromatography. Examples of the use of this method for the

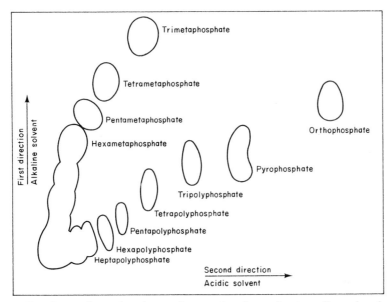

Figure 10. Separation of poly- and metaphosphates by two-dimensional paper chromatography. The basic solvent is isopropanol–isobutanol–water–25% ammonia (40:20:39:1) and the acidic solvent is isopropanol–water–25% ammonia (70:30:5: 0.3)[106]

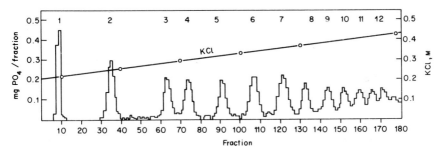

Figure 11. Separation of linear polyphosphates of low molecular weight by ion-exchange chromatography on Dowex 1 X10 in a KCl gradient. The numbers 1–12 represent the number of phosphorus atoms in the polyphosphate molecules constituting the various fractions[497]

investigation of biological material are the work carried out in Penniall's laboratory[194] and that of Janakidevi et al.[273] A method has also been developed in Ebel's laboratory[165] for the separation of inorganic polyphosphates on Sephadex columns (Fig. 12).

The identification of polyphosphates can be avoided by the use of Thilo and Wieker's method of chromatographic analysis of the products of partial hydrolysis.[749] In this method, hydrolysis of condensed phosphates in neutral solution at 60° C affords trimetaphosphate and orthophosphate, which are readily identified chromatographically, especially when Ebel's basic solvent is

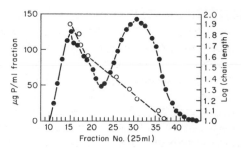

Figure 12. Separation of polyphosphates of high molecular weight on Sephadex G-100[165] ● – P or polyphosphates; ○ – Log of polyphosphates chain length

used.[129] The use of this method for the identification of inorganic polyphosphates in biological subjects has been described in several papers.[47,288,320,422,425 745,769,775]

PHYSICAL METHODS OF DETECTION

Glonek et al.[186] and Salhany et al.[624] have recently developed methods for the identification of linear polyphosphates by nuclear magnetic resonance (NMR) spectroscopy. The first attempt to use this method for the identification of these macromolecules in extracts of living cells have given promising results.[181]

METHODS OF EXTRACTION FROM BIOLOGICAL MATERIAL

In the earlier work on the isolation of condensed polyphosphates from the cells of living organisms, the same methods as those used for the extraction of nucleic acids were usually employed.[15,336,453,454] It was not until 1936 that MacFarlane[477,478] proposed a specific method for the extraction and fractionation of condensed phosphates present in cells. It was found that these condensed phosphates could be divided into two main fractions—one soluble in 5% trichloroacetic acid (TCA) and the other insoluble, and ever since than cellular condensed polyphosphates have been divided into acid-soluble and acid-insoluble fractions. Although most workers have used the same extractant, 5% TCA (or occasionally 0.5 M $HClO_4$) to obtain the acid-soluble fraction, a variety of methods have been used to isolate the acid-insoluble condensed phosphates.

1. The most commonly used method of extraction from cells involves the use of dilute sodium hydroxide solution (pH 8–9).[50,58,335,345,477,478,529]

2. A widely used method for the extraction of acid-insoluble condensed phosphates from cells is some variant of the method of Schmidt and Thannhauser,[643] i.e. the use of 1 N potassium hydroxide at 37 °C for various lengths of time.[21,83,84,193,194,240,297,298,588,837–842]

3. A method differing in principle for the extraction of this condensed

phosphate fraction which involves the use of hot solutions of acids, either 5% TCA (Schneider).[50,58,73–75,324,325,335,805,809] or 10% perchloric acid, at 80–100° C.[121,123,162,213,214,272,316,351] When this method of extraction is employed, the condensed phosphates are hydrolysed to orthophosphate, the amount of which indicates the amount of condensed phosphates present in the acid-insoluble fraction. Hughes and co-workers[89,256,257] used a prolonged (5 h) extraction of acid-insoluble polyphosphates with 10% TCA at 20–22 °C. In the opinion of the author, this method ensures almost complete extraction of the acid-insoluble polyphosphates from the cells of bacteria and other microorganisms.

Later investigations showed that it was possible to carry out further fractionation of the polyphosphates present in biological material according to the lengths of the molecules. Such fractionation has been carried out by Langen, and Liss in the laboratory of Lohmann.[442,444,464,465,468,469] Their method consists in the successive extraction of the cells in the cold with 1% TCA, a saturated solution of a salt such as $NaClO_4$, dilute NaOH solution (pH 10), and a more concentrated solution of alkali (0.05 N NaOH).

This method, either in its original version or modified in various ways, has been used extensively for the fractionation of polyphosphates from different organisms.[158,243,320,357,381,383,390,391,404,405,407,409,413,431,433,510,511,754,794] It possesses the advantage that the fractions obtained were localized at different intracellular sites and showed different physiological activity.

Another, and apparently successful, method for the fractionation of polyphosphates present in biological material is that developed by Miyachi and co-workers in the laboratory of Tamiya.[228,293,520–524] These workers successively extracted the polyphosphates present in the cells of Chlorella and other organisms with 8% TCA in the cold (fraction A), then with a solution of NaOH of pH 9 in the cold (fraction B), and finally with a 2 N solution of potassium hydroxide at 37 °C for 18 h. The polyphosphates extracted with 2 M KOH by the method of Schmidt and Thannhauser[643] were further separated by Miyachi into two fractions, namely that precipitated by neutralization with $HClO_4$ in the presence of $KClO_4$ (fraction C), and that which was not precipitated under these conditions (fraction D). The work of Miyachi et al.[522] showed that this mode of fractionation of Chlorella polyphosphates afforded fractions which differed in their physiological activity, and also apparently in their cellular location.

It should be pointed out that none of the methods described above was aimed at obtaining completely unmodified preparations of cellular condensed polyphosphates. They were used mostly to fractionate this group of compounds to different extents, an essential preliminary to working out the details of their interconversions and the way in which they participate in the cellular metabolism as a whole.

Other, much milder, methods of extraction from cells have been developed in order to obtain samples of condensed phosphates which are as little modified as possible, and which completely avoid the use of strong acids and bases.

The mildest methods for the extraction of condensed phosphates are as follows:

(a) extraction with hot water;[85,106,108,332,468,469]

(b) extraction with dilute sodium carbonate solution;[131,267]

(c) extraction with hot 2 M sodium chloride solution;[288-290] and

(d) extraction with cold distilled water following a preliminary treatment of the material with alcohol and ether[84,106,108,130,469,481,635,640,641] or sodium hypochlorite.[217]

Extraction of polyphosphates with distilled water should, of course, be the best method for obtaining unmodified preparations. In the opinion of Schmidt,[635] this method ensures complete extraction of all the cellular condensed phosphates.

However, in assessing the potential of this method in yeasts and young mycelia of *P. chrysogenum*, we have found that, firstly, a much higher proportion of the condensed phosphates passes into water than into cold 5% TCA under the same conditions, and secondly, the use of cold water alone does not ensure the complete extraction of all the cellular condensed phosphates, a substantial proportion of which remain after extraction in the biological material, the actual amounts depending on the age of the cells undergoing extraction.

On the basis of work carried out in our own laboratory, we consider the fractionation method of Langen and Liss to be one of the best available for the separation and quantitative determination of the different polyphosphate fractions localized at different intracellular sites and apparently displaying specific physiological roles.

The best of the methods available for the isolation of the least modified forms of the inorganic polyphosphates is extraction with cold water following treatment of the cells with alcohol and ether, or with sodium hypochlorite solution.

Chapter 3

The Distribution of Condensed Inorganic Phosphates in Living Organisms

The first report of the presence of condensed inorganic phosphates in living organisms dates back to 1888, when Liebermann[453] found them to be present in yeast 'nuclein'. Soon afterwards, Kossel[336] and Ascoli[15] showed that condensed phosphates formed part of the 'plasminic acids' obtained by partial hydrolysis of yeast nucleic acids. Since that time, the presence of these phosphorus compounds has been demonstrated in many other lower animals and plants, especially protozoa, bacteria, fungi, and algae. They have also been found in some, albeit very few, animals (insects, amphibia, and mammals), mosses, and in various organs and parts of the higher flowering plants. We list below those organisms in which condensed phosphates have been identified by one means or another.

Bacteria

Acetobacter suboxydans;[314] *Agrobacterium radiobacter;*[726] *Aerobacter aerogenes* *;* [216–219,221,222,227,671,812] *Aerobacter cloaceae;*[671] *Azotobacter agile* *;*[842] *A. vinelandii* *;*[160,835–843] *Bacillus brevis;*[257] *B. cereus ;*[726] *B. freidenreihii;*[671] *B. megaterium;*[257] *B. prodigiosus;*[257] *Brevibacterium ammoniagenes ;*[341,342] *Caulobacter vibroides;*[199] *Chlorobium thiosulphatophilium;*[89,162,163,256,659] *Chromatium* sp.;[162,163] *Chr. okenii;*[632] *Clostridium* sp.*;*[709] *Cl. sporogenes;*[257] *Corynebacterium diphtheriae* *;*[57,100,101,128,129,131,335,518,625,626,719] *C. xerosis* *;*[106,110,257,531,812] *Escherichia coli;*[541–543,599,600] *Hydrogenomonas* sp.,[288,289,631,634] *Lactobacillus casei;*[257] *Micrococcus denitrificans;*[290] *Micrococcus lysodeikticus;*[173] *Mycobacterium avium;*[623] *M. cheloni;*[529] *M. phlei;*[117,118,120–123,185,257,316,820,821] *M. smegmatis;*[257,817–822] *M. thramnopheos;*[529] *M. tuberculosis ;*[257,321,362,506–509,820,821] *Myxococcus xanthus;*[780] *Nitrobacter winogradsky;*[758,776] *Nitrobacter* sp.;[80] *Nitrosomonas europaea;*[730] *Propionibacterium shermanii;*[326,437]

Rhodopseudomonas palustris;[162,163] *Rhodopseudomonas spheroides*;[81,794] *Rhodospirillum rubrum*;[425,561,794] *Serratia marescens*;[671] *Spirillum volutans*;[48,49] *Staphylococcus albus*;[257] *Staphylococcus aureus*;[438] *Streptococcus sl-1*;[724] *Thiobacillus thiooxydans.*[39]

Actinomycetes

Actinomyces aureofaciens*;[47,201-203,320,396,606] *A. eritreus*;[204] *A. griseus.*[150,151,208]

Fungi

Agaricus bisporus*;[360,361,390,409] *Aspergillus niger*;[20,21,51,53,54,58,151,263,267,365,367,368, 386,387,457,483,484,562,563] *A. nidulans*;[661] *A. oryzae*;[279,280] *Badihamia utricularis*;[560] *Botryotridium pililiferum, Brethanomyces animalis, Candida tropicalis, C. vulgaris*;[151] *C. guillermondii*;[197,450,538,702,781] *Chaetonium globosum*;[151] *Choanephora cucurbitarumi*;[105] *Colletotrichum lindenmuthianum, Cryptococcus laurentii, Culvularia maculans, Cuninghamella elegans*;[105] *Diplodia natalensis*;[151] *Claviceps paspali**;[430] *C. purpurea*;[782,783] *Fusarium coeruleum, Cliocladium roseum*;[151,534,535] *Giromitra esculenta*;[390] *Endomyces magnusii**;[365,381-384,404,667] *Isaria gelina, Kloeckera mulleri*;[151] *Lenthinus tigrinus**;[352,353] *Morchella hortensis*;[151]*; Mortierella alpina, Mucor javanicum,*[105] *Mucor racemosus*;[272] *Mycelia steria, Mycotorula eisnerosa*;[151] *Mycothecium verrucaria*;[151,534,535] *Necktria rubi*;[151] *Neurospora crassa**;[212-215,225,226, 254,357,403,405,406,408,413,429,510,844] *Penicillium chrysogenum**;[53,88,114,151,365,367,368,392,397, 415,420,421,511,573] *Phycomyces balkesleanus*;[468,469,593] *Physarum polycephalum*;[188,207,629] *Physarum sp.*;[301] *Rhodotorula rubra*;[151] *Saccharomyces cerevisiae*;[15,50,51,53,60,74,75,83, 84,104,108,113,127-133,140-146,166,230,243-247,251,261,286,287,297-300,324,325,330,335,338,367,368,442,444,445,447,453, 454,458,459,464-469,477,478,497,498,609,616,640-642,677-680,681,682,685,690-694,781,801-810,831-834] *S. mellis*;[796,798] *S. carlsbergensis*;[432,433,754] *Schizosaccharomyces pombe*;[407] *Staphylotrichum coccosporum*;[151] *Torula sp.*;[129,130] *Torula utilis*;[85,613] *Torula spora roxi, Vertiollium honigii*;[151] *Zygorhyncus exponens.*[105]

Blue–Green algae

Anabaena variabilis*;[81,91,93,94,125,337] *Anacystis nidulans*;[551] *Cylindrosperum licheniforme*;[125] *Clocothece sp.*;[177] *Lingbyaerogineo coerula, L. amplivaginata*;[72] *Oscillatoria sp.*;[151,301] *O. amoena*;[177] *O. limosa*;[125] *Phormidium ambiguum*;[132] *P. frigidum*;[301] *P. uncinatum*;[301] *Plectonema boryanum.*[278]

Algae

Acetabularia mediterranea*;[200,615,621,622,695-697,745] *Acetabularia crenulata*;[424,552] *Ceramium sp.*;[441,467] *Ankistrodesmus braunii*;[115,292,476,757,759] *Chara sp.*;[301] *Chilomonas sp.*;[151] *Chlamydomonas*;[301] *Chlorella* sp.*;[91,150,250,252,253,786,787] *C. ellipsoides*;[228,293,521-524,558,559,722,723] *C. pyrenoidosa*;[22-24,97,98,240,294,576,636,673,825-827] *C. vulgaris*;[363] *Cladophora sp.,*[441] *Cosmarium sp.*;[301] *Dunaliella viridis**;[771] *Enteromorpha sp.*;[441,]

[467] *Euglena gracilis*;[9,467,670] *Fragillaria* sp.;[301] *Hydrodiction reticulatum*?[363,364,601] *Mongeotia* sp., *Navicula* sp., *Oedogonium* sp.;[301] *Polytomela ceca*;[810] *Rhopalocystis gleifera*;[727] *Scenedesmus* sp.;[187,705] *S. quadrianda**;[42,583,584] *S. obliquus**;[431,585,770,771] *Spirogira* sp., *Ulotrix* sp., *Vaucheria* sp., *Zygnema* sp.,[301] some brown and red algae.[550]

Mosses

Leptobrium, Polytrichum.[301]

Flowering plants

Ripening and mature seeds of cotton;[12,13,773,775] tomato (*Solanum lycopersicum*), leaves[252,253] and hypocotyls;[312] leaves of spinach (*Spinacea oleracea*);[520] roots and stems of *Banksia ornata*, a plant of the Australian wastelands;[277] dodder (*Cuscuta reflexa*);[734] leaves of wheat (*Triticum vulgare*);[785] roots of maize;[769] leaves of apple (*Malus domestica*);[637–639] roots of *Deschampsia flexiosa* and *Urtica dioica*;[537] plants of *Lemna minor.*[268]

Protozoa

Amoeba sp.;[151] *Amoeba chaos chaos*;[502,503] *Tetrahymena pyriformis*;[619] *Crithidia fasciculata.*[273]

Insects

Imago stage of *Blaberus cranifera*;[423] larvae of *Deilephila euphorbiae*;[233–237] *Galeria mellonela*;[555,557] excretions of *Achroca grissela*;[598] *Celleria euphorbiae*;[233] *Galeria melonella.*[553–557,598,828]

Frog embryos

Rana japonica.[662]

Mammals

Brain of rat;[179,181,422] other mammalian tissues;[179,181] nuclei of animal cells;[198] animal liver nuclei;[193,194,588–590] rat liver nuclei.[492]

It is apparent that condensed phosphates have been found in a wide variety of organisms, but it must be pointed out that this list also includes those organisms in which condensed phosphates have been detected cytochemically, and this method, as we have seen, is inaccurate and non-specific. The following may serve as an example. Ebel et al.[151] and Hagedorn[208] demonstrated the presence of condensed phosphates in *Actinomyces griseus* by cytochemical methods. However, both Kokurina et al.[320] and Guberniev and Torbochkina,[203] using more precise chemical methods, were unable to detect them in this organism. As we have stated above, metachromatic staining by toluidine blue, methylene blue, and other similar dyes is a property not only of polyphosphates, but also of other polyanions. For example, granules consisting of poly-β-hydroxybutyric acid and various other ether-soluble lipids have been found in many bacteria.[461,479,495,633,814] These granules undergo metachromatic staining, and are at first sight indistinguishable from volutin. They do not, however, contain polyphosphates, and unlike those which do, the inclusions are surrounded by a clearly defined membrane.[461] There is also the possibility that there are yet other types of granule present in the cells of microorganisms which undergo metachromatic staining even though they do not contain polyphosphates. In any case, the identification of polyphosphates in the cells of a given organism solely on the basis of the presence therein of metachromatic granules must now be regarded as totally unsatisfactory. This has been demonstrated specifically by our observations[320] of the absence of polyphosphates in *Actinomyces streptomicini*, strain LS-3, when grown on a phosphate-rich medium, i.e. under conditions which lead to the formation of a large number of volutin granules. Microscopic observations of a 96-h culture of this organism showed it to contain a large number of volutin granules, but polyphosphates were totally absent, as was shown conclusively by both chemical and chromatographic analyses of the appropriate fractions. It is interesting that both in the alkali extracts and in the residue remaining after the removal of the alkali- and acid-soluble fractions other, extremely labile, phosphorus compounds were found. Following hydrolysis of these compounds by the method of Thilo and Wieker, trimetaphosphate (the characteristic hydrolysis product of inorganic polyphosphates) was not found, but in addition to orthophosphate an unidentified phosphorus compound was present, the mobility of which on paper chromatography in Ebel's acidic solvent was somewhat less than that of pyrophosphate. Visualization with Haynes and Isherwood's reagent followed by treatment with ammonia vapour gave a blue spot of a different colour from those obtained with inorganic ortho- and polyphosphates. We have found similar compounds in association with polyphosphates in other lower organisms. In particular, they were observed in the alkali and salt fractions isolated from the mycelia of the fungus *Neurospora crassa*,[405] and in the acid-soluble fractions from *N. crassa*[405] and *Penicillium chrysogenum*.[392]

These examples provide further evidence that chromatography is the only fully reliable method for the identification of polyphosphates, and the presence of inorganic polyphosphates is most reliably indicated by the presence of trimetaphosphate among the products of partial hydrolysis.

In the lists given above, those species of organisms in which the presence of polyphosphates has been demonstrated by chromatography are marked with an asterisk. As will be seen, the number of such species is very small, particularly among the higher plants and animals, in which until recently there have been few reliable demonstrations of the presence of inorganic polyphosphates.

On the other hand, it is also necessary to treat with caution any conclusions about the absence of polyphosphates from any particular organism. For example, Ebel[129,130] found no polyphosphates in *Escherichia coli* in the stationary growth phase. However, Nesmeyanova *et al.*[541] found them to be present in this organism in the initial stages of logarithmic growth. The compounds were not detected by Hughes and Muhammed[257] in *Aerobacter aerogenes* or *Bacillus cereus* under normal growth conditions, but when the cells of these bacteria were cultured under conditions which were unfavourable to growth, polyphosphates were found to accumulate in significant amounts.[227,600,726] Polyphosphates were not found by Gotovtseva[189] in old mycelia of *Penicillium chrysogenum* when grown on maize extract, but they were present in this organism during all stages of development on synthetic media, and in young mycelia when the fungus was grown on maize extract.[391]

Literature reports of the absence of polyphosphates from various microorganisms (*Streptococcus faecalis*,[220] several algae,[441] some insects,[810] many Actinomycetes,[320] lichens and crayfish,[368] and cells and tissues of higher animals and plants[132,335,467,635]) should therefore not be regarded as proof of their inability to synthesize and accumulate these compounds under suitable conditions of growth. It is likely that improvements in analytical methods and extension of our knowledge of the physiology of these organisms will result in condensed inorganic phosphates being found therein.

The existing data do indicate, however, that inorganic polyphosphates are not invariably to be found among the cell constituents. In some organisms they appear to accumulate only under certain growth conditions and at certain stages of development.

Chapter 4

The Content and Degree of Polymerization
of Polyphosphates in Various Organisms

There can no longer be any doubt that inorganic polyphosphates, including high-molecular-weight polyphosphates, are to be found at certain stages of development and under suitable conditions in most, if not all, lower and higher animals and plants. Their presence has been detected and their amounts determined with varying degrees of reliability in various groups of organisms. It may nevertheless be said that, in general, inorganic polyphosphates occur in relatively large amounts in the lower organisms (bacteria, fungi, and algae), whereas in the higher plants and animals they are usually present in very small, sometimes scarcely detectable, amounts. For this reason, neither the content nor the degree of polymerization of condensed phosphates in higher organisms has yet been investigated with anything like sufficient thoroughness.

An adequate amount of data of this type is now available for microorganisms, however, and this can be summarized to a certain extent, although it is true that this is rendered more difficult by the fact that different methods of fractionation have been employed by different workers in determining the content and degree of polymerization of the polyphosphates present in various lower organisms. For this reason, it is preferable to consider this question by first summarizing the data obtained using each method of fractionation in turn.

Table 2 presents data obtained by Hughes and Muhammed[257] for the contents of acid-soluble and acid-insoluble polyphosphate in various microorganisms. These workers extracted the acid-soluble polyphosphates with 5% trichloroacetic acid (TCA) in the cold, and the acid-insoluble polyphosphates by keeping for extended periods at room temperature in 10% TCA.

These results show that acid-insoluble polyphosphates were present in all of the microorganisms studied, except for *Aerobacter aerogenes* and *Bacillus*

Table 2. Amounts of acid-soluble and acid-insoluble polyphosphates in the cells of various microorganisms[257]

Organism	Polyphosphates (μM/g dry weight)	
	Acid-soluble	Acid-insoluble
Baker's yeast (commercial)	0.00	0.19
Brewer's yeast (commercial)	4.58	3.37
Corynebacterium xerosis	0.00	10.01
Mycobacterium smegmatis	0.05	0.01
Aerobacier aerogenes	0.00	0.00
Mycobacterium phlei	0.08	0.01
Bacillus brevis	0.00	0.02
Bacillus megaterium	0.01	0.01
Bacillus pradigiosus	0.40	0.01
Bacillus ccreus	0.00	0.00
Clostridium sporogenes (grown on glucose, NH_4^+, phosphorus, and broth)	0.00	0.02
Mycobacterium tuberculosis (BCG stain, grown on broth)	0.00	0.01
Loctobacillus casei (grown anaerobically on broth)	0.13	0.01
Staphylococcus albus (grown on broth)	5.9	3.39

cereus, in which they were subsequently found by other workers[227,600,726] when conditions unfavourable to the growth of these organisms were employed. Acid-soluble polyphosphates were found, however, by Hughes and Muhammed in only half of the organisms which they studied.

Table 2 shows that the amount of polyphosphate present in the cells of microorganisms can vary very considerably, occasionally reaching very high values (for example, in brewer's yeast and *Staphylococcus albus*).

Thorough investigations have shown that acid-soluble condensed phosphates are much less highly polymerized than the acid-insoluble fractions. Table 3, taken from a review by Yoshida,[834] gives the mean values for the length of the polyphosphate chains ($\bar{n}$) for acid-soluble and acid-insoluble fractions obtained by different methods. It can be seen that, depending on the species of organism, the acid-soluble fractions contain polyphosphates with n values from 3 to 20, whereas the degree of polymerization of the acid-insoluble polyphosphates is much higher. For material extracted with dilute alkali, the degree of polymerization was 60–85.

Using the method developed by Langen and Liss, it has proved possible to separate the cellular condensed phosphates from brewer's yeast into four fractions having mean chain lengths of 4, 20, 55, and 260 phosphoric acid residues, respectively.[443,464,465,468] The results obtained by these workers are summarized in Table 4, from which it can be seen that polyphosphates account for over a third of the total phosphorus present in the cells of ordinary brewer's yeast.

Using this method, we have determined the polyphosphate content of some

Table 3. Degree of polymerization ($\bar{n}$) of acid-soluble and acid-insoluble polyphosphate fractions in various inicroorganisms[834]

Organism	Method of extraction	Method of determination of mean chain length*	Mean chain length, $\bar{n}$	References
Acid-soluble fraction				
Saccharomyces cerevisiae	H$_2$O	1	13	130, 131
	5% TCA	1	10	833
	5% TCA	2	7	833
	1% TCA	5	$n = 3$ (50%) $n = 4$ (30%) $n = 5$ (15%)	468
	5% TCA	3	16	299
	0.5 M HClO$_4$	6	13	497
Gliocladium roscum	5% TCA	5	10	534
Fusarium coeruleum	5% TCA	5	10	534
Mycothecium verrucaria	5% TCA	5	10	534
Neurospora crassa	0.5 M HClO$_4$	1	20	213
Acid-insoluble fraction				
Saccharomyces cerevisiae	Dilute alkali (pH.9)	1	50	833
	Dilute alkali (pH 9)	2	60	833
	Dilute alkali (pH 9)	3	85	299
Aspergillus niger	Dilute alkali (pH 9)	4	60–80	267

*Methods employed: 1, potentiometric filtration; 2, viscosity determination; 3, determination of rate of diffusion; 4, ultracentrifugation; 5, paper chromatography; 6, chromatography on ion-exchange resin.

Table 4. Degree of polymerization of polyphosphate fractions from brewer's yeast[447]

Fraction No.	Extractant (at 0 °C)	Polyphosphate content mg/g dry weight	% of total phosphorus	Mean chain length, $\bar{n}$	Mean molecular weight of the potassium salt
1	1% TCA	0.54	6	4	530
2	Saturated NaClO$_4$	1.74	21	20	2 400
3	NaOH$_1$ H10	0.33	4	55	6 500
4	0.05 N NaOH	0.54	6	260	30 700
	Total polyphosphates	3.15	37	—	—

fungi. The data presented in Tables 5–7 show that all of the species examined were found to contain greater or lesser amounts of inorganic polyphosphate. To judge from the acid-labile phosphorus of the various fractions, the polyphosphate-phosphorus content of the cells of certain fungi (budding yeasts, mycelia, sclerotia, spores, and fruiting bodies) range from 2 to 10 mg per gram dry weight.

The polyphosphate-phosphorus content of strongly sporulating mycelia and

Table 5. Polyphosphate content of cells of *Endomyces magnusii*,[384] *Neurospora crassa*,[405] and of the fruiting bodies of *Giramitra esculenta*,[390] expressed as mg P/g dry weight

Polyphosphate fraction	Extractant	*E. magnussi,* cells, 12 h growth	*N. crassa,* mycelia, 17 h growth	*G. esculenta,* fruiting bodies*
PP₁	0.5 M HClO₄, 0–4 °C	1.10	0.62	0.00
PP₂	Saturated NaClO₄ solution, 0–4 °C	0.90	1.24	1.52
PP₃	NaOH, pH 9, 0–4 °C	0.20	0.12	0.24
PP₄	NaOH, pH 12, 0–4 °C	0.90	0.82	0.01
PP₅	10% HClO₄, 100 °C	0.40	0.00	—
Total polyphosphates	—	3.50	2.80	1.77
Total phosphorus	—	17.30	15.6	6.03

*Instead of cold 0.5 M HClO₄, extraction was carried out with 1% TCA at 0–4 °C.

Table 6. Polyphosphate content of resting spores and mycelia of a 24-h culture of *Penicillium chrysogenum* under different conditions of growth.[391] Results in mg P/g dry weight

Polyphosphate fraction	Extractant	Spores	Mycelia on synthetic medium	Mycelia on medium containing maize extract
PP₁	1% TCA, 0–4 °C	4.90	0.09	1.07
PP₂	Saturated NaClO₄ solution, 0–4 °C	2.18	1.75	1.01
PP₃ + PP₄ + PP₅	10% HClO₄, 100 °C	0.44	8.18	1.96
Total polyphosphates	—	7.52	10.02	4.04
Total phosphorus	—	16.15	28.28	44.27

of the spore-forming organs (fruiting bodies) is shown by chemical analysis to amount to 30–50% of the total phosphorus present. Most of the polyphosphates in these cells and tissues are lower polyphosphates, extracted by cold acid and saturated salt solutions.

The most highly polymerized, acid-insoluble polyphosphates are present in comparatively large amounts in actively growing mycelia (for example, in *Penicillium chrysogenum* and *N. crassa*). The polyphosphate content is highly dependent on the growth conditions and stage of development of the fungi (see, for example, Table 6). This has been amply demonstrated in the case of *N. crassa*.[86,357,405] The results presented in Fig. 13 show the extent to which the contents of the different polyphosphate fractions may change during the

Table 7. Polyphosphate content of spores and fruiting bodies of the mushroom *Agaricus bisporus* at various stages of development.[390] Results in mg P/g dry weight

Polyphosphate fraction	Extractant	Fruiting bodies			Spores leaving the fruiting bodies
		1.5 day's growth (undifferentiated)	3–4 days' growth (fully differentiated)	8 days' growth (vigorously sporulating)	
PP_1	1% TCA, 0–4 °C	1.69	1.15	2.38	0.24
PP_2	Saturated $NaClO_4$ solution, 0–4 °C	1.76	1.10	2.56	0.97
PP_3	NaOH, pH9, 0–4 °C	0.31	0.27	0.61	0.44
PP_4	NaOH, pH 12, 0–4 °C	0.11	0.10	0.09	0.09
PP_5	10% $HClO_4$, 100 °C	0.18	0.02	0.03	0.24
Total Polyphosphates	—	4.05	2.64	5.67	1.98
Total phosphorus	—	11.44	8.80	11.28	5.03

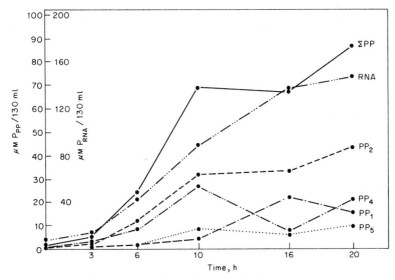

Figure 13. Changes in the content of different polyphosphate fractions and of RNA during the growth of a culture of *N. crassa* (in μM of PO_4^{3-} per 130 ml of medium[88]

growth of this fungus. Chernysheva *et al.*[88] have shown that in *N. crassa*, as in yeasts, these fractions contain polyphosphates of different chain lengths. It can be seen from Table 8 that the mean chain length $(\bar{n})$ of the polyphosphates extracted from *N. crassa* by cold $HClO_4$ is 19 orthophosphoric acid residues. This value is almost the same as that found previously by Harold $(\bar{n} = 20)$ for this polyphosphate fraction in *N. crassa*.[213] The mean chain length of the salt-soluble polyphosphates in *N. crassa*[88] is 33, i.e. it falls within the range of values of $\bar{n}$ found in the comparable fraction from yeasts (Table 4)[447] and the simple flagellate *Crithidia fasciculata* $(\bar{n} = 50–80)$[273]. However, the higher molecular weight polyphosphate fractions obtained by us on extraction of *N. crassa* with 1% dodecyl sulphate solution has a mean chain length of 151 orthophosphate residues. It is noteworthy that the polyphosphate fraction extracted from *N. crassa* with 1% dodecyl sulphate solution is almost the same as the sum of fractions PP_3 and PP_4 extracted by the Langen and Liss method using alkaline solutions from the cells of yeasts and other organisms. This is shown, for

Table 8. Degree of polymerization of polyphosphate fractions extracted from *Neurospora crassa*[88]

Polyphosphate fraction	Extractant	Mean chain length, $\bar{n}$	Mean molecular weight of the potassium salt
PP_1	0.5 M $HClO_4$, 0–4 °C	19	2 300
PP_2	Saturated $NaClO_4$ solution, 0–4 °C	33	5 000
$PP_3 + PP_4$	1% dodecyl sulphate solution in 0.02 M Tris–HCl buffer, pH 7.4	151	18 000

Table 9. Polyphosphate content of sclerotia of ergot, *Claviceps paspali*, at various stages of growth,[430] and in the mycelia and fruiting bodies of the basidial fungus *Lentinus tigrinus*,[354] expressed as mg P/g dry weight

Fractionation of polyphosphates	C. paspali*		L. tigrinus	
	10 days, first growth stage	10 days, second growth stage	Mycelia, 10 days' growth	Fruiting bodies
0.5 M HClO$_4$, 0–4 °C	0.71	2.10	—	—
0.5 M KOH, 37 °C, 18 h (Schmidt and Thannhauser RNA fraction)	0.51	0.45	0.3–1.0	0.3–1.8
0.5 M KOH, 37 °C, 18 h (Schmidt and Thannhauser DNA fraction)	0.05	0.14	—	—
Total polyphosphates	1.27	2.69	—	—
Total phosphorus	14.49	12.20	—	—

*The material was treated with 50% ethanol at 50 °C prior to extraction with cold 0.5 M HClO$_4$.

example, by comparing the degrees of polymerization of the corresponding polyphosphate fractions from *N. crassa* and yeasts (see Tables 4 and 8).

In our work on the detection of polyphosphates and investigations of the variations in their content in various fungi, other methods of fractionation have also been employed. In particular, in an investigation of the sclerotia of ergot[430] and the mycelia and fruiting bodies of the basidial fungus *Lentinus tigrinus*,[354] we made use of the method of Schmidt and Thannhauser, which is normally employed for the separative isolation of RNA and DNA. As shown in Table 9, we found in these organisms approximately the same amounts of polyphosphate as in other fungi. It is also important to note that our results are in good agreement with those of other workers. Thus, Harold[213,215] has shown that young mycelia of *N. crassa* contain 4 mg of polyphosphate phosphorus per gram dry weight of material, i.e 0.4%. Both our own work and that of others[9,272] shows that the proportion of these compounds often amounts to 40–60% of the total phosphorus in the cells of microorganisms grown in the laboratory under normal culture conditions. Under special conditions of growth (culture on a medium containing phosphorus and glucose following phosphorus starvation), brewer's yeast accumulates large amounts of polyphosphates, amounting to 20% of the dry weight of the cells.

It is important to note that in most of the organisms in which the amounts of the various polyphosphate fractions present have been investigated in any detail, both high- and low-polymeric polyphosphates are invariably found. Thus, chromatographic examination of low-polymeric polyphosphates in yeasts[106,108,129,330,332,384,468,497,609,779] and other fungi,[390,391,405,409,430,534] and also in some algae[524] and bacteria,[118,425,709,794] showed them to contain linear polyphosphates with $\bar{n}$ values from 2 to 15, together with the cyclic trimetaphosphate. Examples which may be cited are the chromatograms obtained in the

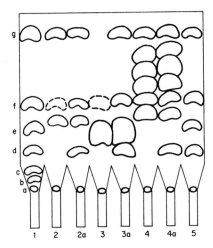

Figure 14. Composite diagram of the chromatographic separation of polyphosphates isolated from *Neurospora crassa*. Solvent: isopropanol–water–TCA–20% ammonia (75:2:5:0.3), pH 1.5. 1, Reference spots, (a) high-molecular-weight polyphosphates ($\bar{n} \geq 10$), (b) hexapolyphosphate, (c) tetrapolyphosphate, (d) trimetaphosphate, (e) tripolyphosphate, (f) pyrophosphate, (g) orthophosphate; 2, polyphosphate fraction obtained by cold perchloric acid extraction (Ba precipitate, pH 4.3); 2a, the same after hydrolysis; 3, polyphosphate fraction from the salt extract (Ba precipitate, pH 4.3); 3a, the same after hydrolysis; 4, polyphosphate fraction from the alkali extract (Ba precipitate, pH 3.0); 4a, the same after hydrolysis, 5, residual material from the above extractions, after hydrolysis[405]

separation of the polyphosphates from the mould fungus *Neurospora crassa* (Fig. 14) and the mushroom *Agaricus bisporus* (Fig. 15). Although there is no longer any doubt as to the presence of linear polyphosphates in the cells, the same cannot be said of trimetaphosphate.

In contrast, most workers[106,108,363,390,506,745] incline to the view that trimetaphosphate is formed during the extraction of high-molecular-weight linear condensed phosphates from the cells, perhaps as a result of the partial hydrolysis of the high-polymeric polyphosphates.[740] However, the contrary view has also been put forward,[609] according to which trimetaphosphate is not only *not* a hydrolysis product of polyphosphates, but also plays a specific part in the metabolism of the living cell.

The presence of pyrophosphate in the acid-soluble polyphosphate fraction has been demonstrated in all of the fungi which we have examined. It is present in particularly large amounts in strongly sporulating mycelia of *Penicillium chrysogenum*, in old, actively sporulating fruiting bodies of mushrooms (Fig. 15), and in old, alkaloid-prodicing cultures of ergot. The presence of pyrophosphate in fungal cells is unexpected. Prior to our work, it was considered to be well established[442,469] that in yeasts, as in all other organisms, pyrophosphate

30

Figure 15. Chromatograms of polyphosphates isolated from the spore-bearing fruiting bodies of the mushroom *Agaricus bisporus*. Chromatogram A: I, orthophosphate (1) and Graham's salt (2); II, tripolyphosphate (3); III, polyphosphates from the 1% TCA extract (6 is the tetrapolyphosphate); IV, pyrophosphate (4) and tetrametaphosphate (5). Chromatogram B: I, orthophosphate (1) and tripolyphosphate (3); II, pyrophosphate (4) and trimetaphosphate (5); III and IV, polyphosphates from the salt extract; V, mixture of orthophosphate and polyphosphates from pyrophosphate to heptapolyphosphate (9); pentapolyphosphate (7), and hexapolyphosphate (8)[390]

was not able to accumulate in any appreciable amounts. It was thought[247,329] that, on being formed in a variety of biosynthetic reactions, it must be hydrolysed immediately to orthophosphate by pyrophosphatase, since it is highly toxic to the cell.[469] Direct determinations of the amounts of pyrophosphate present in the cells of yeasts carried out by Mansurova *et al.*[488] have shown, however, that pyrophosphate accumulates in the cells of these organisms in amounts which are comparable to and which, as is shown in Table 10, in some instances even exceed those of ATP.

The presence of significant amounts of pyrophosphate in the cells of many organisms is also indicated by the results of other workers. For example, Rosenberg[619] isolated from cells of *Tetrahymena pyriformis* specialized granules consisting of Ca–Mg pyrophosphate. Similar granules could probably also occur in other organisms, especially in fungi, since they are valuable

Table 10. Pyrophosphate and ATP contents of the facultative anaerobic yeast *Saccharomyces carlsbergensis*, grown on glucose under aerobic and anaerobic conditions (μM/g dry weight)[488]

Growth conditions	ATP	Pyrophosphate	Pyrophosphate/ATP
Aaerobic, 0.8% glucose	14.10	1.15	0.08
Anaerobic, 0.8% glucose	0.20	1.39	7.0
Anaerobic, 2% glucose	0.28	5.12	18.3

reserves of active phosphate, Ca^{2+}, and Mg^{2+}. The localization of pyrophosphate in such granules may also provide a convenient method of detoxification.

In addition to pyrophosphate and trimetaphosphate, the acid-soluble fraction from mushrooms (Fig. 15) and other organisms has been found to contain tripolyphosphate. In most cases, in addition to these comparatively low-molecular-weight poly- and metaphosphates, this fraction also contains varying amounts of more highly polymerized polyphosphates having $\bar{n} \leq 10$, which do not move from the origin under the chromatographic conditions used (see Figs. 14 and 15). However, the precise amounts of tripolyphosphate have been determined in only one instance. Thus Egorov and Kulaev[156] showed that in *N. crassa* mycelia the amount of tripolyphosphate present is equal to that of pyrophosphate, and is lower by a factor of ten than the ATP content. All of the results obtained for the content and degree of polymerization of condensed phosphates in the various microorganisms were obtained by using fractionation methods which involved at some stage treatment of the biological material with acids and alkalis. From what is known of the chemical properties of polyphosphates, however, (see Chapter 1) such treatment might be expected to result in a considerable amount of depolymerization. Several investigations have shown experimentally that this does, in fact, take place.[106,108,552] Further confirmation of the possibility of some depolymerization occurring during extraction of the cells with dilute acids and alkalis is provided by the fact that the maximum figures for the mean degrees of polymerization of these compounds were obtained by using a method of fractionation which differs basically. Thus, Harold,[217] who extracted the polyphosphates by preliminary lysis of the cells with a 1.8 M solution of hypochlorite followed by extraction with cold water, was able to isolate from *Aerobacter aerogenes* a macromolecular fraction having a mean degree of polymerization $\bar{n} = 500$–600. This fraction was precipitated with alcohol from a 1.5 M aqueous solution of NaCl, and it comprised 50–80% of the total polyphosphates in the cells of this organism. A very high-molecular-weight fraction with mean chain length $\bar{n} = 850$ has also been obtained using this method, from the simple flagellate *Critidia fasciculata*.[273]

It has already been pointed out that information on the content and degree of polymerization of polyphosphates in higher plants and animals is extremely fragmentary, and is to a large extent of a preliminary nature. This is due primarily to the extremely small amounts of these high-energy phosphorus compounds present in relatively highly organized organisms, together with the inadequate sensitivity and reliability of the methods available for their determination.

Although there are a number of reports in the literature of the presence of inorganic polyphosphates in plants and animals (see p. 19), they have been rigorously identified in only a limited number of cases.

There are only four reliable reports of the presence of these compounds in the tissues of plants and animals, i.e. those in which the presence of cyclic trimetaphosphate was detected among the products of partial acid hydrolysis (in maize root tips,[769] cotton seeds,[775] rat brain,[422] and rat liver nuclei,[492] although they have been shown with reasonable reliability to be present in certain other plant[253,520] and animal[179,194] specimens.

In the first investigations in which more or less reliable proof of the presence of inorganic polyphosphates in animal[194] and plant[520] tissues was obtained, quantitative estimates were also made of the amounts present in these subjects.

Thus, according to Miyachi,[520] 1 g dry weight of spinach leaves contains 40 μg of polyphosphate phosphorus, while Griffin et al.[194] found rat liver to contain only 1–2 μg of phosphorus per gram wet weight which could be ascribed to inorganic polyphosphate. It has recently been found in our laboratory that the high-molecular-weight polyphosphate content of rat liver nuclei amounts to 100–200 μg per gram dry weight of nuclei.

Table 11 summarizes the data available in the literature on the inorganic polyphosphate content of animal tissues. As a consequence of the small amounts of inorganic polyphosphate present in plant and animal tissues, their content is usually determined in the total acid-insoluble fraction extracted from the cells after removal of the acid-soluble and alcohol-soluble phosphorus compounds with 0.5–1.0 N KOH by Schmidt and Thannhauser's method.[643] In particular, this is the method used in most of the investigations the results of which are given in Table 11. It can be seen from Table 11 that the inorganic polyphosphate content of whole cells varies over an order of magnitude. This variation is probably due to differences in the conditions under which the tissues were fixed after the animals had been killed.

The investigations of Gabel and Thomas[181] and Kulaev and Rozhanets[422] have shown that the highest figures for the polyphosphate content of rat brain are obtained when the brain is frozen in liquid nitrogen immediately following

Table 11. Content of high-molecular-weight polyphosphates in some animal tissues

Tissue	Total phosphorus (mg/g)	High-mol.-wt. polyphosphate phosphorus (μg/g wet wt.)	Reference
Rat			
Liver nuclei	—	100–200	492
Liver	—	1–2	194
Liver	—	3–5	181
Brain	—	12.8–15	181
Brain	—	10–15	422
Ox			
Brain	1.16	2.95	181
Liver	1.18	2.24	181
Pancreas	1.19	2.10	181
Kidney	1.18	1.89	181
Spleen	1.35	1.98	181
Rabbit			
Erythrocytes	0.74	7.48	181

decapitation of the animal. In all other cases, lower figures were obtained. The lower figures found by Gabel and Thomas for various tissues of the ox in comparison with the corresponding values for the same tissues in the rat appear to be due to the less favourable conditions for the freezing of the larger animal. The most interesting conclusion to be drawn from the figures given in Table 11, from our point of view, is the relatively high polyphosphate content of the cerebral tissues and erythrocytes of the animals.

It is important to note that approximately the same (very small) amounts of polyphosphate have been found by Kulaev et al.[423] in certain insects, not only in the intact insects (tropical cockroaches), but also after removal of the intestines (Table 12). Hence, their presence in these subjects does not appear to be related to the intestinal microflora. It is interesting to note that the amounts present in insects are less than those of ATP.

It is particularly noteworthy that the polyphosphate content of rat liver nuclei is about 100 times greater than that of whole liver tissue.[194,492] It is possible that in animals these compounds are mainly present in the nuclei. It is also interesting that when the proteins of rat liver nuclei are fractionated, polyphosphates are found in the globulin, histone, and acidic protein fractions.

The amounts of inorganic polyphosphate normally present in the tissues of higher plants also appear to be very small. In addition to the previously cited work of Miyachi,[520] attention is drawn to the findings of Valikhanov and Asamov[12,773–775] who determined the inorganic polyphosphate content of cotton seeds, both when ripe and during the ripening process. Some of their results are given in Table 13.

It can be seen from these results that fairly large amounts of polyphosphate accumulate at an early stage of ripening (10 days). At this time, polyphosphate-phosphorus accounts for more than 10% of the total phosphorus present in the seed, and it exceeds the phytin phosphorus by a factor of more than two. Later, as the seeds ripen, their polyphosphate content falls to low, scarcely detectable, levels, while the phytin content steadily increases. These findings will be discussed in greater detail below, in considering current views of the physiological functions of high-molecular-weight polyphosphates. At this point, we merely note that appreciable amounts of inorganic polyphosphates may accumulate

Table 12. Inorganic polyphosphate content of certain insects (μM P/g body weight)

Phosphorus compounds	Mulberry moth larva	Mosquito larva	Image of granary weevil	Tropical cockroach	
				Intact insect	Insect without intestines
Orthophosphate	13.30	9.05	13.40	8.40	3.26
Acid-soluble polyphosphates	0.04	0.09	0.06	0.45	0.51
Acid-insoluble polyphosphates	0.17	0.01	0.005	0.03	0.03
ATP	2.05	2.00	2.46	—	—

Table 13. Inorganic polyphosphate and phytin content of ripening seeds of cotton[12] (μg P per 100 seeds)

Phosphorus compounds	Age of seeds (days)						
	10			40			70
	Whole seeds	Embryos	Cotyledons	Whole seeds	Embryos	Cotyledons	Whole seeds
Polyphosphates	120	20	60	80	0	30	30
Phytin	50	1740	11 200	12 940	3200	32 600	35 800
Total phosphorus	1120	4300	39 500	43 800	6200	56 200	62 400

in higher plants at certain stages of development and in specific organs and tissues, the amounts present being comparable to those of other phosphorus compounds which are characteristic of these organisms.

However, the amounts of polyphosphates which have so far been found in higher plants, as in animals, are far from sufficient to permit their degree of polymerization to be established with acceptable accuracy. The available literature data on this point are extremely scanty, and appear to be unreliable. They are all concerned with polyphosphates isolated from animal tissues.[179,181,194] In all of these investigations, the degree of polymerization was determined by the method of Ohashi and Van Wazer,[567] which involves chromatographic separation of the polyphosphates. This method permits some assessment to be made of the degree of polymerization of the polyphosphates from their mobility in the chromatograms. It is, however, strictly applicable only to the pure sodium salts of the polyphosphoric acids.[567] The assessment of the degree of polymerization of polyphosphates isolated from biological specimens from the results of chromatographic separation is wholly inadmissible. It is likewise completely erroneous to draw any conclusions as to the presence of high-molecular-weight polyphosphates in biological materials and extracts solely from the observation that when the phosphorus compounds are subjected to paper chromatography, some phosphates remain at the origin. This is due principally to the fact that it is not possible to remove completely alkaline earth and heavy metals from preparations of biogenic polyphosphates.[745] In their presence, for example, the usually highly mobile tripolyphosphate remains at the origin in the form of a polymeric, metal-containing compound. Further, work by Okorokov and co-workers[418,419,568,572] has shown that magnesium, calcium, manganese, cobalt, and iron orthophosphates are present in biological materials in substantial amounts, in the form of polymers with molecular weights in excess of 400 000. These phosphate complexes also remain for the most part at the origin in paper chromatography. It follows that the determination of the mean chain lengths of polyphosphates by the method of Ohashi and Van Wazer,[567] i.e. by estimating the degree of polymerization from chromatographic mobility, is clearly invalid.

It follows that the very high values found first by Griffin *et al.*[194] and later by Gabel[179,181] for the mean chain lengths of polyphosphates in animal tissues using Ohashi and Van Wazer's method[567] are, in our view, of little validity, and require careful checking and, it would appear, substantial correction.

In order to obtain an accurate estimate of the degree of polymerization of polyphosphates in higher organisms, it will be necessary either to develop a sufficiently accurate micro-method for the determination of mean chain lengths, or to carry out large-scale preparative work in order to obtain sufficient material from the tissues of higher plants and animals to determine the degree of polymerization from viscosity, or by means of potentiometric titration.

Chapter 5

The Form in which High-Molecular-Weight Polyphosphates are Present in the Cell. The Problem of the in vivo Existence of Polyphosphate–Ribonucleic Complexes

Research workers have long been interested in the question of whether poly-phosphates are present in the cell in the free or bound state.

As far as the acid-soluble polyphosphate fraction is concerned, there is usually no disagreement.[145,388,443] Since the classical work of MacFarlane,[477] it has been regarded as firmly established that this comparatively low-mole-cular-weight polyphosphate fraction is present in the cell in the free state. However, the question as to the state of the more highly polymerized, acid-insoluble polyphosphates within the cell cannot yet be considered as having been finally resolved. It seems very likely that this polyphosphate fraction is bound in some way to other cellular components. What these components may be is at present difficult to say, despite the many attempts which have been made over the last 30–40 years to resolve this question.

Some workers[15,48,50,51,60,85,91–94,143,286,335,336,365,477,785] consider the acid-insoluble polyphosphates to be present in the cell in combination with nucleic acids, pri-marily RNA. Others have suggested the possibility that the more highly poly-merized polyphosphates are bonded within the cell to proteins[445,465,469,809] or to polysaccharides of the polyhexosamine type.[214,220,226] Ashmarin et al.[14] and others[846] have suggested that temporary, or even permanent, bonding of a cer-tain polyphosphate fraction to histones and desoxyribonucleoproteins of the cell nucleus may take place.

However, most attempts to settle the question of the state of the higher poly-meric acid-insoluble polyphosphates within the cell have involved model ex-periments, and an examination of the complexing ability of the polyphosphates with respect to various possible cell partners.[83,84,214,300,475] This type of approach has not proved fruitful, and is unlikely to contribute to the understanding of the true state of the highly polymerized polyphosphates within the cell.

The greatest efforts of the workers who have attempted to resolve this problem have been concentrated on examining the polyphosphate–ribonucleic acid complexes which have frequently been isolated from the cells of a variety of organisms. It appeared at one time to be very likely that such complexes were actually present in the cell. This question has, in fact, been discussed in the literature over a very long period, virtually from the time when condensed phosphates were first found in biological materials. It became the subject of especially intensive investigations in the 1940s and 1950s, in connection with the physiological functions of the nucleic acids. At that time many workers, including Brachet, Spiegelman, Belozersky, Ebel, and Bresler, considered it likely that RNA phosphorylated by polyphosphates was a specific energy donor in protein biosynthesis (a unique 'macro-ATP'). Subsequent developments in the understanding of the physiological functions of the nucleic acids, in particular the elucidation of the principal ways in which they are involved in protein synthesis, resulted, however, in a decrease in interest in the possibility of the actual existence of polyphosphate–ribonucleic acid complexes. Nevertheless, if the physiological significance of inorganic polyphosphates in the metabolism is to be explained, a solution of this question remains important.

What are the principal facts on which the hypothesis of the actual existence and functioning of polyphosphate–ribonucleic acid complexes in the cell is based? Firstly, whatever the method used to isolate acid-insoluble polyphosphates from the cell, they are always accompanied by nucleic acids, particularly RNA. Acid-insoluble polyphosphate fractions always contain a certain amount of RNA, and conversely inorganic polyphosphates have been found to be present in many RNA fractions derived from microorganisms. The proportion of polyphosphate to RNA in these materials depends to some extent on the proportions in which they are present in the cells from which they were extracted. This is well demonstrated by the results of Ebel et al.,[153] shown in Table 14.

It can be seen from Table 14 that RNA fractions obtained from cells which were almost completely devoid of polyphosphates contained only traces of polyphosphates. From yeasts which were harvested in the stationary growth phase, when polyphosphates had begin to accumulate in the cells, fractions were obtained which contained 0.2 polyphosphate phosphorus atom per nucleotide phosphorus atom. Finally, fractions were obtained from yeasts enriched with polyphosphates by Jeener and Brachet's method,[275] in which each RNA nucleotide phosphorus atom was accompanied by 4–6 polyphosphate phosphorus atoms.

These polyphosphate-nucleic acid complexes have been thoroughly investigated in the laboratories of Belozersky[50,51,53,60,335,365,368,386,387] and other workers.[83–85,91–94,111,113,141–144,442,445,447,464,465,536,676,682,685,785,795].

All of these investigations sought to resolve the question of whether the RNA was combined with the polyphosphates, or whether they were simply coprecipitated during extraction and separation from the cells as a result of the similarities in their chemical and physicochemical properties (RNA and poly-

38

Table 14. Phosphorus content of RNA fractions obtained from yeast containing different amounts of polyphosphate, depending on the conditions of growth[153]

Experiment No.	Phosphorus-deficient (very little polyphosphate)		Normal conditions, logarithmic phase		Normal conditions, stationary phase		Phosphorus-enriched medium	
	RNA-P (% found)	$\frac{\text{RNA-P}_{found}}{\text{RNA-P}_{calc.}}$ *	RNA-P (% found)	$\frac{\text{RNA-P}_{found}}{\text{RNA-P}_{calc.}}$ *	RNA-P (% found)	$\frac{\text{RNA-P}_{found}}{\text{RNA-P}_{calc.}}$ *	RNA-P (% found)	$\frac{\text{RNA-P}_{found}}{\text{RNA-P}_{calc.}}$ *
1	7.49	1.06	7.35	1.07	9.00	1.24	18.80	5.78
2	7.14	1.06	7.29	—	8.25	1.18	16.60	4.46

*Phosphous content calculated from the ribose content.

phosphates are both polymeric phosphorus-containing compounds, bearing on their surfaces large numbers of negatively charged groups). The solution of this problem has been found to present great difficulties. It was found impossible to separate these compounds completely by precipitation and re-precipitation in the presence of Ba^{2+}, Ca^{2+}, Mg^{2+}, or other metal ions.[51,52,58,83,113,335] The use of widely different conditions for the separation of RNA (by the use of dodecyl sulphate,[111,113,139,143,153] by the phenol method,[91,93,94,143,682,785] or by a combination of these methods[85,795] from yeasts and other sources, the cells of which contained substantial amounts of polyphosphates, failed to afford pure RNA fractions. In all cases, polyphosphates were also present.

It has proved especially difficult to separate polyphosphates from such RNA fractions when they contain relatively small amounts of polyphosphate. For example, in our own work,[387] carried out with a similar RNA–polyphosphate complex obtained from A. niger, electrophoretic analysis of the fraction in a Tiselius apparatus did not result in its separation into its constituent parts. The same conclusion was reached when paper chromatography was used in an attempt to separate polyphosphate from RNA in a polyphosphate–nuclein complex from yeast, in which RNA predominated.[85] When fractions which contained large amounts of polyphosphate were examined, however, it was found possible to separate the latter from RNA either by electrophoresis,[60,85,113,143,368,] or by paper chromatography.[111,143,153] In particular, we have carried out an electrophoretic examination of three polyphosphate-nuclein complex fractions in a Tiselius–Svensson apparatus.[60] This series of experiments was carried out on one polyphosphate–RNA complex from brewer's yeast (after preliminary purification by electrophoresis on a cellulose column) which had a polyphosphate to RNA ratio of 1:7, and on two fractions obtained from baker's yeast (without further purification) having polyphosphate to RNA ratios of 1:4 and 1:9.

Parallel experiments were also carried out using artificial mixtures of polyphosphates and RNA, with various ratios of the components. Electrophoresis was carried out in an acetate buffer at pH 4.5–4.7, ionic strength 0.04, temperature 2 °C, current 9 mA, and potential gradient 6–7 V/cm. In all of the experiments, the mean electrophoretic mobilities of the polyphosphates and RNA were calculated, together with the approximate ratios of polyphosphates to RNA, from the areas under the peaks on the electrophoretograms (Figs. 16 and 17). The results obtained by electrophoresis in a Tiselius apparatus are given in Table 15, and in Figs. 16 and 17.

It can be seen from Table 15 and Fig. 16A that electrophoresis of the polyphosphate–RNA complex from brewer's yeast, following preliminary purification by electrophoresis on a cellulose column, gave only a single, symmetrical peak on the electrophoretogram. This would appear to indicate the presence in this fraction of a homogeneous polyphosphate–RNA complex, but this particular experiment had the disadvantages that the fraction was present in a very low concentration, and that only a small amount of polyphosphate was present therein.

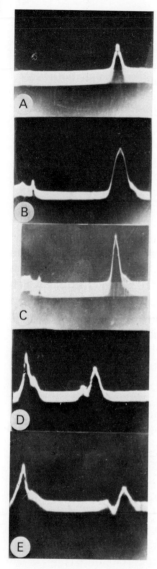

Figure 16. Electrophoretograms of the descending boundary obtained in a Tiselius–Svensson apparatus in an acetate buffer at pH 4.5–4.7 and ionic strength 0.04, at 2 °C. A, polyphosphate–RNA complex from brewer's yeast (fraction 1, Table 15) after 95 min, $F = 6.25$ V/cm; B, polyphosphate–RNA complex from baker's yeast (fraction II, Table 15) after 100 min, $F = 6.33$ V/cm; C, the same complex after preliminary dialysis in the presence of versen, after 106 min, $F = 6.40$ V/cm; D, polyphosphate–RNA complex from baker's yeast (fraction III, Table 15) after 77 min, $F = 6.62$ V/cm; E, the same fraction after preliminary dialysis with versen, after 102 min, $F = 6.70$ V/cm[60]

In view of these deficiencies, experiments were subsequently carried out using polyphosphate–RNA complexes from baker's yeast containing larger amounts of polyphosphate. Further, in each of these experiments a several times greater amount of the fraction was taken. It can be seen from Fig. 16B and Table 15 that the electrophoretograms from these experiments showed two fairly mobile peaks. When the fractions contained more RNA and less polyphosphate (cf. Fig. 16B and C), the first of these two peaks was found to be small. Electrophoresis of a fraction in which the polyphosphate content was much greater than that of RNA gave a first peak which was several times larger

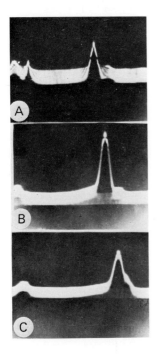

Figure 17. Electrophoretograms of the descending boundary obtained in a Tiselius–Svensson apparatus in an acetate buffer at pH 4.65–4.70 and ionic strength 0.04, at 2 °C: A, 3:1 mixture of yeast RNA (Merck) and polyphosphate from brewer's yeast ($\bar{n} = 30$) after 78 min, $F = 6.48$ V/cm; B, the same yeast RNA–brewer's yeast polyphosphate ($\bar{n} = 30$) mixture in proportions of 6:1, after 90 min, $F = 6.56$ V/cm; C, a mixture of yeast RNA with synthetic polyphosphate, ($\bar{n} = 75$), in proportions of 3:1, after 105 min, $F = 6.21$ V/cm[60]

than the second. This showed that the more mobile component was polyphosphate, and the following peak was attributed to RNA.

Comparison of the peak areas on the electrophoretograms showed, however, that the polyphosphate component was much smaller than it should have been had it contained all the polyphosphate present in the complex. Thus, in the less polyphosphate-rich fraction, instead of a polyphosphate to RNA ratio of 1:4, the electrophoretogram showed a ratio of 1:11, and electrophoretic examination of a fraction with a polyphosphate to RNA ratio of 9:1 gave peaks in the ratio of 4:1, i.e. this suggested that part of the polyphosphate present in these fractions was combined with the RNA, and part was in the free state.

In order to establish whether the polyphosphates were bonded to the RNA, even if only in part, by divalent metal ions, in some experiments the polyphosphate-RNA fractions were dialysed against 10^{-3} M versen, (EDTA) a well known complexing agent, before electrophoresis. Table 15 and Fig. 16 show that this preliminary treatment with versen resulted in some increase in the polyphosphate peaks, although it did not afford complete separation of all the polyphosphate from the RNA.

These results suggested that divalent metal ions played some part in the formation of the polyphosphate–RNA complexes. In order to ascertain whether this was so, a series of experiments was carried out in which artificial mixtures of polyphosphate and RNA were subjected to electrophoretic separation. In some of these experiments, a polyphosphate obtained from a yeast acid-soluble

Table 15. Analytical results from the electrophoresis of polyphosphate–RNA complexes from yeast in a Tiselius apparatus (pH 4.6–4.7)

Fraction under examination	Weight (mg)	Poly phosphate to RNA ratio	Further treatment	mean electrophoretic mobility (cm^2/V.S. $\times 10^{-5}$)		Ratio of areas under peaks due to polyphosphate and RNA to total peak area (%)		Ratio of poly-phosphate to RNA peak areas
				Poly-phosphate	RNA	Poly-phosphate	RNA	
PP–RNA brewer's yeast (I)	21	1:7	Purified by electrophoresis on a cellulose column and treatment with Ku-2 (H² form)	—	14.6	—	—	—
PP–RNA brewer's yeast (II)	150	1:4	Treated with Ku-2 cation-exchange resin (H⁺ form)	14.8	13.6	8.4	91.6	1:11
PP–RNA brewer's yeast (II)	75	1:4	Dialysis in presence of 10^{-3} M versen (EDTA)	15.8	14.2	10.5	89.5	1:9
PP–RNA brewer's yeast (III)	70	9:1	Treated with Ku-2 cation exchange resin (H⁺ form)	16.3	13.7	80.5	19.5	4:1
PP–RNA brewer's yeast (III)	70	9:1	Dialysis in presence of 10^{-3} M versen (EDTA)	17.2	14.4	83.4	16.6	5:1
RNA* + PP†	40	1:3	No further treatment	17.5	14.2	12.3	87.7	1:7
RNA* + PP†	70	1:6	No further treatment	17.4	14.2	5.3	94.7	1:18
RNA* + PP‡	80	1:3	No further treatment	15.2	14.1	22.5	77.5	1:3,5

* Merck yeast RNA.
† Polyphosphate isolated from the acid-soluble fraction of brewer's yeast ($\bar{n}$ = 30).
‡ Synthetic sodium polyphosphate ($\bar{n}$ = 200).

fraction ($\bar{n} = 30$) which contained more than 3% of Ca^{2+} was used, and in the remainder a synthetic sodium polyphosphate ($\bar{n} = 75$) was employed.

The pure yeast RNA (Merck) which was used was the same throughout the experiments. The results are shown in Fig. 17 and Table 15. Examination of the results showed that those mixtures which contained yeast polyphosphate (in which Ca^{2+} was present) displayed marked discrepancies in the ratios of poly-phosphate to RNA present in the initial mixture, and after separation in the Tiselius apparatus. The area under the early running peak was substantially smaller than it should have been, bearing in mind the initial polyphosphate content of the material. On the other hand, in those experiments in which syn-thetic polyphosphate was used, the ratio of polyphosphate to RNA remained nearly the same as in the initial mixture.

These results therefore demonstrated firstly, the absence of covalent bonds between polyphosphate and RNA in the polyphosphate–RNA complexes isolated from biological material, and secondly, the strong likelihood of the presence in the complexes of bonding by means of divalent metal ions.

With respect to the chemical nature of the bonds present in the polyphos-phate–nucleic acid complexes isolated from biological materials, it should be added that in parallel with our own work, which was carried out in the labora-tory of Belozersky, investigations were intensively pursued by Ebel and his co-workers.[111,113,141–144,153,536] In particular, methods were developed in his labo-ratory for the preparative separation of polyphosphates and RNA present in the polyphosphate–nucleic acid complex from yeast, based on the use of acti-vated carbon[143,536] and Sephadex G-200.[111] We have shown[60,368] that these com-plexes can be separated into their components, i.e. polyphosphates and RNA, by precipitation of the polyphosphates in the presence of high concentrations of barium.

It is therefore unlikely that stable covalent bonds between polyphosphates and RNA are present in these complexes, although the possibility remains that electrostatic or hydrogen bonding, or bonding via divalent metal ions occurs. Such bonds would be expected to be broken readily during separation proce-dures.

Investigations into the possible existence of such bonds in yeast polyphos-phate–nucleic acid complexes have shown that hydrogen bonds are absent, but electrostatic interactions mediated by Ca^{2+}, Mg^{2+}, and other metal ions are possibly present.[60,111,143,368] The actual presence in these complexes of signi-ficant amounts of divalent metal ions, primarily Ca^{2+} and Mg^{2+}, has been demonstrated in the laboratories of both Belozersky[60,368] and Ebel.[144,145,682] These investigations furnished substantial proof of the involvement of divalent metal cations in the formation of complexes between RNA and polyphos-phates. Possible modes of linkage of the polyphosphate and RNA chains through divalent metal ions, as proposed by Ebel and his co-workers,[145] are shown in Fig. 18.

Following the work of Belozersky and Kulaev,[60,368] and of Ebel and his co-workers,[144,145,682] it appeared that the problem of the nature of the bonds be-

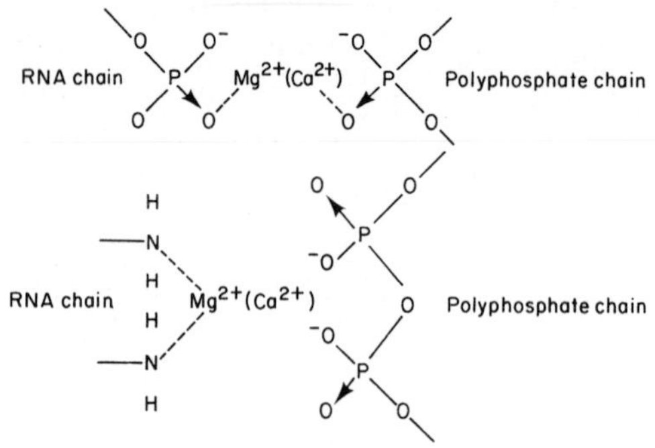

Figure 18. Possible modes of linkage of the chains of high-molecular-weight polyphosphates and RNA through divalent metal ions[145]

tween polyphosphates and RNA had been resolved in principle. However, soon after the principal results of these workers had been published, a series of papers by Correll[91,93,94] and a communication by Wang and Mancini[785] appeared in which, following the isolation of polyphosphate-nucleic acid complexes from algae and higher plants, respectively, the question was again raised of the possible existence in the cells of the organisms which they investigated of stable polyphosphate–nucleic acid complexes in which the polyphosphates and RNA were convalently bonded. In their investigations, these workers attempted to separate the polyphosphate–nucleic acid complexes from the algae *Anabaena* and *Chlorella*[91,93,94] and from the leaves of wheat[785] by means of chromatography on DEAE-cellulose, and on columns of methylated albumin on Kieselguhr (MAK). They were unable, however, to effect a separation of RNA from polyphosphate by the use of these methods. Using various methods of chemical and enzymatic hydrolysis of the polyphosphate–nucleic acid complexes from *Chlorella*, Corell[91,94] showed that the RNA and polyphosphate fragments formed also remained linked together. Correll[91] obtained a pure polyphosphate fraction from this complex only after complete digestion of the RNA by snake venom diesterase. However, precise and unambiguous evidence for the existence of covalent bonds in the polyphosphate–nucleic acid complexes was again not obtained. Their results could all be explained in terms of the existence alone of divalent cation linkages between RNA and polyphosphate. Belozersky and Kulaev[60,368] and Stahl and Ebel[682] showed that Ca^{2+} and Mg^{2+} ions were capable of forming very stable and difficult to separate polyphosphate–RNA complexes. The relatively high stability of the complexes obtained by Correll from *Chlorella* is apparently due to the presence of a certain number of imide groups in the polyphosphates which were isolated from these complexes, alternating with polyphosphate groups.[92] These latter findings of Correll are, however, in need of careful checking.

Whether complexes of polyphosphates with other cell components are present in the living organism is as yet even less clear. Langen and co-workers,[447,464,465] for example, showed that the most highly polymerized yeast polyphosphate fraction, extractable only with comparatively strong alkali (0.05 M) or by keeping for a long period with dilute $CaCl_2$ solution, is apparently firmly bonded to some cell component other than RNA. They concluded that this fraction was actually present in the cell in a combined form from the fact that its extraction from the cell followed a first-order rate law, i.e. it involved fission of some covalent bonds.[464] The partner of polyphosphate in this case was not RNA, since removal of the latter with RNAase had no effect on the rate of extraction of polyphosphate. They considered that in this case the polyphosphates were bonded to certain protein cell components. It is also possible, however, that if some part of the polyphosphates present in the cell is in fact combined with other compounds, the latter could include polysaccharides, especially polyhexosamines such as chitin.

Polysaccharides have been shown repeatedly to be present in polyphosphate-ribonucleic complexes.[365,785] Further, direct experiments by Harold[214, 225] on the formation of complexes between polysaccharides which occur in the cell walls of mycelia of *N. crassa* and polyphosphates have shown that polysaccharides, particularly fractions containing polygalactosamine, combine readily with polyphosphates.

Figure 19 shows the results of Harold and Miller[226] on the formation of complexes between polyphosphates of different degrees of polymerization and the cell wall polysaccharides of *N. crassa* at different pH values. It will be seen that inorganic polyphosphates are capable of complexing with polysaccharides *in vitro*, the reaction being dependent on pH and the chain length of the polyphosphates. However, as has already been pointed out, in our opinion evidence of the reaction of polyphosphates with cell components *in vitro* has only a very indirect bearing on the form in which these polyphosphate fractions are present *in vivo*.

In concluding this chapter, it has to be said that our current ideas on the mode of occurrence of polyphosphates within the cell are extremely vague. The only thing which is clear is that, in contrast to what was previously believed, stable, covalently bonded complexes of polyphosphates with RNA are not present in

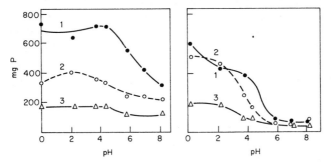

Figure 19. Effect of polyphosphate chain length and pH on the combination of polyphosphates with cell walls.[226] Ordinate: mg P in the cell walls per 1.5 g of mycelium. (1) $\bar{n} = 50$; (2) $\bar{n} = 6$; (3) $\bar{n} = 3$

the cell. It is possible that inorganic polyphosphates, or at any rate the major proportion of them, are not bonded in any way to nucleic acids, and the polyphosphate–nucleic acid complexes which are frequently isolated from organisms of different types may well be secondary products. However, the mode of bonding of polyphosphates which has been demonstrated in these complexes, i.e. interaction with RNA via Ca^{2+} and Mg^{2+} ions, may occur *in vivo*. Polyphosphates could combine in the cell, either permanently or temporarily, via such chelate bridges both with nucleic acids and with other cell components such as proteins, phospholipids, and acidic polysaccharides.

The precise form in which the various polyphosphate fractions are present in the cell remains to be clarified in the future. Some light may be shed on this problem, however, by currently available information on the localization of the various polyphosphate fractions within the cell. The following chapter is devoted to this subject.

Chapter 6

The Localization of Inorganic Polyphosphates Within the Cell Volutin Granules of Microorganisms—The Site of Localization of Inorganic Polyphosphates

Since the earliest work of Wiame,[802–805,808,811] it has been known that polyphosphates, or at least a part of them, are present in the cells of yeasts and other microorganisms as constituents of the cellular inclusions long known as Babesh-Ernst bodies, metachromatic granules, or volutin granules (for reviews, see refs. 49, 50, 114, 123, 124, 204, 335, 363–365, 506, 515, 606, 777, 812, and 815). The presence of polyphosphates in these granules is indicated by the following indirect observations:

1. Volutin granules undergo metachromatic staining by basic dyes such as toluidine blue and methylene blue, i.e. they possess properties which are shared by inorganic polyphosphates, as we have already noted. Furthermore, metachromatic granules can be detected by means of other reactions which are fairly specific for polyphosphates.[133,150,151]

2. In the electron microscopic examination of ultrathin sections of bacteria and other organisms, volutin granules absorb electrons strongly, in exactly the same way as polyphosphates.[124,151,257,780]

3. The accumulation of volutin granules is nearly always well correlated with the accumulation of specific polyphosphate fractions.[47,118,124,128–130,133,201,202,335,365,529,626,808,809,815]

4. Bacterial mutants which are unable to synthesize polyphosphates do not accumulate volutin granules in their cells.[221]

5. Utilization of the polyphosphates present in the cells of certain bacteria for the phosphorylation of glucose, mediated by the specific enzyme polyphosphate glucokinase, is accompanied by the disappearance of volutin granules.[712,719]

There can no longer be any doubt that polyphosphate-containing granules are actually present in cells, and are not artefacts formed during the fixation and staining of cells by particular dyes. That this is so may be concluded from the fact that they are readily visible without staining in living cells of microorganisms by phase-contrast microscopy.[812,815]

Another question which arises when considering the literature data is whether the terms 'voluting granules' and 'metachromatic granules' always denote the same cell inclusions.

It has already been pointed out that it is not only polyphosphates which undergo metachromatic staining within the cell. For example, in one of our papers[320] it was shown that in various actinomycetes, volutin-like granules were present when the cells contained no polyphosphates. Further, a number of papers have appeared in which metachromatic granules containing poly-β-hydroxybutyric acid rather than polyphosphates have been found in various microorganisms.[461,479,495,633,814] Like the polyphosphates, poly-β-hydroxybutyric acid is a polyanion, and like most polyanions it is stained by basic dyes such as toluidine blue.

Metachromatic granules containing poly-β-hydroxybutyric acid constitute reserves of readily mobilizable plastic substances. This store of carbon compounds is separated from the surrounding cytoplasm by a lipid membrane.[814] Lipids constitute about 10% of the dry weight of these granules. Polyphosphate granules, in contrast to those containing poly-β-hydroxybutyric acid, do not appear to possess a boundary membrane,[124,780,814] and they occur most frequently in other parts of the cell.[814] However, the identical behaviour of these two types of granule towards various stains, and towards changes in the growth conditions of microorganisms (accumulation under conditions which retard growth, and utilization when cell growth is stimulated), appear to have resulted in these cell inclusions frequently being confused. This has considerably hindered the investigation of their chemical behaviour by cytochemical methods.

The only reliable method for the investigation of the chemical nature of the polyphosphate-containing granules involves the isolation of their components in an uncontaminated condition, followed by detailed analysis. This approach, however, is very laborious. Early attempts to obtain volutin in which the usual methods of mechanical disintegration and differential centrifugation were employed were unsuccessful. The isolation of volutin in aqueous media resulted in the degradation of the polyphosphates by polyphosphatases, which made it impossible to obtain them in an unmodified state.[124,257,495] Of the investigations using this method, only the work of Rosenberg, who attempted a chemical analysis of volutin granules from *Tetrahymena periformis*,[619] was relatively successful. His results are shown in Table 16. Rosenberg found the principal constituent of *Tetrahymena* granules to be calcium magnesium pyrophosphate. Polyphosphates were not found in the volutin from *Tetrahymena* and this appears to have been the reason for the isolation of these inclusions in a relatively intact condition.

The work of Rosenberg is of particular interest in that it provided the first demonstration of the possibility that substantial amounts of pyrophosphate could accumulate in the form of discrete granules within the cell. Another

Table 16. Chemical composition of granules isolated from *Tetrahymena*[619]

Component	% of total	Component	% of total
Carbon	2.50	Calcium	13.50
Hydrogen	2.61	Magnesium	8.1
Phosphorus		Sodium	0.6
(as orthophosphate)	17.25	Potassium	0.9
Phosphorus		Total ash	69.94
(as pyrophosphate)	48.50	Moisture	26.20

successful attempt to obtain unchanged volutin granules was that of Hase et al.[228] in which the isolation of these inclusions from the cells of *Chlorella ellipsoidea* was effected in a nonaqueous medium consisting of a mixture of cyclohexane and carbon tetrachloride. The results obtained by these workers for the materials which they isolated are shown in Table 17.

The main component of volutin from *Chlorella* is polyphosphates which are extracted by cold 8% TCA, although other, alkali-soluble polyphosphates are also present.

The results of the Japanese workers are in good agreement with those of other investigators who attempted to establish a relationship between the accumulation of volutin and the occurrence of various polyphosphate fractions in certain microorganisms.[47,118,124,133,204] For example, Belousova and Popova,[47] using the Langen and Liss method for the fractionation of polyphosphates[442,445] showed that, in actinomycetes, an increase in volutin was accompanied by accumulation of salt-soluble (in the investigation of Hase et al.,[228] those extracted by 8% TCA in the cold) and alkali-soluble polyphosphates. It is possible however, that, depending on the species of organism and even on the age of the cells taken for examination, the chain length may vary, and hence also the extractability of the components of the volutin. The results of Drews[118,124] are in good

Table 17. Amounts of polyphosphates and other phosphorus compounds present in volutin granules from *Chlorella*[228]

Acid-soluble phosphorus compounds	μM P per 100 mg of cells	Acid-insoluble phosphorus compounds	μM P per 100 mg of cells
Extraction with cold 8% TCA:		*Extraction with cold NaOH, pH9:*	
Orthophosphate	11.5	Polyphosphates	2.5
Nucleotides	22.0	RNA	0.7
Polyphosphates	93.2	DNA	0.0
		Extraction with cold 2n KOH at 37° C for 18 h:	
		Polyphosphates precipitated on neutralization of the extract	0.9
		RNA	Trace

agreement with this assumption, showing that an increase in the content of volutin granules was associated with increases in polyphosphate fractions which differed at different stages of development in mycobacteria. It must be stressed, however, that an increase in the more highly polymerized polyphosphates, which are normally extractable only by relatively strong alkali, is never associated with an increase in volutin. Rather, this fraction is responsible for the extremely powerful basophilicity of young, actively growing cells from which volutin is known to be usually completely absent.[363] It appears to be this polyphosphate fraction which is responsible for the accumulation of substantial amounts of acid-insoluble, labile phosphorus in *Chlorella* when this alga is grown in a medium devoid of Mg^{2+}, when volutin granules are totally absent.[363]

It follows that in all of the organisms examined, only part of the polyphosphate is found in volutin granules, the remainder apparently not being associated with volutin. It is also apparent that volutin granules are not obligatory components of the cells in which they are found. Further, there have been suggestions that volutin granules may be regarded as equivalent to bacterial microsomes.[529,625,626] or their nuclear equivalents.[72,317] On the basis of careful electron microscopic work, however, Drews[117,122,124] has shown that polyphosphate granules may be found in any region of the bacterial cell, including the nuclear zone and the neighbourhood of the mesosomes. His results show that, in mycobacteria, polyphosphate granules are laid down in the protein matrix of the cytoplasm, and do not possess an internal structure or a boundary membrane. Under certain conditions of culture of mycobacteria, for instance when grown on a medium containing oleic acid, the polyphosphate granules were surrounded by a lipid layer.[630]

Voelz *et al.*[780] described a detailed investigation into the formation of polyphosphate granules in *Myxococcus xanthus* under various growth conditions. It was found that the polyphosphate granules in this organism are closely associated with glycogen inclusions, and are either distributed throughout the cytoplasm, or are localized within the nucleoids (the nuclear equivalents) (Fig. 20). Similar localization of polyphosphate granules has been found in *Micrococcus lysodeikticus*.[173]

The course of the formation of polyphosphate granules in *Myxococcus xanthus*[780] involves gradual deposition of polyphosphates on strands of cytoplasmic material. Such localization of volutin granules is particularly characteristic of bacterial cells.

In large cells, particularly those of fungi and algae, some of the polyphosphate granules are also found in the vacuoles.[259–261,811] The work of Indge[259–261] on the localization of polyphosphates in fungi, especially yeasts, is of great interest. He isolated cellular vacuoles from yeast, and found them to contain high-molecular-weight polyphosphates. He further showed that within the yeast protoplasts, part of the polyphosphates were found outside the vacuoles, apparently being concentrated, as in bacteria, on the cytoplasmic strands. These careful investigations established that the vacuolar polyphosphates were on high molecular weight, and were much less readily exchangeable with external ortho-

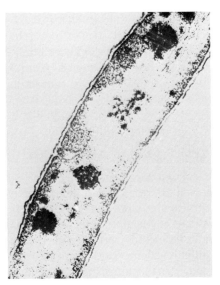

Figure 20. Polyphosphate granules in *Myxococcus xanthus* (magnification ×110 000)

phosphate than were the cytoplasmic polyphosphates. It thus appears that two types of polyphosphate granules are present in yeast, one vacuolar and the other cytoplasmic.

The question of whether only polyphosphates are present as such in the volutin-like polyphosphate-containing granules found in many organisms, or whether they occur therein in combination with other, probably organic, components, remains totally unresolved.

The presence of RNA in the volutin granules which contain polyphosphates has not yet received strict experimental proof, although its presence cannot be categorically denied. Many workers have long believed that polyphosphate–RNA complexes are present in volutin granules.[48–50,118,124,317,335,365,606,812] In view of the high probability of the formation of such complexes during the process of isolation,[106,135,215,220] however, the question cannot yet be regarded as having been settled.

The detection of small amounts of RNA in volutin from *Chlorella*, obtained by Hase *et al.*[228] using a non-aqueous method of extraction, apparently cannot be regarded as a conclusive argument for the presence of RNA in these granules *in situ*, since, as was pointed out by the authors, the material obtained was contaminated with cytoplasmic material.

Perhaps the only indication of the presence of RNA in volutin granules is the fact that the latter are broken down by RNAase. In the presence of this enzyme, the granules either disappear completely, or their numbers are reduced considerably.[123,518,812] However, it was not proved conclusively that the RNAase acted directly on the volutin, rather than causing destruction of the cellular structure *in toto*, thereby affecting the volutin. Evidence against the existence of RNA in volutin is provided by the fact that it is frequently impossible to detect it therein by spectrophotometry.[114,244,695] It was shown by Dmitrieva and Bekker,[114] who investigated this problem in the volutin-like granules present in *Penicillium chrysogenum*, that the detection of RNA in these inclusions at certain stages of development of the fungus was an artefact arising during the vital staining of the cells by basic dyes.

THE LOCALIZATION OF INORGANIC POLYPHOSPHATES IN VARIOUS CELLULAR STRUCTURES

In discussing the intracellular localization of polyphosphates, mention must be made of the fact that this topic is no longer solely concerned with the location of the volutin granules since, as has already been pointed out, part of the polyphosphates may be localized outside these granules.

The earliest investigations aimed at establishing the localization of the whole of the intracellular polyphosphate were those of Harold and Miller[226] and Hughes and Muhammed.[257] In these investigations, the cells of microorganisms were ruptured mechanically, and their contents subjected to differential centrifugation. Hughes and Muhammed, working with the cells of the bacterium *Corynebacterium xerosis*, in determining the amount of polyphosphate in the various fractions, did not at the same time identify the cellular structures associated with these fractions. The work of Harold and Miller, using young mycelia of *N. crassa*, is of greater interest in this respect, since an attempt was made to carry out such an identification.

Table 18 shows the results obtained by Harold and Miller[226] for the distribution of polyphosphates and certain other compounds in various cellular structures in *N. crassa*, following disruption of the mycelia in a Nossal's apparatus in 0.05 M Tris buffer at pH 7.0, containing 0.25 M sucrose. It will be seen that following disruption of the cells and separation of the various cellular structures

Table 18. Distribution of polyphosphates and other compounds between fractions of subcellular structures in 24-h mycelia of *Neurospora crassa* enriched in phosphorus (per 1.5 g of fresh weight)[226]

Fraction	Acid-insoluble polyphosphates		Phospholipids		Nucleic acids		Protein	
	P (μg)	% of initial value	P (μg)	% of initial value	P (μg)	% of initial value	P (μg)	% of initial value
Whole homogenate	740	100	480	100	15 500	100	82	100
Cell walls	465	63	190	40	1 100	7	17.5	21
Mitochondria	145	20	275	57	1 200	8	20.2	25
Ribosomes + supernatant	68	9	48	10	11 000	71	34.0	42
Total	—	92	—	107	—	86	—	88

by differential centrifugation, the acid-insoluble polyphosphates appear for the most part in the cell wall material, and to some extent in the mitochondrial and ribosomal fractions. However, in ribosomes which were isolated in a more or less pure state (these results are not shown in Table 18), these workers found only traces of acid-insoluble polyphosphates, together with approximately 80% of the total cellular RNA.

It must again be emphasized, however, that the results obtained by these workers, and also those obtained in other similar investigations, must be regarded only as preliminary. This follows from the considerations presented above concerning the very rapid degradation of polyphosphates following mechanical disruption of the cells, and the possibility of secondary sorption and desorption of the polyphosphates during fractionation. Harold and Miller reported the occurrence of both of these processes. For example, the acid-soluble polyphosphates (and therefore probably part of the acid-insoluble polyphosphates) decomposed to orthophosphate in the course of this work. Further, Harold showed in separate experiments that the presence of large amounts of acid-insoluble polyphosphates in cell-wall material from N. crassa could be due to secondary sorption by polysaccharides such as polygalactosamine (chitin), which forms part of the cell walls of this organism.[214] Figure 19 shows the results of experiments which confirm this hypothesis.

Much more promising are those investigations into the intracellular location of polyphosphates which do not have recourse to mechanical disruption of the cell. In particular, the work of Weimberg and Orton[798] on the localization of polyphosphates involved treatment of the cells of the yeast Saccharomyces mellis with an enzyme preparation from the snail Helix pomatia. Treatment with this preparation is known[798] to cause lysis of the fungal polysaccharide cell wall, resulting in the formation of spheroplasts devoid of cell walls. Weimberg and Orton showed that approximately one quarter to one third of the total cellular polyphosphate was localized in the immediate vicinity of the external cytoplasmic membrane. This polyphosphate fraction was removed from the cells of S. mellis on lysis of the cell walls with the snail enzyme preparation, and was hydrolysed to orthophosphate in the process. These results were later confirmed by Weimberg.[796]

The results of Souzu[674,675] show that a significant portion of the polyphosphates in yeasts are localized in the region of the cytoplasmic membrane, possible in the membrane itself. He showed that yeast cells, during freezing and thawing, underwent a conformational change of the cytoplasmic membrane, accompanied by rapid hydrolysis of the cellular polyphosphates to orthophosphate.

The localization of a portion of the polyphosphates in yeast cells at the surface of the cells is also shown by the work of Van Steveninck and Booij.[693] They obtained substantial evidence for the participation of polyphosphates in the transport of glucose through the external cytoplasmic membrane in Saccharomyces cerevisiae, and on these grounds they proposed also that part of the polyphosphate occurs on the outside of the external cell membrane of the yeast cells, and is involved in the phosphorylation of the carriers which transport glucose from the medium into the cell.

With respect to the presence of polyphosphates in specific intracellular structures in various organisms, very little information has been available until recently. It is interesting that the literature on this question, up to 3 or 4 years ago, was concerned mainly with cellular components isolated from the cells of higher animals.

We cite here only the completely reliable work of Griffin and Penniall[193,588] and Grossman and Lang,[198] which demonstrated the presence of very small amounts of inorganic polyphosphates in nuclei from rat liver cells and other animals.

A preliminary communication by Lynn and Brown[474] on the presence of inorganic polyphosphates in rat liver mitochondria aroused interest, but subsequent work by Heldt and Klingenberg,[232] Kulaev et al.,[403,410] and by Lynn and Brown themselves[220] failed to confirm these observations.

Preliminary indications of the presence of very small amounts of polyphosphates in Chlorella chloroplasts were obtained by Hase et al.,[228] but these workers did not adduce rigorous proof of the presence of polyphosphates in the chloroplasts themselves, rather than in impurities present in the material.

In considering the localization of inorganic polyphosphates in the cell, other studies by Miyachi and co-workers should be borne in mind in which, on the basis of indirect findings from an examination of the physiological functions and the metabolism of various polyphosphate fractions from Chlorella, it was suggested that their intracellular localization was variable.[293,521–524] In particular, they showed that there exists a polyphosphate fraction (fraction A) which is a constituent of volutin granules and is closely involved in nuclear metabolism. It participates actively in nuclear fission of cells of Chlorella. There is also another fraction (fraction C) which is sited adjacent to the chloroplasts, and is closely involved in the metabolism of these cellular structures, especially with photosynthesis. Fractions B and D appear to be localized at other cellular sites, and fulfil other functions in the metabolism of the cells.

THE INTRACELLULAR LOCALIZATION OF POLYPHOSPHATES OF DIFFERENT MOLECULAR WEIGHTS IN THE LOWER EUKARIOTES

The different intracellular localizations of the various inorganic polyphosphate fractions, first observed in Chlorella by Miyachi and co-workers,[293,521–524] were clearly demonstrated by the author in collaboration with Krasheninnikov, Afanas'eva, and others, in the cells of heterotrophic microorganisms, namely the fungi Neurospora crassa and Endomyces magnusii.[343–345,381–384,403–406,408,667] The initial investigations carried out in our laboratory of the fungus Neurospora crassa showed that a solution of the problem of the intracellular location of the polyphosphates by the use of differential centrifugation following mechanical disruption of the mycelia was not possible.[405] When this method was used, it was found to be impossible to obtain the individual subcellular structures in a state which was at the same time both uncontaminated and sufficiently intact, and furthermore, secondary sorption of the major part of the cellular polyphosphates by the cell walls occurred. Control experiments designed to obtain pure preparations of the mycelial cell walls showed them to be completely devoid of inorganic polyphosphates.[343] Model experiments carried out by Krasheninnikov et al.,[343] together with the work of Harold using the same material,[214] demonstrated the strong tendency of certain polysaccharide–protein fractions, constituents of the cell walls of N. crassa, to combine with the inorganic polyphosphates which were originally located in the cytoplasm of this fungus.

For these reasons, in our work on the elucidation of the intracellular localization of polyphosphates in mycelia of N. crassa[405] and in the cells of Endomyces magnusii, we employed a method which had been used with success by Weimberg and Orton[798] in their work with Saccharomyces mellis. An attempt was made to isolate protoplasts without their cell walls (i.e. spheroplasts) from mycelia of N. crassa and cells of E. magnusii, and to compare the amounts of various polyphosphate fractions present in the whole cells with those in the protoplasts. This comparison, together with the known absence of polyphosphates from the cell walls, made it possible to decide whether any particular polyphosphate fraction was present between the cell wall and the external cytoplasmic membrane.

Microscopic examination of protoplasts from N. crassa and E. magnusii showed them to resemble protoplasts and spheroplasts from fungi examined in other laboratories,[218,753,778] being spherical bodies of varying sizes and high osmotic sensitivity. Under favourable conditions (isotonic medium, 30 °C) they were capable of regenerating new cell walls at their surfaces, and eventually of giving normal mycelia.

Electron microscopic examination showed the protoplasts to be, from the morphological point of view, structures fully equivalent to the entire cells, the only significant difference being that they were entirely devoid of cell walls.

The principal cellular structures of the protoplasts (cytoplasmic membrane, nucleus, mitochondria, etc.) were much the same as those in the entire cells, indicating the integrity of the protoplasts, and the possibility of isolating from them organelles of interest to us in an unchanged condition. That the protoplasts obtained by us retained their structure intact, and had not undergone a change in their chemical composition, was also shown by the fact that both the protoplasts and the entire cells contained almost identical amounts of nucleic acids and phospholids, the principal phosphorus-containing polymeric compounds. This is shown in Table 19, in which are summarized the analytical results obtained in our laboratory for the principal phosphorus compounds present in the entire cells, protoplasts, nuclei, and mitochondria of 18-h mycelia of N. crassa. It can be seen from Table 19 that, in addition to the nucleic acid and phospholipid contents, the total phosphorus contents in the cells and protoplasts were virtually identical. This provides support for the view that phosphorus compounds, in particular inorganic polyphosphates, are absent not

Table 19. Content of inorganic polyphosphates and other phosphorus compounds in the cells, protoplasts, nuclei, and mitochondria of mycelia of *Neurospora crassa* (expressed as mg P/g of dry mycelium)[345,405,406]

Phosphorus compounds	Entire cells	Protoplasts	Nuclei	Mitochondria
High-molecular-weight poly-phosphates (total)	5.6	3.0	0.2	0.0
Acid-soluble (PP_1)	1.8	1.8	0.0	0.0
Salt-soluble (PP_2)	1.0	1.2	0.2	0.0
Alkali-soluble (PP_4)	2.0	0.0	0.0	0.0
Hot $HClO_4$ extract (PP_5)	0.8	0.0	0.0	0.0
Orthophosphate	1.1	3.0	0.0	0.1
Total orthophosphate and inorganic polyphosphates	6.7	6.0	0.2	0.1
Nucleotides	1.2	1.0	0.1	0.3
Nucleic acids	6.7	6.9	0.6	0.9
Phospholipids	2.1	2.1	0.4	0.9
Sugar phosphates	0.4	0.5	0.2	0.1
Total phosphorus	17.3	17.2	1.5	2.4

only from the cell walls of *N. crassa*, but also from the region between the cell wall and the external cytoplasmic membrane.

Although the total amount of phosphorus present in the entire cells and the protoplasts was the same, nevertheless the amounts of the different polyphosphate fractions differed considerably. Table 19 shows that when the protoplasts are produced, the most highly polymeric fractions (the alkali and hot perchloric acid extracts) disappear completely, while the acid-soluble and salt-soluble fractions remain unchanged. The disappearance of part of the polyphosphates when the protoplasts are formed is accompanied by a corresponding increase in the amount of orthophosphate. This leads to the conclusion that removal of the cell wall results in hydrolysis of the above-mentioned highly polymerized polyphosphate fractions to orthophosphate. It is possible that these polyphosphate fractions are in some way bound to the outer cytoplasmic membrane of *N. crassa* cells, and that the formation of the protoplasts may be accompanied by some conformational change in the membrane (extension or compression) which is also the reason for such rapid hydrolysis of the more highly polymerized polyphosphate fractions to orthophosphate. However this may be, these polyphosphate fractions are far more sensitive to the integrity of the cellular structure than the acid-soluble and salt-soluble fractions. This suggests in turn that the different polyphosphate fractions occur at different sites within the *N. crassa* cell. Further, the alkali-soluble polyphosphates, and those extractable from the cell with hot perchloric acid, which occur in this organism adjacent to the outer cytoplasmic membrane, appear to be sited in *N. crassa* at the inner surface of the membrane, since the orthophosphate which results from their hydrolysis accumulates within the protoplast, and is not lost from the cell on removal of the cell wall.

An examination of protoplasts obtained by more prolonged incubation of *N. crassa* mycelia with the snail enzyme gave basically similar results, although they differed somewhat in detail from those described above. Under these conditions, all of the more highly polymerized polyphosphates, present originally in the mycelium in the alkali-soluble fraction only, also disappeared from the protoplast. However, in contrast to the previous experiment, when the time of contact of the protoplast with the snail enzyme was increased, substantial amounts of orthophosphate and acid-soluble polyphosphates were also lost. Further, salt-soluble polyphosphates were also found in this experiment in the protoplasts in the same amounts as in the intact cells. This suggests that this polyphosphate fraction differs in its localization, and perhaps also in its state, within the *N. crassa* cell, both from the more highly polymerized polyphosphates (alkali-soluble, and extracted with hot perchloric acid), and also from the less highly polymerized, acid-soluble polyphosphates.

On the basis of these findings, it is likely that the fractions isolated by the Langen and Liss method from *N. crassa* cells differ from each other not only in their molecular weights, but also in their intracellular localization and their state within the cell. Thus, the most highly polymerized fractions are apparently located at the periphery of the *N. crassa* cell, and removal of the cell wall results in their rapid hydrolysis. The less polymerized, acid-soluble polyphosphates are evidently located within the cell, and it therefore appears very likely that they are present for the most part in the free state. In contrast to these fractions, the salt-soluble polyphosphates in *N. crassa* are localized in such a way that the formation of protoplasts from entire cells has no effect on the amounts present. It is possible that this polyphosphate fraction (perhaps together with some of the acid-soluble polyphosphates) occurs in the cellular vacuoles of *N. crassa*, and is protected from the hydrolytic effects of the polyphosphatase by the tonoplast. One gains the impression from these results that this polyphosphate fraction may be a constituent of the volutin granules which were described in this organism by Zalokar.[844] Our cytochemical results,[381] which showed that the more highly polymerized polyphosphates were absent from the volutin granules present in *N. crassa* protoplasts, may be cited in support of this hypothesis.

In concluding this comparison of the amounts of phosphorus compounds present in the entire cells and protoplasts of *N. crassa*, it only remains to say that removal of the cell wall results not only in the loss of certain polyphosphate fractions, but also part of the nucleoside di- and triphosphates. A detailed examination of the acid-soluble nucleotides of the entire cells and protoplasts of *N. crassa* showed[405,408] that the formation of the protoplasts was accompanied by substantial losses of ADP and ATP in addition to GTP and UTP, but the nucleoside monophosphates and nucleoside diphosphate sugar remained entirely within the protoplasts. It can therefore be concluded that a substantial proportion of the ADP and ATP, like the other nucleoside di- and triphosphates, is localized at the periphery of the cell, adjacent to the outer cytoplasmic membrane.

Summarising briefly, the most significant observations arising from these investigations are perhaps the following. Different polyphosphate fractions are localized in mycelial cells of *N. crassa* at different sites. The more highly polymerized fractions are located at the cell periphery, and the less highly polymerized fractions occur within the cell, in part as constituents of volutin granules. In order to test these highly significant conclusions, we investigated the localizations of the polyphosphates within the cells of another fungus, *Endomyces magnusii*.

A comparative investigation of the amounts of the various inorganic polyphosphate fractions in entire cells and protoplasts of *E. magnusii* was carried out in conjunction with Afanas'eva, using a normal 12-h culture of this yeast, and also a culture enriched with phosphorus after preliminary phosphorus starvation. In the latter case, *E. magnusii* was cultured for 12 h on a normal medium, followed by 6 h on a phosphorus-free medium, and finally for 4 h on a medium rich in phosphorus and sucrose.

Yeasts enriched with phosphorus in this manner are known[471] to produce large amounts of polyphosphates, which accumulate in the cell for the most part as volutin granules. It was of interest to determine precisely which polyphosphate fractions were associated with the accumulation of volutin, and whether this accumulation occurred in the protoplasts of *E. magnusii*.

The amounts of the various polyphosphate fractions present in the intact cells and protpolasts of the normal 12-h and phosphorus-enriched cultures of *E. magnusii* are shown in Tables 20 and 21, respectively. Table 20 also shows the amounts of some other phosphorus compounds present in the cells and and protoplasts of this organism, together with analyses for phosphorus compounds in the nuclei and mitochondria.

Table 20. Amounts of inorganic polyphosphate and other phosphorus compounds present in cells of *Endomyces magnusii*, and in the protoplasts, mitochondria, and nuclei obtained from them (expressed as mg P/g dry weight of cells)[4,384,667]

Phosphorus compounds	Entire cells	Protoplasts	Mitochondria	Nuclei
High-molecular-weight polyphosphates (total)	2.0	1.5	0.0	0.4
Acid-soluble (PP_1)	0.2	0.7	0.0	—
Salt-soluble (PP_2)	0.7	0.4	0.0	0.4
Alkali-soluble ($PP_2 + PP_4$)	0.9	0.4	0.0	0.0
Hot $HClO_4$ extract (PP_5)	0.2	0.0	0.0	0.0
Orthophosphate	3.2	1.1	0.3	0.1
Total orthophosphate and inorganic polyphosphates	5.2	2.6	0.3	0.5
Nucleotides	1.0	0.6	0.3	—
Nucleic acids	6.2	5.4	1.1	—
Phospholipids	1.2	1.2	0.9	0.2
Sugar phosphates	0.9	0.5	Trace	—
Total phosphorus	14.4	10.3	2.6	2.3

Table 21. Amounts of inorganic polyphosphates present in cells of *Endomyces magnusii* enriched in phosphorus (expressed as mg P/g dry weight of cells)[384]

Phosphorus compounds	Entire cells	Protoplasts
Inorganic polyphosphates (total)	20.6	14.3
Acid-soluble (PP_1)	11.2	12.1
Salt-soluble (PP_2)	3.4	1.4
Alkali-soluble ($PP_3 + PP_4$)	3.3	0.8
Hot $HClO_4$ extract (PP_5)	2.5	0.0

Comparison of the entire cells and protoplasts from a 12-h culture of *E. magnusii* (Table 20) showed that removal of the cell wall resulted in a reduction of 30% in total phosphorus. This was due primarily to a reduction in the amounts of orthophosphate and inorganic polyphosphate. As in *N. crassa*, the formation of protoplasts from cells of *E. magnusii* resulted in complete loss of polyphosphates extractable by hot perchloric acid. In contrast to *N. crassa*, however, the alkali-soluble polyphosphates present in *E. magnusii* are not completely hydrolysed on incubation with the active enzyme, less than half of this fraction remaining in the *E. magnusii* protoplasts. From our point of view, the most important difference in the behaviour of the various poly-phosphate fractions when protoplasts are formed from *E. magnusii* cells, as compared with *N. crassa*, is the marked reduction in the amount of salt-soluble polyphosphates present in the protoplasts as compared with the entire cells. This suggests that in *E. magnusii* this fraction is not homogeneous. Part of the salt-soluble polyphosphate fraction, and also the more highly polymerized fractions, is sensitive to the disturbance of the cellular struc-ture which takes place when the cell wall is removed. This part of the salt-soluble polyphosphate fraction appears to be localized differently in *E. magnusii* than in *N. crassa*.

The behaviour of the different polyphosphate fractions when protoplasts are formed from phosphorus-enriched fungal cells is essentially no different from that observed in normal *E. magnusii* cells. Table 21 shows that this yeast contains about ten times more polyphosphate than that grown in a normal 12-h culture. Despite this, however, when protoplasts are formed from these cells, once again those polyphosphates which are extractable by hot perchloric acid disappear completely, whereas the alkali-soluble and salt soluble polyphos-phates, as in normal cells of this fungus, remain partially in the protoplasts.

Consequently, as in *N. crassa* mycelia, in *E. magnusii* the most highly polymerized inorganic polyphosphates are localized outside the cell, adjacent to the outer cytoplasmic membrane, while the less highly polymerized acid-soluble and part of the salt-soluble polyphosphate fractions accumulate within the cell.

These results suggest that the behaviour which we have observed is general for all organisms, or at least for fungi. This conclusion is supported by a comparison of our findings with those described in the literature, for example by Weimberg and Orton[798] and Souzu,[674,675] obtained in parallel with our own on two species of yeasts. Indirect support is also provided by the work of Van Steveninck and co-workers.[104,616,690,693,694] Consideration of the general features of the distribution of polyphosphates in the cells of a variety of organisms reveals certain differences, however. In the two species of fungus with which we have worked, there are at least two differences. The first has already been mentioned, viz. the heterogeneity of the salt-soluble and also perhaps the alkali-soluble polyphosphate fractions in E. magnusii with respect to their localization and intracellular condition. The second difference is the more rapid disappearance of a proportion of the polyphosphates and orthophosphate from the cells of E. magnusii when protoplasts are formed therefrom. In fact, when protoplasts are formed from cells of N. crassa by incubation for 4 h with snail enzyme, the whole of the phosphorus compounds, including both orthophosphate and polyphosphates, remain within the protoplast, whereas in the case of E. magnusii cells, only 2 h treatment with snail enzyme results in substantial losses of both polyphosphates and orthophosphate. This latter difference is perhaps due both to the different localization of the polyphosphates in these fungi (for example, at the outside of the cytoplasmic membrane in E. magnusii, or in the inside, in N. crassa), and to the characteristics of the ultrastructure of the cells. A very marked common feature of both species of fungus is the occurrence of the less highly polymerized polyphosphate (the salt-soluble and particularly the acid soluble fractions) in the so-called volutin granules. As can be seen from Table 21, most of the phosphorus in phosphorus-enriched cells is found in the acid-soluble polyphosphate fraction. This fraction is also completely retained within the protoplasts of E. magnusii. On the other hand, a cytochemical investigation of the protoplasts from phosphorus-enriched yeasts has shown them to be completely filled with volutin granules. These two observations lead to the conclusion that probably, in both fungal species, the principal volutin-forming component is acid-soluble polyphosphate. This hypothesis becomes even more plausible when it is recalled that in the other organisms which have been investigated in this respect, i.e. protozoa (Table 16) and algae (Table 17), acid-soluble polyphosphates also form the principal constituents of the voluntinoid granules. Nevertheless, the above results do not exclude the possibility that salt-soluble polyphosphates may also be present in fungal volutin.

A final solution of this question will not be possible until these inclusions have been isolated in the pure state, and their chemical composition examined. Our own efforts in this direction will be described.

Finally, in this examination of the behaviour of the various phosphorus compounds when protoplasts are formed by these fungi, attention must certainly be drawn to results obtained with the acid-soluble nucleotides. It appears that in both E. magnusii and N. crassa, when protoplasts are formed, substantial amounts of nucleoside di- and triphosphates (mainly ADP and ATP) disappear.[384] Thus, in both organisms, substantial amounts of nucleoside polyphosphates behave in a manner similar to the most highly polymerized inorganic polyphosphates, when protoplasts are formed. These results suggest the possibility of the simultaneous occurrence of these high-energy compounds at the periphery of the cell, adjacent to the outer cytoplasmic membrane.

Following the demonstration in two different speciments (mycelia of *N. crassa* and cells of *E. magnusii*) that the most highly polymerized polyphosphates were localized adjacent to the cell periphery, in the region of the cytoplasmic membrane, while the less polymerized polyphosphates were found within the cell, it appeared to us important to carry out a more detailed investigation into the location of the polyphosphates which remained within the protoplasts. We first attempted to isolate the cell nuclei from the protoplasts obtained from these organisms. The results of an examination of the phosphorus compounds present in intact, uncontaminated *N. crassa* nuclei carried out by the author in collaboration with Krasheninnikov and Polyakov,[406] and of a similar analysis of highly purified *E. magnusii* nuclei carried out by Skryabin *et al.*,[667] are given in Tables 19 and 20. The results of analyses of the phosphorus compounds present in the cell nuclei of *N. crassa* (Table 19) show that the total nuclear phosphorus comprises only 8% of the phosphorus content of the entire cell. It is interesting that the nuclei were completely devoid of any trace of orthophosphate, and contained comparatively small amounts of nucleotides. Their phosphorus content was slightly over 5% of the nucleotide phosphorus of the entire cells. In addition to nucleotides, the cell nuclei of *N. crassa* also contained sugar phosphates, the phosphorus content of which accounted for over 10% of the total phosphorus of these compounds, and 30% of that of the sugar phosphates found in the entire cells. The main sugar phosphate fraction from the cell nuclei contained ribose-5-phosphate, and phosphate esters of glucose and glycerol. The proportion of nucleic acid phosphorus in the nuclei was approximately half the total phosphorus present in the material, and only one tenth of that of the nucleic acid in the entire cells. The phospholipid phosphorus of the cell nuclei represented approximately 15% of the phosphorus present in these compounds in the entire cell.

Of the greatest interest to us, however, was the detection in *N. crassa* nuclei of inorganic polyphosphates which were extractable by saturated salt solution.[406] The presence of polyphosphates in this fraction was proved conclusively by chromatography. Among the products of partial hydrolysis of this fraction by the method of Thilo and Wieker, cyclic trimetaphosphate (a specific degradation product of inorganic polyphosphates) was found. Polyphosphates were not found in the other fractions. The phosphorus content of the salt-soluble nuclear fraction amounted to 15% of the total polyphosphate phosphorus of this fraction in the entire cells. Further purification of the nuclear material by centrifugation in a sucrose concentration gradient, or by other methods, did not result in the removal of the polyphosphates from the nuclei. These findings suggest that some part of the salt-soluble polyphosphate fraction is actually present as such within the nucleus of *N. crassa*. Similar results were obtained in our laboratory by Skryabin *et al.*,[667] who investigated the phosphorus compounds of cell nuclei from *E. magnusii*. In this case, as can be seen from Table 20, approximately 16% of the total cell phosphorus was present in nuclear phosphorus compounds, from which orthophosphate was almost entirely absent. A most notable finding was, however, that the polyphosphates as detected by a Langen and Liss fractionation were present in the nuclei of both *E. magnusii* and *N. crassa* in the salt-soluble fraction only. The polyphosphates of *E. magnusii* nuclei represented approximately 20% of the total high-molecular-weight polyphosphates, and over 50% of the total salt-soluble polyphosphates present in entire cells of this fungus. Of interest also was the finding that the high-molecular-weight polyphosphate fraction present in the nuclei of *E. magnusii* represented approximately 15% of the total phosphorus content (approximately 13% in the case of *N. crassa*). Such high concentrations of high-molecular-weight polyphosphate in fungal cell nuclei suggest that these compounds play a significant part in the metabolism of these important cell structures.

Following the detection of high-molecular-weight polyphosphates in nuclei of *N. crassa* and *E. magnusii*, it was of particular interest to resolve the question of whether or not these high-energy phosphorus compounds were present in the mitochondria of these same organisms. This was all the more important, since in mitochondira of animal origin, in particular those from rat liver and beef heart, neither we[403,410] nor others[220,232,684] have been able to demonstrate conclusively the presence of inorganic polyphosphates. As has already been mentioned, the preliminary finding of Lynn and Brown[474] that these compounds were present in rat liver mitochondria has not been confirmed.

Mitochondria isolated from *N. crassa* mycelia and *E. magnusii* by the author in collaboration with Krasheninnikiv and Afanas'eva appeared from electron microscopy and the appropriate biochemical tests to be substantially intact.[4,345] Unpurified mitochondrial fractions from both fungi were found to contain compounds which, from their chemical properties, were identical with polyphosphates. Acid-labile phosphorus compounds, present as their barium salts at pH 4.3, were detected in both cases in the cold perchloric acid extract, in amounts corresponding to 0.5 mg of phosphorus per gram dry weight of cells. Further, the presence of these compounds was also detected in the salt extract from the mitochondria of *E. magnusii*, in which the acid-labile phosphorus compounds sometimes amounted to as much as 0.7 mg per gram dry weight of cells. In order to establish definitely the presence of inorganic polyphosphates in mitochondrial preparations from *N. crassa* and *E. magnusii*, the barium salts of the acid-labile phosphorus compounds were subjected to paper chromatographic analysis. The results proved conclusively that high-molecular-weight polyphosphates were actually present in the mitochondrial preparations from these organisms.

The same conclusion can also be drawn from the detection of the characteristic cyclic trimetaphosphate, together with low-molecular-weight polyphosphates, among the products of partial hydrolysis of the acid-labile phosphorus compounds precipitated by barium from the above-mentioned fractions at pH 4.3.

However, in an electron microscopic examination of the mitochondrial fractions, in addition to the mitochondria we observed the presence of electron-dense granules similar to volutin granules. The concurrent presence of volutin-like granules in the mitochondrial fractions is also indicated by the observation of metachromatic granules in these materials on staining with toluidine blue. In order to establish whether or not high-molecular-weight polyphosphates are present in *N. crassa* and *E. magnusii* in the mitochondria themselves, or whether they occur in the mitochondrial preparations as a result of contamination with polyphosphate-containing volutin granules, we attempted to purify these cellular structures by centrifugation in a sucrose concentration gradient. For this purpose, suspensions of mitochondria from *N. crassa* and *E. magnusii* immediately after isolation from the cells were applied to a linear sucrose concentration gradient (1.2 to 2 M) in a 0.0005 M Tris–HCl buffer of pH 7.4 containing 0.001 M EDTA. Centrifugation was carried out at 100 000 g for 3 h with *N. crassa* mitochondria, and for 1 h with the mitochondrial preparation from *E. magnusii*. Protein and acid-labile phosphorus was determined in the centrifuged fractions from each preparation. In addition, following centrifugation of *N. crassa* material, the distribution of cytochromes in the fractions was determined, and the metachromatic granule content in the fractions obtained by purification of the mitochondria from *E. magnusii* was measured using toluidine blue.

Figure 21 shows that the distribution of acid-labile phosphorus when the mitochondria were centrifuged in a sucrose concentration gradient in neither case corresponds with the distribution of protein and cytochromes. When each preparation is purified, the peak due to acid-labile phosphorus runs in front of the protein and cytochrome peaks, which correspond with the distribution of mitochondria in the fractions.

A similar analysis of the fractions in the presence of volutin granules showed that their distribution corresponded with that of the acid-labile phosphorus, and that they were almost completely absent from fractions containing the greatest amounts of mitochondrial protein.

The most important conclusion to be drawn from this work is that the mitochondria from *N. crassa* and *E. magnusii*, like those from beef heart and rat liver, contain no polyphosphates. The high-molecular-weight polyphosphates found in incompletely purified mitochondrial preparations from *N. crassa* and *E. magnusii* are probably present in volutin-like granules, which provides yet further confirmation of the polyphosphate nature of volutin in fungi. It was of interest to isolate preparative amounts of volutin granules from *N. crassa* and *E. magnusii* so that it could be examined more closely. Our attempts to do this have however not yet been successful. The use of Rosenberg's method,[619] which he used to obtain volutin from *Tetrahymena pyriformis*,

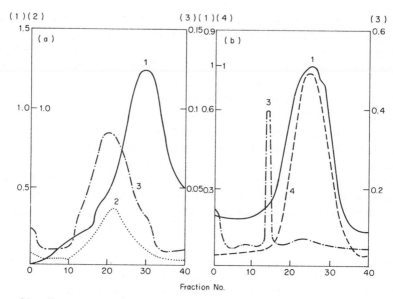

Figure 21. Separation of mitochondria from polyphosphate-containing granules by centrifugation in a sucrose concentration gradient (1.2 to 2 M).[403] (a) Preparation from *E. magnusii*, 60 min at 100 000 *g*; (b) preparation from *N. crassa*, 180 min at 100 000 *g*. 1, Protein; 2, volutin granules; 3, acid-labile phosphorus; 4, cytochrome a (All data expressed as conditional units).

and the method of Hase *et al.*,[228] which was developed for the isolation of pre-parative amounts of volutin granules from chlorella cells, failed in our hands. Both methods were completely unsatisfactory for the isolation of volutin-like granules from fungi. This is indicative of the marked dissimilarities between granules containing inorganic polyphosphates from different organisms, and of the need in each case to develop a method of isolation which is specific for the particular organism.

It is possible that the substantial difficulties encountered in obtaining intact volutin granules from bacteria and fungi is due to the presence in these organisms of highly active polyphosphatases[166,222,377,404,531] which rapidly cleave the polyphosphate during the mechanical disruption of the cells, even at very low temperatures. It is thus apparent that, in any attempt to obtain volutin from the cells of bacteria and fungi, the isolation of the granules should be carried out in the presence of adequately powerful and specific inhibitors of polyphosphatase.

Despite the lack of success in attempts to isolate volutin granules in preparative amounts from fungi, the analytical data which we obtained during the purification of the crude mitochondrial fraction demonstrate unambiguously the polyphosphate nature of these granules in *N. crassa* and *E. magnusii*. Further, the results provide evidence that the granules in these fungi may contain both acid-soluble and salt-soluble polyphosphates.

On the basis of the overall results obtained in our laboratory, it can be concluded that at least three different pools of inorganic polyphosphate are present in *N. crassa* and *E. magnusii* (and perhaps in other fungi).

The most highly polymerized polyphosphates (those extractable by hot perchloric acid and weakly alkaline solutions) are localized in these organisms at the cell periphery, in the region of the cytoplasmic membrane and perhaps bonded to it. The less highly polymerized, salt-soluble polyphosphates are localized mainly within the cell, partly as components of the cell nucleus and partly in volutin granules. The least highly polymerized, acid-soluble polyphosphates are apparently wholly localized in the cytoplasm, in volutin granules. This general picture of the localization of polyphosphates in fungi, based on our own findings, is in good agreement with results obtained by other workers, both with fungi and other lower eukaryotes.

Firstly, the previously mentioned investigations of Weimberg,[796,798] Souzu,[674, 675] and Van Steveninck and co-workers,[690–694] in which a wide variety of methods were used, demonstrated that in yeasts a proportion of the polyphosphates is found at the cell periphery, in the region of the cytoplasmic membrane.

Secondly, Goodman *et al.*,[188] working in the laboratory of Rusch, obtained results which showed the presence of high-molecular-weight polyphosphates in the nuclei of the mycomycete *Physarum polycephalum*.

Thirdly, the work of Indge[261] has shown that the polyphosphates which occur in yeast protoplasts are present in both the vacuoles and the cytoplasm proper. It was also shown that the more highly polymerized polyphosphate fractions found in the protoplasts were localized in the vacuoles. A comparison of these results with our own, obtained with *N. crassa* and *E. magnusii*, suggests that the vacuolar polyphosphates found by Indge in yeasts are identical with the salt-soluble fractions which were isolated from *N. crassa* and *E. magnusii*, and the yeast cytoplasmic polyphosphates are identical with the acid-soluble polyphosphates of *N. crassa* and *E. magnusii*.

It may be that the pattern of the intracellular distribution of polyphosphates observed by us and other workers in fungi is a general one for eukaryotic organisms, and even for higher plants and animals. In fact, such scanty and fragmentary data as are to be found in the literature on the intracellular localization of high-molecular-weight polyphosphates in higher organisms are also in good agreement with the general observed in the lower eukaryotes.

The previously cited work of Griffin and Penniall,[193,588] in which clear proof was obtained of the presence of high-molecular-weight polyphosphates in rat liver nuclei, should be borne in mind. We also draw attention to the work of Bashirelashi and Dallam,[41] which demonstrated the presence of polyphosphates in calf thymus nuclei, and in which the metabolism of these compounds was also investigated to some extent.

In addition, the work of Gabel and Thomas[179,181] supports the view that in brain tissue cells, the greatest amounts of polyphosphate occur in the myelin membrane fraction. This worker further showed the presence of comparatively large amounts of high-molecular-weight polyphosphates in pigeon erythrocyte membranes.

In conclusion, it is appropriate to mention here results obtained by Vagabov and Kulaev[769] during an attempt to detect polyphosphates in maize root tips. It was found that high-molecular-weight polyphosphates are present in maize roots only at their extremities where, according to Ivanov,[270] volutin-like metachromatic granules are present in the root epidermis cells between the meristematic and extension zones.

These findings lead to the conclusion that high-molecular-weight polyphosphates can accumulate in volutin-like granules in the cells of higher plants, as well as in those of microorganisms.

SOME COMMENTS ON THE POSSIBLE LINK BETWEEN THE INTRACELLULAR LOCALIZATION OF HIGH-MOLECULAR-WEIGHT POLYPHOSPHATES AND THE MODE OF NUTRITION OF THE ORGANISM

Information currently available in the literature suggests that the location and metabolism of high-molecular-weight polyphosphates in the cells of various organisms must be dependent on their mode of nutrition. Specifically, the localization of these compounds in heterotrophs and autotrophs may be completely different. The previously mentioned work of Miyachi and co-workers[293,521-524] in particular shows this. These workers, using *Chlorella*, demonstrated the presence of specific polyphosphate fractions which were very closely associated with the chloroplasts, and which accumulated in the cells of this alga only under conditions of photosynthesis.

Evidence in support of this hypothesis is provided by the results obtained by Kulaev et al.[424] It was shown in *Acetabularia* (a giant single-celled alga), during the period of vegetative growth, the most high polymerized, alkali-soluble polyphosphates, extractable by hot perchloric acid, were usually completely absent from the cells. In the cells of this photosynthesizing organism, 80–90% of the total high-molecular-weight polyphosphate is found in the acid-soluble fraction. It may be concluded that, in *Acetabularia*, there is no peripherally-localized fraction which is bonded to the cytoplasmic membrane. On the other hand, it appears likely that the bulk of the (acid-soluble) polyphosphates are localized within the cells of this alga. How valid this hypothesis may be, further investigations will show, but it has now been established beyond doubt that any consideration of the localization of high-molecular-weight polyphosphates in the cells of different organisms must be approached on an individual basis, taking into account above all the structural organization of the cells and their mode of nutrition.

Chapter 7

The Biosynthesis of Inorganic Polyphosphates

RELATIONSHIP BETWEEN THE BIOSYNTHESIS OF INORGANIC POLYPHOSPHATES AND ENERGETIC PROCESSES

Since condensed inorganic phosphates are high-energy phosphorus compounds,[514,832] it is obvious that sufficient amounts of phosphorus and energy must be present in the cells of the organism if they are to be formed. It has in fact been shown repeatedly that the presence of large amounts of polyphosphate and polyphosphate-containing granules is dependent in the first instance on the presence of large amounts of phosphorus[11,50,120,126,129,132,201–204,227,239,272,275,276,335,529,604,614,625,640,673,745,824,847] and readily oxidizable substrates[73,121,123,244,246,314,529,625,671,794,812,834,847] in the surrounding medium. It has further been shown that in yeasts and other microorganisms, the formation of polyphosphates is highly dependent on the presence of ions of potassium[117,126,363,388,529,625,635,641,796,804,821,834,847] and other metals (beryllium,[834] magnesium,[117,126,821] manganese, zinc, and iron,[199,348,351,834] and, in the case of *Azotobacter*, calcium.[160,835a,839] The action of these ions is, of course, not restricted to the biosynthesis of polyphosphates, each having its own mechanism which is apparently specific and often complex. This is, moreover, an indication that the formation of polyphosphates is an extremely complex process which is intimately involved in a wide range of vital processes taking place in the organisms in which they are found.

It was shown in the earliest investigations of Wiame,[801,811] and has subsequently been confirmed on numerous occasions,[74,75,98,100,121,123,126,244,245,288,290,786,826,834] that the metabolic systems on which the formation of polyphosphates in microorganisms depend are energy-providing processes such as fermentation, respiration, and photo- and chemosynthesis. Examples of substrates for biological oxidation which provide the energy for the formation of polyphosphates are glucose,[50,74,244,245,335,671,768,769,801,811,826,834,844] and occasionally fructose,[314] and various components of the Krebs cycle.[527,529,625,834] Further, in chemoautotrophic organisms, of which apart from the photosynthesizing sulphur bacteria[162,163,425,794] only hydrogen-oxidizing[288,289,631,634] and nitrifying bacteria[80,290,730,758,776] have been investigated from the point of view of polyphosphate metabolism, the formation of these phosphorus compounds takes place as a result of the oxidation of inorganic compounds. As far as the photosynthesizing organisms are concerned, it is not yet clear which is the energy donor for the synthesis of polyphosphates in these organisms, the photochemical reactions of photosynthesis,[240,558,695,696,745,794,825,826] or the oxidation of glucose formed during photosynthesis.[162,163,292,363,520,524,753,759]

With reference to the latter, experiments carried out by Kulaev and Vagabov,[431] using a culture of *Scenedesmus obliquus* in the middle of its logarithmic growth phase, are of some significance. The results of these experiments, presented in Fig. 22, provide evidence in support of the view that the biosynthesis of polyphosphates in this green alga in the presence of glucose proceeds at substantially the same rate in the light and in the dark (at least when phosphate is absent from the medium).

However, when the culture is grown in the light in the absence of glucose and phosphate, no significant synthesis of polyphosphate takes place in the medium initially (during the first hour). Thereafter (perhaps following the accumulation of a certain amount of carbohydrate as a result of photosynthesis), the polyphosphate content increases rapidly. These findings suggest that the biosynthesis of high-molecular-weight polyphosphates in *S. obliquus* depends in the first instance on the presence of glucose in the medium, and that it is connected indirectly with photosynthesis, although these experiments do not exclude the possibility that the biosynthesis of polyphosphates under conditions of photosynthesis might be mediated by some intermediate phosphorylated, high-energy compound which is formed directly, utilizing the energy of photochemical reactions. In particular, it is possible that, in addition to ATP, such a high-energy intermediate phosphorylated compound could be pyrophosphate which, as has been shown by the experiments of Baltchevskii *et al.*[28] and ourselves,[425] can be formed during photosynthetic phosphorylation in the purple bacterium *Rhodospirillum rubrum*.

The numerous investigations which have been carried out into the effects on the formation of polyphosphates of such uncoupling poisons as 2,4-dinitrophenol, iodoacetic acid, azide, and sodium fluoride suggest that this process is

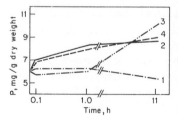

Figure 22. Change in high-molecular-weight polyphosphate content extractable by NaOH at 37°C and $HClO_4$ at 90–100°C, when *Scenedesmus obliquus* is grown (after preliminary incorporation of ^{32}P) in the absence of phosphorus in a medium under different conditions of carbon nutrition.[431] 1, Darkness − glucose − P; 2, darkness + glucose − P; 3, light − glucose − P; 4 light + glucose − P

in some way connected with either oxidative phosphorylation at the level of the respiratory chain, or glycolytic phosphorylation.[51,74,75,121,122,140,201,230,248,431,494, 529,549,671,709,821,835,840,847]

Information on the action of inhibitors of oxidative phosphorylation, in particular 2,4-DNP, on the biosynthesis of polyphosphates must however be integrated with caution. Their effects on the formation of polyphosphates may be indirect rather than direct. The results of investigations carried out with yeasts,[73-75] *Mycobacteria*,[119,122] mould fungi and algae[421,431] show that the main effect of 2,4-DNP on intact cells of microorganisms is not disruption of the link between respiration and phosphorylation, but entry into the cell of orthophosphate from the surrounding medium, i.e. the effect occurs at a stage preceding the formation of polyphosphate. Since the mode of entry of phosphate into the cell may differ markedly in different organisms as a result of the substantial differences in the structure of the cell envelopes and membranes which may be present, the action of inhibitors of oxidative phosphorylation on phosphorus metabolism, and in particular on polyphosphate biosynthesis, may vary markedly from one organism to another. The possibility also exists that the action of oxidative phosphorylation inhibitors such as 2,4-DNP and iodoacetic acid on polyphosphate biosynthesis in any given organism may vary according to whether or not an oxidizable substrate (glucose) is present in the medium. For example, investigations carried out by Kulaev and Vagabov[431] using cell cultures of *S. obliquus* grown in the presence of glucose have shown (Fig. 23a) that 2,4-DNP at a concentration of 10^{-3} M is much more effective in reducing the uptake of exogenous phosphate than is iodoacetic acid at a concentration of 5.10^{-5} M. On the other hand, iodoacetic acid at this concentration greatly

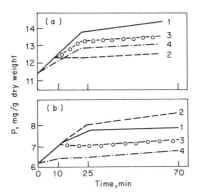

Figure 23. Change in total phosphorus (a) and high-molecular-weight polyphosphates extractable by NaOH at 37 °C and $HClO_4$ at 100 °C (b), when *S. obliquus* is grown in the dark after preliminary 10-min incorporation of ^{32}P, on a medium containing non-radioactive phosphate.[431] 1, Darkness + glucose + P; 2, darkness + glucose + P + 2,4-DNP (10^{-3} M); 3, darkness + glucose + P + ICH_2COOH (5.10^{-4} M); 4, darkness − glucose + P

reduces polyphosphate synthesis under these growth conditions, whereas 2,4-DNP even stimulates their biosynthesis (Fig. 23b).

Contrary results were obtained by Markarova and Baslavskaya,[494] working with cells from the same culture of *S. obliquus*, but growing on a medium which did not contain glucose. In this case, 2,4-DNP inhibited very strongly the biosynthesis of polyphosphates, whereas iodoacetic acid had relatively little effect on this process.

On the basis of these findings, it is assumed that in the presence of glucose, glycolysis of glycolytic phosphorylation are of greater importance in the biosynthesis of polyphosphates than respiratory phosphorylation. However, when the formation of polyphosphates is brought about solely by endogenous energy sources, processes associated with the functioning of the electron chain, which are inhibited by 2,4-DNP, assume greater importance. However, at present it is difficult to say what precisely is the mechanism of the inhibitory effect of 2,4-DNP in this case and others described in the literature.

It has been shown[669] that the effect of 2,4-DNP on mitochondria (and possibly on intact cells) is related to its action on the proton permeability of the membranes. In its presence, a reduction in the membrane potential takes place as a result of the equilibration of proton concentration on either side of the biomembranes, and since the generation of a membrane potential is, according to current ideas,[669] the major 'energy trap' of the cell, preceding the formation of ATP in the functioning of the electron transport chain, its removal by whatever means must affect polyphosphate biosynthesis. The actual mechanism by which polyphosphates are biosynthesized cannot, however, be established solely on the basis of inhibition analysis, although its use certainly provides some preliminary information. For example, our results on the effects of 2,4-DNP and iodoacetic acid on certain aspects of the phosphorus metabolism of *S. obliquus* (see Fig. 23), although they did not provide a detailed elucidation of the mechanism of polyphosphate biosynthesis in this protococcal alga, nevertheless led to the two important conclusions concerning its general features.

Firstly, they showed that polyphosphate biosynthesis is not directly related to the uptake of external phosphate into the cells of *S. obliquus*. This finding is in good agreement with those obtained using other lower eukaryotes,[369,371,397,409,412,415,421] and this appears to be a general rule which holds for the cells of all microorganisms.

Secondly, they demonstrate the possibility that, at least in some microorganisms, enzymatic reactions take place which link the biosynthesis of polyphosphates with glycolytic phosphorylation. As we shall see below, this conclusion has been amply confirmed in work with other microorganisms.[394,395,426,429,437,541]

ATP AND POLYPHOSPHATE BIOSYNTHESIS

As yet, very little is known about the actual mechanism (or mechanisms) of polyphosphate biosynthesis, and current hypotheses are of a highly speculative nature. For example, most investigators who have studied the metabilism of polyphosphates have assumed, without adequate grounds, that the biosynthesis of polyphosphates in the cell involves ATP alone.[220,243,257,333,443] This totally unproved hypothesis was first put forward by Hoffman-Ostenhof and Weigert in 1952 on theoretical grounds,[248] when they suggested the possibility of the reversible transfer of phosphoric acid groups from ATP to polyphosphates, according to the equation

$$ATP + PP_n \rightleftharpoons ADP + PP_{n+1}$$

The first demonstration of the existence of such a reaction in living organisms was provided by the work of Yoshida,[831,834] who synthesized polyphosphates from ATP with the aid of cell-free extracts from yeast, followed by the work of Hoffman-Ostenhof et al.,[245] in which an unpurified enzyme extract, obtained from yeast, catalysed the reaction mentioned above, at least in the direction of the transfer of phosphate groups from polyphosphate to ADP. Especially convincing, however, is the existence of an enzyme which is capable of catalysing the reversible transfer of phosphate residues to and from ATP and ADP, which was established by Kornberg and co-workers.[331,333] These workers succeeded in isolating this enzyme from E. coli, and in purifying it 100-fold. It was termed polyphosphate kinase. The purified enzyme requires the presence of Mg^{2+} ions for the development of its activity, and it is strongly inhibited by fluoride. Only the terminal group of ATP is transferred in the synthesis of polyphosphates. The biosynthesis of polyphosphates by Kornberg's enzyme is strongly inhibited by ADP. It was shown that this takes place as a result of a shift in the reaction towards the left. In fact, the K_m value for ADP was very small ($4.7.10^{-5}$ M), for polyphosphate it was of the same order, $2.6.10^{-5}$ M, and for ATP it was much greater ($1.4.10^{-3}$ M). Consequently, when very low concentrations of ATP are present in the cell, the reaction must proceed in the reverse direction, resulting in the synthesis of ATP from polyphosphate.

It appears that the same enzyme was isolated from Corynebacterium xerosis by Hughes and Muhammed,[257,530] who purified it 1500 times. However, in contrast to Kornberg's enzyme, it catalysed only the formation of polyphosphate from ATP, and not the reverse reaction. This appears strange when it is recalled that in Corynebacterium diphtheriae, i.e. in another species of the same family of bacteria, this enzyme exhibits reversibility of action.[331] A similar enzyme was obtained from cells of Azotobacter by Zaitseva and Belozerskii,[835,836] who demonstrated the high specificity of Azotobacter polyphosphate kinase. This enzyme effected the synthesis of high-molecular-weight polyphosphate from the terminal phosphate of ATP alone, and was unable to utilize other nucleoside triphosphates, in particular UTP and ITP, as phosphate donors.

At the time of writing, polyphosphate kinase has been detected in the following microorganisms: Escherichia coli,[331,333,540,541,542] Corynebacterium xerosis,[257,530] Corynebacterium diphtheriae,[331] Azotobacter vinelandii,[836] Aerobacter aerogenes,[218,221] Clostridium sp.,[709] Salmonella minnesota,[533] Micrococcus *lysodeikticus*,[395] *Propionibacterium shermanii*,[395,437] *Arthrobacter atrocyaneus*,[451] *Mycobacterium smegmatis*,[819] *Actinomyces aureofaciens*,[396] *Chlorobium thiosulphatophillum*,[89,256] *Chlorella* sp.[271] *Aspergillus niger*,[562] and *Saccharomyces cerevisiae*.[166–168,245,683] In *Neurospora crassa* and *Penicillium chrysogenum* the synthesis of polyphosphate by means of this enzyme has not been observed.[395] It is as yet difficult to interpret this result. Either polyphosphate kinase is totally absent from these organisms, or it is present, but functions, as has been shown for instance in yeasts,[166–168] mainly in the direction of the synthesis of ATP from polyphosphates.

It should be pointed out that the properties of polyphosphate kinase in these organisms, or at least the affinity of this enzyme for the substrates of the reaction which it promotes, are very diverse. Also, the direction of action of the enzyme *in vivo* may be very different in different organisms. We have already reported that, in *Corynebacterium xerosis*, polyphosphate kinase operates only in the direction of the formation of polyphosphates from ATP,[530] whereas in yeasts it operates mainly in the direction of the synthesis of ATP from polyphosphates. Nevertheless, in most of the organisms which have been studied, to judge from the values for K_m for ADP, ATP, and polyphosphate, the enzyme can in principle function in either direction, depending on the conditions. The principal factor controlling the mode of action of this enzyme appears to be the intracellular concentration of ADP, ATP, orthophosphate, and possibly polyphosphate. Further, on the basis of fragmentary information in the literature, it appears that the composition of the culture medium, in particular its content of glucose, phosphate, and ammonium salts, must have a marked effect on the direction of operation of polyphosphate kinase. It would appear from some investigations[533] that polyphosphate kinase in the cell is firmly bonded to certain subcellular structures. In order to isolate it, it is often necessary to treat the cell homogenate or lysate with RNAase.[333,533] The view has been expressed[533] that the cellular polyphosphate kinase may be bonded to the cytoplasmic membrane, but the possibility cannot be discounted that the direction of the operation of any given enzyme in the cell may depend to a great extent on its intracellular state and environment, and on the strength of binding with cellular structures and components.

For example, Shabalin *et al.*[656] have recently shown that two isoforms of polyphosphate kinase are present in yeast cells. The principal isoform is localized in the yeast cell cytoplasm, and appears to perform the sole function of synthesizing ATP from high-molecular-weight polyphosphates. This isoform is analogous to the enzyme previously detected and purified by Felter and Stahl.[167,168] In a purified yeast vacuolar fraction, however, another isoform of this enzyme was found which catalysed both the synthesis of ATP from high-molecular-weight polyphosphates, and the reverse reaction (the synthesis of polyphosphates from ATP).[656]

It appears that the direction in which this enzyme operates may change at different stages of development of cultures of microorganisms. This is shown in particular by the work of Nesmeyanova *et al.*[540, 541] on high-molecular-weight polyphosphate content and the activity of the enzymes which are involved in their metabolism, during the growth of a culture of *E. coli*. The results of this investigation (Fig. 24) show that during the development of a culture of *E. coli*, polyphosphate kinase displays three maxima of activity. The first, which is small, occurs during the latent phase; the second, which is also small, occurs during the first third of the logarithmic growth phase, i.e. when the accumulation of high-molecular-weight phosphate is at a maximum. The third maximum of polyphosphate kinase activity, which is of much greater magnitude than the preceding maxima, is observed at the end of the logarithmic and the beginning of the stationary growth phase of *E. coli*, i.e. when polyphosphate which has been formed previously in the cell is being rapidly consumed.

In considering these results, it is important to note two points: firstly, the very low activity of this enzyme at the time when the rate of polyphosphate synthesis is at its greatest, and secondly, the great increase in polyphosphate kinase activity at the growth stage at which the rate of utilization of polyphosphate exceeds its biosynthesis. These observations do not, of course, exclude the possibility that this enzyme also participates in polyphosphate biosynthesis in the early stages of the growth of *E. coli* cultures. However, the activity so far observed is scarcely sufficient to account for the whole, or even a significant part, of the rate of biosynthesis of the polyphosphates.

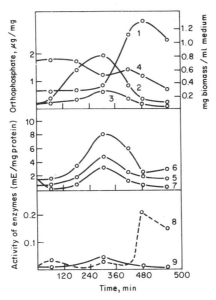

Figure 24. Polyphosphates and polyphosphate-metabolizing enzymes during the growth of *E. coli* K-12.[541] 1, Increase in biomass; 2, polyphosphates; 3, intracellular orthophosphate; 4, exogenous orthophosphate; 5, polyphosphatase; 6, tripolyphosphatase; 7, alkaline phosphatase; 8, polyphosphate kinase; 9, 1,3-DPGA: polyphosphate phosphotransferase

As we shall show later, other systems are present in *E. coli* and other organisms which are capable of effecting the biosynthesis of polyphosphates.

That at least in some organisms polyphosphate kinase is capable of functioning *in vivo* to a large extent in the direction of the utilization of polyphosphate is shown by results obtained by us[437] and other workers[166]. Convincing evidence in support of this hypothesis was provided by the findings of Kulaev *et al.*[395], who were unable to detect any polyphosphate-synthesizing activity of the polyphosphate kinase in mycelia of *N. crassa* and *P. chrysogenum* during the period in which rapid polyphosphate biosynthesis is occurring. It appears that there are at least some organisms in which polyphosphate kinase is wholly or largely responsible for polyphosphate biosynthesis. This is obviously the case in *Corynebacterium xerosis*. Further evidence has been provided by Harold and Harold[221], who isolated mutants of *Aerobacter aerogenes* which were unable to synthesize polyphosphate kinase and to accumulate polyphosphate within their cells.

It may well be that in other bacteria, for instance in *C. xerosis*[257,530] or *Azotobacter*[840], that the biosynthesis of polyphosphates is mainly effected by polyphosphate kinase.

However, the more we learn of the biosynthesis of these interesting polymers in the cells of different organisms, the more convinced we become that in most organisms polyphosphates are formed not so much by polyphosphate kinase as by other enzyme systems.

DETECTION OF 1,3-DIPHOSPHOGLYCERATE–POLYPHOSPHATE PHOSPHOTRANSFERASE

We first succeeded in detecting a biosynthetic route for inorganic polyphosphates other than the polyphosphate kinase route in *N. crassa*[393–395,400] and subsequently in other microorganisms.[395,396,437,540,541] The enzyme involved was first detected by Kulaev *et al.*[426] in the adenine-deficient mutant of *N. crassa* as-6 28610. Kulaev and Urbanek[429] had previously shown that when this mutant is grown for short periods on a medium containing small amounts of adenine, and in particular in the presence of an adenine antimetabolite (8-azaadenine), the biosynthesis of ATP from ADP in the mycelium was inhibited, whereas the biosynthesis of polyphosphate was stimulated. The more strongly the biosynthesis of ATP was inhibited, the more rapidly polyphosphate biosynthesis proceeded.

This increase in the rate of biosynthesis of inorganic polyphosphates under conditions in which the phosphorylation of ADP to ATP was strongly inhibited is evidence against the obligatory participation of the ADP–ATP system in the biosynthesis of polyphosphate in *N. crassa*. The results obtained suggest rather the possible existence in this organism of an enzyme system which mediates the synthesis of polyphosphate without the involvement of ATP.

In our first experiment designed to detect new polyphosphate-synthesizing activity using the incubation medium developed earlier by Kornberg *et al.*[333] for the detection of polyphosphate kinase, we obtained a very low incorporation of [^{32}P]orthophosphate into high-molecular-weight polyphosphate in cell-free extracts of *N. crassa*.

However, on the basis of observations derived from our work with *Scenedesmus obliquus* [431] that the biosynthesis of polyphosphates in the cells of microorganisms might relate to glycolytic phosphorylation, we added to the incubation medium certain components of the glycolytic system which were able to effect esterification of phosphate to form the intermediate high-energy compound 1,3-diphosphoglyceric acid. These components were fructose 1,6-diphosphate, aldolase, 3-phosphoglyceraldehyde dehydrogenase (PGAD), and NAD$^+$.

In the earliest experiments with these mixtures, it was found that the addition of components of the glycolytic system increased very substantially (by 3–4 times) the incorporation of [^{32}P]orthophosphate into polyphosphate. That the incorporation of [^{32}P]orthophosphate into polyphosphate was mediated by some specific enzyme present in the *N. crassa* extract, rather than by one of the newly added components of the glycolytic system, followed from control experiments in which *N. crassa* extracts were introduced into the reaction mixture after incubation and the addition of HClO$_4$. In the absence of the enzyme preparations from *N. crassa*, incorporation of [^{32}P]orthophosphate into polyphosphate in no case differed from that in the controls.

All the components of the glycolytic system which were added proved to be essential for the attainment of a high rate of incorporation of [^{32}P]orthophosphate into polyphosphate. Incorporation of ^{32}P was reduced to a particularly great extent (nearly to the level of the control) following removal of NAD$^+$ from the mixture. It was retarded to a slightly lesser extent by the absence of fructose-1,6-diphosphate. Incorporation of ^{32}P was reduced to an even lesser extent in the absence of PGAD (by 35%) and aldolase (by 20%) from the incubation mixture. These enzymes are apparently present to a certain extent in the *N. crassa* extract. Following numerous experiments aimed at determining the optimal concentrations of each component, an incubation mixture was obtained which afforded the maximum rate of incorporation of [^{32}P]orthophosphate into inorganic polyphosphate.

As can be seen from Table 22, in this incubation mixture the specific activity of this enzyme system is 0.92 mE per milligram of protein in the cell-free extract. In this case, incorporation of [^{32}P]orthophosphate into polyphosphate was 23 000 times greater than in the control experiment.

In further experiments to determine the glycolysis-dependent polyphosphate synthesizing activity, we employed the same incubation mixture. It was shown conclusively that this enzyme system resulted in the incorporation of ^{32}P into high-molecular-weight polyphosphate only. The radioactive product obtained was undialysable, and was almost completely (80%) hydrolysed to orthophosphate by treatment with 1 N HCl for 10 min at 100 °C. Hydrolysis by Thilo and Wiecker's method afforded trimetaphosphate among the products of incomplete hydrolysis. Finally, it was able to function as a substrate for

Table 22. Determination of the polyphosphate-synthesizing activity of cell-free extracts of mycelia of an adenine-deficient mutant of *Neurospora crassa* ad-6, grown for 18 h on a medium containing adenine, and for 6 h on a medium containing 8-azaadenine[426]

Components of the incubation mixture	Control	Experiment 1	Experiment 2
7% $HClO_4$ (ml)	0.58	0.58	0.58
Sample volume (ml)	0.58	0.58	0.58
Glycylglycine buffer, pH 7.4 (ml)	15	15	15
$MgCl_2$ (μM)	6	6	6
Polyphosphate, Graham's salt, $\bar{n} = 75$ (μM)	0.015	0.015	0.015
Fructose-1, 6-diphosphate (μM)	5.2	5.2	—
Aldolase (μg)	12	12	—
3-Phosphoglyceraldihyde dehydrogenase (PGAD) (μg)	14.4	14.4	—
NAD^+ (μM)	0.6	0.6	—
$Na_2H^{32}PO_4$ (μM)	8	8	8
Cell-free extract from *N. crassa* (mg protein)	2	2	2
H_2O (ml)	0.03	0.03	—
Specific activity (mE/mg protein in the extract)	0.00004	0.92	0.0143

the specific enzyme polyphosphate glucokinase. On incubation with this enzyme, the ^{32}P from the radioactive product was transferred to glucose, with the formation of radioactive glucose-6-phosphate.

These experiments leave no room for doubt that the enzymatic reaction which we have established results in the formation of high-molecular-weight polyphosphates. In order to prove that the synthesis of polyphosphates by this system is a result of glycolytic phosphorylation, the effects thereon of inhibitors of glycolytic phosphorylation and of phosphorylation at the level of the respiratory chain were examined with care. It was found that iodoacetic acid (12×10^{-3} M) and a mixture of sodium arsenate (5×10^{-2} M) and sodium fluoride (2×10^{-3} M) inhibited polyphosphate biosynthesis in our system by 96 and 95%, respectively. Inhibitors of oxidative phosphorylation in the respiratory chain (2,4-DNP at 1.4×10^{-5} M and sodium azide at 3×10^{-5} M) had no effect on the incorporation of [^{32}P]orthophosphate into polyphosphate, but an increase in the concentration to 10^{-3} M retarded the process by 25%.

These results therefore confirm the hypothesis that, in our enzyme system, polyphosphate biosynthesis is associated with glycolytic phosphorylation reactions.

Since the high-energy compounds 1,3-diphosphoglyceric acid (1,3-DPGA) and phosphoenol pyruvate (PEP) are direct donors of phosphate in the synthesis of polyphosphates, we decided to carry out direct experiments using these compounds as constituents of the incubation medium. For this purpose, 1,3-[^{32}P]DPGA amd [^{32}P]PEP labelled at the high-energy phosphate group were synthesized.

The results of work carried out by Kulaev and Bobyk,[394] in which these radio-active substrates were used in the biosynthesis of polyphosphates, are presented in Table 23. It can be seen that in those experiments in which the incubation medium consisted of a glycylglycine buffer, Mg^{2+}, polyphosphate ($\bar{n} = 180$), 1,3-[^{32}P]DPGA, and cell-free *N. crassa* extract, incorporation of the radio-active label into inorganic polyphosphate was three times greater than in experiments in which the incubation mixture contained components of the glycolytic system and [^{32}P]orthophosphate. This shows that transfer of high-energy phosphate took place from 1,3-DPGA to polyphosphate. This was also indicated by the absence of incorporation of the label into polyphosphate in control experiments in which the substrate used was hydrolysed 1,3-[^{32}P] DPGA (10 min at 100°C in 1 N HCl).

However, introduction of 1,3-[^{32}P]DPGA into the incubation mixture, and the use of a cell-free *N. crassa* extract as the enzyme preparation, did not preclude the possibility that ATP labelled at the terminal phosphate had been formed, which could function as a phosphate donor. In order to settle this question, a large number of experiments were carried out in which the place of 1,3-[^{32}P]DPGA in the incubation mixture was taken by ATP labelled at the terminal phosphate. The results of these experiments are given in Table 23. It can be seen that no incorporation of radioactive phosphate into polyphosphate occurred in those experiments in which [^{32}P] ATP was used as substrate. These experiments showed in their entirety that at least one substrate in the biosynthesis of polyphosphates by our enzyme was 1,3-DPGA.

When [^{32}P]PEP or its precursors (fructose-1, 6-[^{32}P] diphosphate and 3-[^{32}P]PGA) were used as substrates, no radioactive lable was detected in the high-molecular-weight polyphosphates (see Table 23). This group of experiments therefore provided direct proof that the phosphate donor in the biosynthesis of high-molecular-weight polyphosphates in the adenine-deficient mutant of *N. crassa* is 1,3-DPGA. Consequently, the enzyme which we have found to catalyse the transfer of phosphate from 1,3-DPGA to polyphosphate may be termed 1,3-DPGA–polyphosphate phosphotransferase, and it is considered established that it participates in the following reactions:

$$
\begin{array}{ccc}
\text{COO} \sim \text{P} & & \text{COOH} \\
| & & | \\
\text{CHOH} \quad + \text{PP}_n & \underset{\text{phosphotransferase}}{\overset{\text{1,3-DPGA-polyphosphate-}}{\rightleftharpoons}} & \text{CHOH} + \text{PP}_{n+1} \\
| & & | \\
\text{CH}_2\text{O} - \text{P} & & \text{CH}_2\text{O} - \text{P}
\end{array}
$$

Using an unpurified enzyme preparation (a cell-free extract of the adenine-deficient mutant of *N. crassa*), experiments were carried out to determine the effect on its activity of adding ATP and ADP to the incubation medium. It transpired that ATP in the concentration range tested (3.4×10^{-3} to 3.4×10^{-2} M) strongly inhibits the activity of 1,3-diphosphoglycerate–polyphosphate phosphotransferase (in a concentration of 3.4×10^{-2} M by a factor of two), while ADP in a concentration of $1.9.10^{-3}$ to 3.8×10^{-3} M increased the activity of the enzyme by 15–20%, but at higher concentrations (0.9×10^{-3} to 5.7×10^{-2}

Table 23. Identification of the phosphate donor in the biosynthesis of polyphosphates in *Neurospora crassa* by means of coupling with glycolytic reactions[394]

Constituents of the incubation mixture	Modifications											
	1	2	3	4	5	6	7	8	9	10	11	12
Glycylglycine buffer, pH 7.2 (μM)	15	15	15	15	15	15	15	15	15	15	15	15
$MgCl_2 \cdot 2H_2O$ (μM)	6	6	6	6	6	6	6	6	6	6	6	6
Polyphosphate, $\bar{n} = 180$ (μM)	0.005	0.005	0.005	0.005	0.005	0.005	0.005	0.005	0.005	0.005	0.05	0.05
Fructose-1, 6-diphosphate (μM)	5	—	—	—	5	5	—	—	—	—	—	—
Aldolase (μg)	12	—	—	—	12	12	—	12	—	—	—	—
PGAD (μg)	14.4	—	—	—	1.4	14.4	14.4	14.4	—	—	—	—
NAD^+ (μM)	0.6	—	—	—	.6	0.6	0.6	0.6	—	—	—	—
$Na_2H^{32}PO_4$ (μM)	8	8	—	8	—	8	8	8	8	—	—	—
$Na_3H^{31}PO_4$ (μM)	—	3	8	—	3	3	8	8	—	—	—	—
[^{32}P] PEP, $3.7.10^4$ counts/min.μM (μM)	—	—	—	—	—	—	—	—	—	—	—	—
3-[^{32}P] PGA, 7.10^5 counts/min-μM (μM)	—	—	2.6	—	—	—	2.6	—	—	—	—	—
Fructose-1, 6-[^{32}P] diphosphate, $5.1.10^5$ counts/min.μM (μM)	—	—	—	—	—	—	—	4.1	—	—	—	—
1,3-[^{32}P] DPGA, $4.8.10^4$ counts/min.μM (μM)	—	—	—	4.1	—	—	—	—	—	—	—	—
1,3-[^{32}P] DPGA, hydrolysed, $4.5\text{-}10^5$ counts/min. μM (μM)	—	—	—	—	—	—	—	—	—	6	—	—
[^{32}P] ATP, 6.10^6 counts/min.μM (μM)	—	—	—	—	—	—	—	—	—	—	6	—
Cell-free extract from *N. crassa* (mg of protein)	2	2	2	2	2	2	2	2	2	—	—	0,3
H_2O (ml)	—	—	—	—	—	—	0.10	0.05	0.40	1.8	1.8	1.8
Volume of sample (ml)	0.61	0.61	0.61	0.61	0.81	0.81	0.81	0.81	0.81	0.76	0.76	0.76
7% $HClO_4$ (ml)	0.60	0.60	0.60	0.60	0.80	0.80	0.80	0.80	0.80	0.76	0.76	0.76
Specific activity (mE)	0.55	0.00	0.00	0.00	0.00	0.00	0.00	0.00	0.05	1.56	0.002	0.00

M) markedly inhibited it. It has already been pointed out that in mycelia of
N. crassa, in particular in the same auxotrophic adenine-deficient mutant ad-6,
we have been unable to detect polyphosphate kinase activity, even in mycelia
which were rapidly accumulating polyphosphates. It follows that the biosynthesis of polyphosphates in this strain of *N. crassa* involves some polyphosphate-synthesizing enzyme other than polyphosphate kinase. It may be that
more than one system is present in this organism for the biosynthesis of polyphosphates, but it may be said with a reasonable degree of confidence from our
results that one system for the biosynthesis of polyphosphates in this *N. crassa*
mutant is 1,3-diphosphoglycerate–polyphosphate phosphotransferase. Following the detection of this enzyme in the adenine-deficient *N. crassa* mutant
ad-6, it was important to discover whether the enzyme was present in this
mutant only when it was grown in the presence of 8-azaadenine, or whether it
was also present in the cells of a wild parent strain grown when under normal
conditions. From the results of Kulaev et al.,[395] shown in Table 24, cell-free
extracts from the mycelia of a wild strain of *N. crassa* (Abbot-4A) grown for 24 h
on a normal Beadle and Tatum medium, also possessed 1,3-DPGA–polyphosphate phosphotransferase activity, although to a much lesser degree than that
found in the ad-6 mutant grown in the presence of 8-azaadenine, whereas polyphosphate kinase was also absent from the wild strain. The question further
arose as to the extent of the distribution of 1,3-DPGA–polyphosphate phosphotransferase among different microorganisms. To provide an answer to

Table 24. Specific activity of polyphosphate-synthesizing enzymes in some microorganisms (mE/mg protein)[395,396]

Microorganism investigated	ATP–polyphosphate phosphotransferase	1,3-DPGA–polyphosphate phosphotransferase
Neurospora crassa Abbot-4A (wild strain) (1-day-old culture)	0.00	0.53
N. crassa, adenine-deficient mutant ad-6 (grown for 18 h on a medium containing adenine, and 6h in the presence of 8-azaadenine)	0.00	1.45
Penicillium chrysogenum Q-176 (1.5-day-old culture)	0.00	0.28
Propionibacterium schermanii (4-day-old culture)	0.16	0.04
Micrococcus lysodeikticus (2-day-old culture)	0.15	0.06
Escherichia coli, strain 3 (18-h culture)	1.60	0.14
Actinomyces aureofaciens, strain 8425 (chlortetracycline producer) (24-h culture)	1.80	0.05
Actinomyces aureofaciens, strain NRRL 2209 (non-productive strain) (24-h culture)	1.90	0.03

this question, the following microorganisms were studied: the fungus *Penicillium chrysogenum* the bacteria *Escherichia coli*, *Micrococcus lysodeikticus*, and *Propionibacterium shermanii*, and the actinomycere *Actinomyces aureofaciens*.

From the data presented in Table 24, it can be seen that 1,3-diphosphoglycerate–polyphosphate phosphotransferase activity was found in the fungus *P. chrysogenum* at the stage of rapid formation of polyphosphate, and, as in *N. crassa*, polyphosphate kinase activity was not observed.

On the other hand, both types of polyphosphate-synthesizing activity were found in the bacteria and the actinomycete (Table 24). It should be pointed out that the absolute values obtained in these experiments for the specific activities of the enzymes may be by no means the maximum values, since each culture may have its own optimum conditions for the development of enzyme activity. For instance, in some experiments using cell-free extracts from fresh (but not frozen) material, we were able to obtain an ATP–polyphosphate phosphotransferase activity in *Micrococcus lysodeikticus* of 0.80 mE, and in *Propionibacterium shermanii* of 0.87 mE. The activity of 1,3-DPGA–polyphosphate phosphotransferase in these experiments was 0.22 and 0.19 mE, respectively.

It is significant however, that two different polyphosphate-synthesizing enzymes have been found in the same organism. As can be seen from Fig. 24, the activity maxima for these two enzymes during the growth of *E. coli* are not identical. The optimum activity for 1,3-diphosphoglycerate–polyphosphate phosphotransferase coincides with the maximum accumulation of polyphosphate in the cell, whereas the optimum activity for polyphosphate kinase does not occur until the rate of consumption of polyphosphate exceeds its rate of biosynthesis, although during the period of maximum accumulation of polyphosphate in the cells of *E. coli* some increase in polyphosphate kinase activity is also found. During the growth of the cultures, the activities of the polyphosphate kinase and the 1,3-DPGA–polyphosphate phosphotransferase vary in a different manner, as we have shown in *Propionibacterium shermanii* and *Actinomyces aureofaciens*.

Above all, these results show that, in these bacteria, we are dealing with two different enzyme systems for the biosynthesis of polyphosphates. This has been confirmed by the direct separation of these two enzyme activities which are present in cell-free extracts from *E. coli*, by means of gel filtration on Sephadex G-150[395]. The results obtained are shown in Fig. 25. Examination of these results showed that, using this technique, it was possible to obtain fractions which possessed either 1,3-diphosphoglycerate–polyphosphate phosphotransferase activity alone, or polyphosphate kinase activity alone. Thus, taking into account all of these findings, no doubt remains that in some organisms at least there is more than one pathway for the biosynthesis of polyphosphates. It is important to stress the existence of pathways for high-molecular-weight polyphosphate biosynthesis which do not involve ATP, but other, totally different high-energy phosphorus compounds.

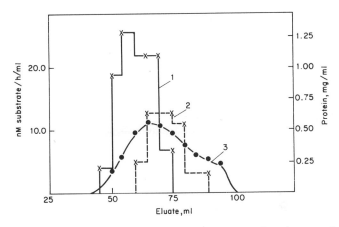

Figure 25. Separation of polyphosphate kinase and polyphosphate synthetase from *E. coli* on Sephadex G-150.[395] A cell-free extract was applied to the column, and elution was carried out with an 0.05 M Tris–HCl buffer at pH 7.2, in the cold. Fractions of 5 ml were collected automatically at a rate of 30 ml/h. 1, Polyphosphate kinase; 2, polyphosphate synthetase; 3, protein content

POSSIBLE PATHWAYS FOR THE FORMATION OF DIFFERENT POLYPHOSPHATE FRACTIONS

Up to now, we have considered the biosynthesis of polyphosphates as a whole, and we have not taken into account the available literature information on the synthesis of individual high-molecular-weight polyphosphate fractions. Nevertheless, the problem exists and deserves the closest attention.

From the earliest experiments of Wiame[809] and Juni and co-workers,[286,287] from which a general picture of high-molecular-weight polyphosphate could be obtained, it was noted that ^{32}P orthophosphate entering the yeast cell from the environment made its appearance in the cell first of all in the acid-insoluble polyphosphate fraction, and only subsequently in the acid-soluble fraction. The authors drew the conclusion that the acid-insoluble polyphosphates possessed greater physiological activity and exchangeability. However, on the basis of results obtained with mould fungi, Kulaev and Belozersky[58,365,386] and Krishnan et al.[351] have suggested that this could be explained at least to some extent as being due to the first stage of polyphosphate biosynthesis involving the acid-insoluble fraction, acid-soluble polyphosphates being formed only secondarily by degradation of the acid-insoluble polyphosphates. This hypothesis was in good agreement with the simultaneous discovery by Lohmann and Langen[468,469] that the more highly polymerized polyphosphates are formed in yeast cells first, before the formation of the less highly polymerized compounds.

These findings were subsequently confirmed on a number of occasions, and investigated in greater detail by Belozersky and co-workers[51-53,73-75,409,420,837-839]

and by other groups.[215,442,444,447,464,465,467,794] All of these investigations showed that radioactive [^{32}P]phosphate entering the cells of microorganisms from their environment appeared first in the most highly polymerized polyphosphate fraction PP$_5$ (extracted by hot perchloric acid), followed by PP$_4$, which is less highly polymerized, then by PP$_3$, PP$_2$, and finally PP$_1$, which is the least highly polymerized of the polyphosphate fractions.

On the basis of these results, it was hypothesized that the most highly polymerized fraction, PP$_5$, was formed first, the other fractions (PP$_4$, PP$_3$, PP$_2$, and PP$_1$) being formed as secondary products only, by stepwise degradation of the more highly polymerized forms by means of the appropriate enzymes (polyphosphate depolymerases), which will be discussed in detail below.

However, subsequent to the determination of the localization of the various polyphosphate fractions in the cells of yeasts and fungi, which enabled a general picture of the distribution of inorganic polyphosphates in the cells of these organisms to be built up, it became possible to address the question of the incorporation of radioactive [^{32}P] orthophosphate from the environment in a different way.

In fact, as was pointed out in the last chapter, the most highly polymerized polyphosphate fractions (PP$_5$, PP$_4$, and PP$_3$) are localized in fungi at the cell periphery, in the neighbourhood of the cytoplasmic membrane, whereas fractions PP$_1$, and PP$_2$ occur within the cell, in the cytoplasm, vacuoles, and nucleus. Consequently, exogenous orthophosphate must first be taken up into the peripherally-sited polyphosphate fractions PP$_5$, PP$_4$, and PP$_3$, and only then enter fractions PP$_2$ and PP$_1$, which are localized within the cytoplasmic membrane. It may be that this is the main reason for the initial appearance of exogenous [^{32}P] orthophosphate in fractions PP$_5$, PP$_4$, and PP$_3$, followed by PP$_2$ and PP$_1$, although it is of course possible that some proportion of the less highly polymerized polyphosphates may also be formed in fungi and other organisms by degradation of more highly polymerized forms. It can, however, by no means be assumed that all polyphosphate fractions are genetically related to each other, and cannot be formed independently of the other fractions directly from external or intracellular orthophosphate. That the formation of individual polyphosphate fractions does in fact take place independently of other fractions is indicated by a substantial body of indirect evidence.

For example, Mel'gunov and Kulaev[510] have shown that, depending on the conditions of growth and the physiological state of the cells of *N. crassa*, the biosynthetic pathways for individual polyphosphate fractions may differ. Under some conditions, for the synthesis of some polyphosphate fractions exogenous orthophosphate can be used, while for the synthesis of others reserves of internal orthophosphate can be employed. It is interesting that there is now some indication, from limited data, that in the biosynthesis of inorganic polyphosphates in higher plants (in particular, the work of Khomlyak and Grodzinskii[252,253]) [^{32}P]orthophosphate entering the leaves from the root system first passes into the acid-soluble polyphosphates, and only subsequently appears in the acid-insoluble fraction. That is to say, in these cases acid-soluble polyphosphates are not synthesized by degradation of acid-insoluble forms, but by their own independent route, which is perhaps related in some way to the photosynthetic process.

It is interesting that Bashirelashi and Dallam[41] have adduced convincing arguments in favour of the occurrence of polyphosphate biosynthesis in the nuclei of the cells of higher animals, independently of other subcellular structures. It appears that the biosynthesis of inorganic polyphosphates occurs with great intensity in these structures.

We have shown, in a series of investigations devoted to calculating the correlation coefficients (r) between the rates of formation of individual polyphosphate fractions and the rates of formation of the more important cell components, that it is possible for the biosynthesis of different polyphosphate fractions to take place independently of each other. This approach to the evaluation of the behaviour of the different polyphosphate fractions and the nature of their relationship with various cell components was first employed with success in an examination by Kritsky and Belozerskaya[352] of the relationship of phosphate metabolism to that of RNA in certain fungi. This investigation demonstrated the existence of a direct correlation between the acid-insoluble polyphosphate and the RNA content of the fruiting bodies of the fungus *Lentinus tigrinus*. Their results are shown in Fig. 26.

Using the same microorganism, it was established that the suppression of RNA production by actinomycin D brought about a decrease in the acid-insoluble polyphosphate content, with a simultaneous increase in the acid-soluble polyphosphates. A similar relationship between the formation of certain polyphosphate fractions and RNA in fungal cells was subsequently confirmed by Kritsky and co-workers, working with other microorganisms. [87,354,355,357,510,511] In these investigations, the correlation coefficients (r) between the rates of formation of different polyphosphate fractions and the total RNA in *N. crassa* and certain other fungi were calculated. In calculating the correla-

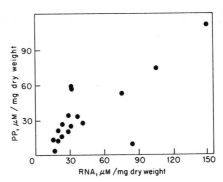

Figure 26. Acid-insoluble polyphosphate content plotted against RNA content of mycelia and fruiting bodies of *Lentinus tigrinus*[354]

tion coefficients between the rates of formation of the various polyphosphate fractions and RNA, it was established that the acid-insoluble fractions PP_5, PP_4, and PP_3 were in no way connected in their metabolism with RNA (cf. Fig. 27). Also, under normal growth conditions, no correlation was found to exist between the rate of formation of the acid-soluble polyphosphate fraction PP_1 and RNA. The only polyphosphate fraction found to be closely associated in its metabolism with that of RNA (as shown by the calculated correlation coefficients, Fig. 27) in *N. crassa* was fraction PP_2 ($r = 0.88 \pm 0.06$). This is particularly interesting, in view of the observation of Kulaev *et al.*[406] that it is this salt-soluble polyphosphate fraction which is localized in the nucleus, where the synthesis of RNA is known to take place. These results force one to the conclusion that the biosynthesis of the salt-soluble polyphosphates which are found in the nucleus is in some way intimately connected with the biosynthesis of nucleic acids in the nucleus, and it is possible that the relationship between these two nuclear processes may be that proposed in one of our papers[407], which is shown in Fig. 28. This assumes that the biosynthesis of RNA and polyphosphates may be linked by the intermediate formation of pyrophosphate. We do not exclude the possibility that our postulated mechanism for the synthesis of polyphosphates from pyrophosphate may be mediated by a phosphotransferase reaction brought about by pyrophosphatase[377]. It has been shown by several workers[564,689] that pyrophosphatase may exhibit both hydrolase and phosphotransferase functions.

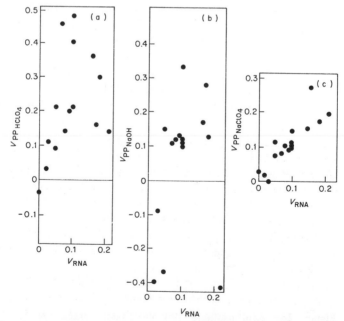

Figure 27. Correlation between the rates of formation of RNA and different polyphosphate fractions in mycelia of *N. crassa*: (a) PP_{HClO_4}; (b) PP_{NaOH}; (c) PP_{NaClO_4}.[357]

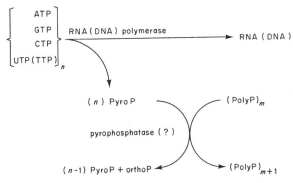

Figure 28. Proposed scheme for the interrelation of the biosynthesis of salt-soluble polyphosphates and nucleic acids[407]

Support for this hypothesis has been provided by the work of Kulaev *et al.*,[407] using a synchronous culture of the yeast *Schizosaccharomyces pombe*, during the development of which two maxima in pyrophosphatase activity were observed, corresponding to the maxima of RNA and DNA synthesis in the cell (cf. Fig. 56). Using the same culture, it was shown that of all the high-molecular-weight polyphosphate fractions, the only one to show a direct correlation between its formation and that of the nucleic acids in the cell was the salt-soluble fraction PP_2. In the same investigation[407] as in an earlier one dealing with the metabolism of polyphosphates and RNA in various adenine-auxotrophic mutants of *N. crassa* grown in the presence of 8-azaadenine,[413,510,511] an inverse correlation was found between the formation of RNA and that of the least highly polymerized, acid-soluble fraction (PP_1). It will be apparent from the data given in Fig. 29 that the lower the rate of formation of RNA in the different *N. crassa* mutants growing in the presence of 8-azaadenine, the higher is the rate of formation of acid-soluble polyphosphates. It is of interest that this inverse relationship is only observed in these *N. crassa* mutants in the presence of the antimetabolite, and not under normal growth conditions. Acid-soluble polyphosphates are almost completely absent from mycelia grown under normal conditions. It is likely that the formation of this polyphosphate fraction, which as we have seen is localized in *N. crassa* in the cytoplasm, in the volutin granules[381], is primarily related to the detoxification of intracellular orthophosphate formed by the antimetabolite-induced breakdown of RNA. The synthesis of polyphosphates by the breakdown of RNA under unfavourable growth conditions has already been demonstrated by Harold and Sylvan,[227] working with *Aerobacter aerogenes*.

Results obtained by Kulaev and Konoshenko[400] show that volutin granules from *N. crassa* contained, in addition to acid-soluble polyphosphates, 1,3-phosphoglycerate–polyphosphate phosphotransferase. These results are shown in Fig. 30. By centrifuging an unpurified mitochondrial fraction in a sucrose concentration gradient, we were able to obtain a fraction which, in

86

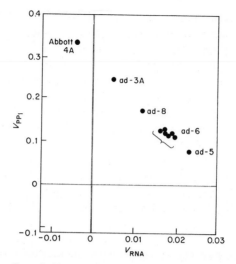

Figure 29. Correlation between the rate of synthesis of RNA and that of polyphosphates in adenine-auxotrophic mutants of *N. crassa*: V_{PP} is the rate of synthesis of acid soluble polyphosphates, and V_{RNA} is the rate of synthesis of RNA[510]

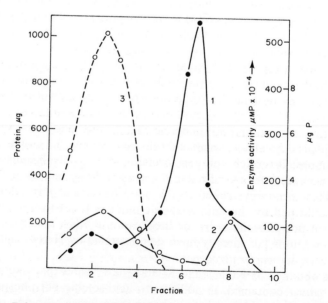

Figure 30. Behaviour of 1,3-DPGA-PP phosphotransferase during purification of a mitochondrial preparation in a sucrose concentration gradient.[400] 1, Protein; P_{lab}; 3, enzyme activity

addition to polyphosphates, contained 1,3-DPGA–PP phosphotransferase. Calculation showed that in *N. crassa* these granules contained not less than one third of the total activity of this enzyme. It is therefore possible that the biosynthesis of acid-soluble polyphosphates in particular is mediated in this organism by the enzyme system the existence of which we have demonstrated.

This is in very good agreement with the observation that 8-azaadenine on the one hand inhibits RNA biosynthesis and accelerates its breakdown,[413,510] and on the other hand greatly increases the activity of 1,3-diphosphoglycerate–polyphosphate phosphotransferase. It appears highly probable that when *N. crassa* is grown in the presence of this antimetabolite, the glucose in the medium, not being required for the biosynthesis of RNA, is rapidly utilized by glycolysis. The excess of phosphate resulting from the breakdown of RNA is used to form 1,3-diphosphoglyceric acid by glycolytic phosphorylation, and the high-energy phosphate is transferred from this to inorganic polyphosphate by means of 1,3-DPGA–polyphosphate phosphotransferase. Thus, the main features of the biosynthesis of acid-soluble polyphosphates in *N. crassa*, and perhaps in other organisms, are becoming clear.

Although reasonably plausible hypotheses concerning the biosynthesis of acid-soluble and salt-soluble polyphosphates in *N. crassa* may be advanced, very little can yet be said concerning the mode of formation of the more highly polymerized fractions (PP_5, PP_4, and PP_3) in this organism.

As has already been mentioned, polyphosphate kinase has not been detected in this organism. The 1,3-DPGA–PP phosphotransferase is wholly contained within the cytoplasmic membrane, mainly in the volutin granules.[400]

The only enzymes involved in phosphate metabolism which have been detected by Kulaev and co-workers in the periplasmic region of *N. crassa* are polyphosphatase[376,377,401] and polyphosphate depolymerase.[376,377,402] Both of these enzymes appear to be primarily involved in the utilization of polyphosphate, and this will be the subject of detailed discussion in the following chapter. It may well be true the polyphosphatase, like ATPases[611,668] and pyrophosphatases,[25,31–33,327,487,491] may participate in both the utilization and synthesis of their substrates. However, as yet little is reliably known apart from the fact, which is difficult to explain, that the changes which occur in the content of high-molecular-weight polyphosphate and polyphosphatase during the development of some organisms are almost completely coincident.[166,398,540,541,781] This is illustrated, for example, by the data given in Fig. 24, although these results could perhaps be explained as being due simply to the activation of polyphosphatase by the high intracellular concentration of its substrate.

In the meantime, the data at our disposal in no way help us to decide just how the more highly polymerized polyphosphates are synthesized in the cells of *N. crassa* and other organisms. Hope that a solution of this problem may be at hand has arisen recently as a result of work carried out in our laboratory.[432–435,656,754–756]

These investigations, carried out with a culture of *Saccharomyces carlsbergensis*, demonstrated above all a high degree of correlation between the rates of formation of acid-insoluble polyphosphates and cell-wall polysaccharides, especially between mannan and the PP_4 polyphosphate fraction ($r = 0.813 \pm 0.098$). This will be apparent from the data given in Table 25.

There is also a good correlation between the rates of formation of glucan and the PP_3 polyphosphate fraction in this yeast ($r = 0.615 \pm 0.122$). It is interesting that the formation of glycogen is not correlated with that of any of the polyphosphate fractions in this organism. The close correlation in the case of mannan and the PP_4 fraction follows from the clear similarities in the behaviour of these compounds under different growth conditions in *Saccharomyces carlsbergensis* (Fig. 31). This obviously constitutes powerful evidence in support of

Table 25. Correlation coefficients (r) between rates of formation of various polyphosphate fractions and polysaccharides in *Saccharomyces carlsbergensie*.[433] Results of 36 experiments

Polysaccharides	Polyphosphates	r
Σ(polysaccharides)	Σ(polyphosphates)	0.806 ± 0.068
Glycogen	PP_1	0.77 ± 0.02
Glycogen	PP_2	0.141 ± 0.008
Glycogen	$\Sigma(PP_2, PP_3, PP_4, PP_5)$	0.173 ± 0.018
Glycogen + mannan	$\Sigma(PP_2, PP_3, PP_4, PP_5)$	0.750 ± 0.087
Glucan	PP_2	0.291 ± 0.180
Glucan	PP_3	0.615 ± 0.122
Mannan	PP_2	0.136 ± 0.192
Mannan	PP_3	0.035 ± 0.196
Mannan	PP_4	0.813 ± 0.098

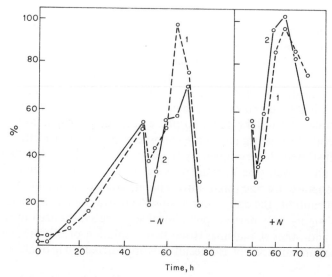

Figure 31. Changes in the mannan (1) and polyphosphate fraction PP_4 (2) contents under different growth conditions[433]

the existence of a special relationship between the biosynthetic processes. It is interesting that fractions PP_4 and PP_3 are localized in this yeast, as in *E. magnusii* and *N. crassa*, at the periphery of the cell, in the immediate proximity of and perhaps on the cytoplasmic membrane itself,[754] i.e. where part of the cell-wall polysaccharides are synthesized.

To summarize, these results suggest the existence of linked mechanisms for the biosynthesis of cell-wall polysaccharides and the more highly polymerized polyphosphates. A hypothetical scheme for such a linked biosynthetic pathway for mannan and the PP_4 polyphosphate fraction is shown in Fig. 32. In contrast to the now widely accepted view that pyrophosphate, formed in the synthesis of nucleoside diphosphate sugars, is cleaved to two molecules of orthophosphate,[247,329] our proposed scheme[432–435] envisages the transfer of one, and possibly both, of the phosphoric acid residues of the pyrophosphate molecule to polyphosphate. This would avoid the possible loss of the energy of the phosphoric anhydride bond of the pyrophosphate, and at the same time, as a result of its removal from the reaction zone by incorporation into polyphosphate, the required shift of the reaction towards the formation of GDP–mannose would be secured. Inspection of Fig. 32 shows that alternative mechanisms for polyphosphate biosynthesis linked with that of mannan may also be their formation from GDP and dolichol pyrophosphate (Dol-pp), which are products formed on the synthetic pathway to yeast mannan.[434]

Data obtained recently in our laboratory[434,435] indicate that the formation of polyphosphates linked with mannan biosynthesis is accomplished in yeasts from GDP–mannose rather than from pyrophosphate, but at which precise stage in the subsequent metabolism of GDP–mannose direct linkage of mannan biosynthesis with that of polyphosphate occurs it is as yet difficult to say.

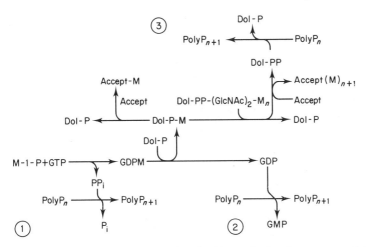

Figure 32. Possible linked mechanism for the biosynthesis of mannan and the PP_4 polyphosphate fraction[433]

It has also been shown that the formation of this polyphosphate fraction from GDP-mannose proceeds in precisely those vesicles of the endoplasmatic reticulum in which mannan biosynthesis takes place in yeasts.[435,440,654]

It is also possible that the formation of this polyphosphate fraction may occur at the cytoplasmic membrane, or at least in association with it. This is shown in particular by the work of Stahl on the biosynthesis of polyphosphates in yeasts,[680] in which he detected some as yet unknown system of polyphosphate biosynthesis which is localized in the membrane and cell wall fractions. That this enzyme system is associated with the cytoplasmic membrane was shown by experiments on the effects of detergents on its activity. It transpired that treatment of the yeast membrane and cell wall fractions with 2% desoxycholate solution resulted in the complete cessation of polyphosphate synthesis mediated by this enzyme. It is also interesting to note that this polyphosphate biosynthetic system is specially activated when the yeasts are subjected to phosphorus starvation, and it may be that the functioning of this enzyme is the prime reason for the occurrence of the phenomenon of 'polyphosphate super-synthesis' following preliminary phosphorus starvation. It should be mentioned that Mühlradt,[533] who investigated this process in *Salmonella*, agreed with the view that polyphosphate biosynthesis in microorganisms takes place at the cytoplasmic membrane. Mühlradt, however, based this conclusion on another polyphosphate biosynthetic enzyme, polyphosphate kinase.

Whether the *in vivo* biosynthesis of polyphosphates involves known or unknown enzymes, the process perhaps occurs at certain subcellular structures, primarily at various cell membranes.

WHAT IS THE ORGANIC BASIS FOR THE BIOSYNTHESIS OF POLYPHOSPHATES?

In the opinion of many workers, the formation of high-molecular-weight polyphosphates must involve some organic basis, but there is as yet no precise information on the nature of the organic components of the cell in which the synthesis of polyphosphates takes place.

In any case, it is likely[465] that the initial formation of inorganic polyphosphates does not involve RNA, since removal of the latter by means of RNAase has no effect on the process. This is also indicated by the experiments of Ebel *et al.*[143] which showed the findings of Dounce and Kay[116] and Bresler and co-workers[69-71] on the enzymatic preparation of phosphorylated RNA to be in error. Ebel *et al.* showed that the phosphorylated RNA obtained by Dounce and Kay and by Bresler and co-workers, supposedly by enzymatic transfer of phosphate from ATP to RNA, was ordinary RNA containing a small amount of absorbed ATP.

Nevertheless, RNA may be involved in some way in the functioning of certain enzyme systems which mediate the biosynthesis of inorganic polyphosphates. For example, Kornberg *et al.*[333] showed at an early date that polyphosphate kinase is extracted from the cell in association with RNA, and that purification of the former required treatment of the material with RNAase. These findings were later confirmed by Mühlradt,[532] who stressed the fact that the properties of a similar enzyme isolated from *Salmonella minnesota*, especially its solubility, changed on treatment of the unpurified enzyme preparation with RNAase. Mühlradt considered that RNA participates in some manner in the binding of polyphosphate kinase to the cellular structures of *Salmonella*, perhaps for example to the cytoplasmic membrane of these bacteria.

On the other hand, Liss and Langen,[465] on the basis of their own work, considered that the biosynthesis of high-molecular-weight polyphosphate involved a protein. It is interesting that Hughes and Muhammed,[257] investigating the properties of polyphosphate kinase from *Corynebacterium xerosis*, showed that this polyphosphate-synthesizing enzyme contained some phosphorus which was determined as orthophosphate. It was apparently bound to the enzyme protein by very labile bonds which underwent hydrolysis to orthophosphate in the presence of ammonium molybdate in acidic solution.[369,419] When this phosphorus had been removed from the enzyme preparation, the ability of the latter to synthesize polyphosphate from ATP was completely lost. The authors considered that the action of this enzyme on the phosphate resulted in the terminal growth of the polyphosphate chain. It is particularly emphasized that, in experiments on the synthesis of high-molecular-weight polyphosphates *in vitro*, oligomeric forms were never found. This problem was investigated in greater depth by Kornberg *et al.*[333] They showed that, using an enzyme isolated from *E. coli*, high-molecular-weight, non-dialysable polyphosphate was synthesized directly from ATP. This was further confirmed by Hughes and Muhammed,[257,530] who showed that the polyphosphate synthesized by an enzyme from *Corynebacterium xerosis* had a mean molecular weight of the order of 15 000 ($\bar{n} = 150$). This was further confirmed by Zaitseva and co-workers,[835, 836] in an investigation of the properties of polyphosphate kinase from *Azotobacter*. Especially impressive results concerning the degree of polymerization of the product synthesized by polyphosphate kinase have been obtained by Mühlradt,[533] who investigated this enzyme in *Salmonella minnesota*. On the basis of sedimentation and gel filtration data, he considered that the molecular weight of the inorganic polyphosphate synthesized by means of purified polyphosphate kinase was of the order of 300 000–400 000 daltons ($\bar{n} = 3000–4000$). These figures are of great interest *per se*, since they give an indication of the values of the molecular weights of the polyphosphates actually present in the cell. The question as to whether an 'matrice' is needed for the operation of the enzymes involved in the biosynthesis of polyphosphate has a direct bearing on this problem. In an investigation of the properties of polyphosphate kinase from *E. coli*, Kornberg *et al.*[333] came to the conclusion that some 'matrice' was necessary for the functioning of the enzyme, but they were unable to determine its nature. Mühlradt,[533] in studying this problem, confirmed that the presence of neither polyphosphates nor RNA were required for the functioning of polyphosphate kinase from *Salmonella*. It is therefore as yet difficult to say whether an 'matrice' is actually required for the functioning of this enzyme.

However, it appears that phosphate (which has been found, as has already been pointed out, in this enzyme from *Corynebacterium xerosis*), loosely bonded to polyphosphate kinase,[257] could be such an 'matrice'. Hoffman-Ostenhof and Slechta[243,247] on the other hand, have suggested that oligomeric polyphosphates may fulfil this role in the cell, in particular tri- and terapolyphosphates, although as has already been pointed out they have never been detected amongst the components involved in the polyphosphate kinase reaction.

In concluding this discussion of the literature information on the mechanism of the biosynthesis of high-molecular-weight polyphosphates in various organisms, and consideration of the various aspects of this problem, it is pointed out that a complete solution remains remote. Nevertheless, the undoubted progress which has been made recently in the understanding of many of the problems mentioned gives confidence that this complex problem will be resolved, even if only in general outline, within the next few years.

SOME COMMENTS ON THE BIOSYNTHESIS OF OLIGOMERIC INORGANIC POLYPHOSPHATES

The biosynthesis of oligomeric polyphosphates may be brought about, firstly, by degradation of high-molecular-weight polyphosphates, and secondly, they may occur as products of certain other enzymatic reactions.

For example, tripolyphosphate may be formed from pyrophosphate and ATP according to the equation

$$\text{ATP} + \text{pyrophosphate} \rightleftharpoons \text{tripolyphosphate} + \text{ADP}$$

That this reaction actually takes place has been demonstrated in a series of investigations.[332,334,455] It may also arise by fission of nucleoside triphosphates, with the formation of the corresponding nucleoside (or its derivative) and tripolyphosphate, such as has been demonstrated in particular by Kornberg *et al.*[334] for guanosine and desoxyguanosine triphosphates, hydrolysis of which by the corresponding enzymes takes place according to the equation

$$\text{GTP} \rightleftharpoons \text{tripolyphosphate} + \text{guanosine (or desoxyguanosine)}$$

In addition, the formation of tripolyphosphate occurs, for example, in the formation of the activated methionine derivative S-adenosylmethionine:[528]

$$\text{ATP} + \text{methionine} \rightleftharpoons S\text{-adenosylmethionine} + \text{tripolyphosphate}$$

and in the synthesis of vitamin B_{12}:[591]

$$\text{ATP} + \text{cyanocobalamin} \xrightleftharpoons{\text{reduced flavine}} \text{5-desoxyadenosylcobalamin} + \text{CN}^- + \text{tripolyphosphate}$$

Formation of tetrapolyphosphate may also occur, for example by the reaction discovered by Rafter[609] in yeasts, which proceeds according to the equation

$$\text{trimetaphosphate} + \text{orthophosphate} \longrightarrow \text{tetrapolyphosphate}$$

This reaction is of special interest, as it shows that cyclic trimetaphosphate, by ring fission in the presence of orthophosphate, may form linear polyphosphates without the participation of ATP. The role of this enzyme is not yet clear, however, since it has not been established that trimetaphosphate is present in the cell.

The matabolism of pyrophosphate in the organism comprises a separate area of contemporary biochemistry which is highly important and extensive. Since, however, this book is devoted to the biochemistry of the polyphosphates, we shall not dwell in detail on pyrophosphate metabolism. Detailed information concerning the metabolism, and in particular the biosynthesis, of pyrophosphate is given in the reviews by Kornberg[329] and Hoffman-Ostenhof. and Slechta.[243,247] Pyrophosphate is formed in a very large number of biosynthetic reactions, which may be summarized in the form of the following equation:

$$\text{pyrophosphoryl–A} + \text{B} \rightleftharpoons \text{A–B} + \text{pyrophosphate}$$

The enzymes which mediate this type of reaction may be divided broadly into two classes:

1. nucleotidyl transferases, which catalyse the reaction

$$\text{nucleoside triphosphate} + \text{X} \rightleftharpoons \text{nucleoside monophosphate–X} +$$
$$\text{pyrophosphate}$$

where X may be an amino acid, fatty acid, sugar, etc., and

2. phosphoribosyl transferases, catalysing the reaction

$$\text{5'-phosphoribosyl pyrophosphate} + \text{Y} \rightleftharpoons$$
$$\text{Y–ribosyl-5'-phosphate} + \text{pyrophosphate}$$

where Y may be in particular a nitrogenous base.

Ipata and Felicioli[269] described the isolation from yeast of an enzyme which catalyses the phosphorolysis of polyphosphates according to the equation

$$\text{orthophosphate} + \text{PP}_n \xrightarrow{E} \text{pyrophosphate} + \text{PP}_{n-1}$$

These workers postulated the occurrence in yeast of a reaction leading to the formation of pyrophosphate by stepwise dephosphorylation of inorganic polyphosphates. In view of the fundamental importance of this reaction, we checked their work.[493] We sought to detect the phosphorolysis of polyphosphate in cell-free extracts of the yeast *Saccharomyces cerevisiae* and the fungi *Endomyces magnusii* and *Neurospora crassa*, and in the nuclei and mitochondria of the yeast. In no instance were we able to detect this enzymatic reaction. It appears that Ipata and Felicioli were observing a non-enzymatic transphosphorylation of orthophosphate by polyphosphate. We have observed such a reaction under the same conditions as those used by the Italian workers.

It appears that, in the presence of the divalent ions Mg^{2+}, Ca^{2+}, Mn^{2+}, Cd^{2+}, and Ba^{2+}, the following reaction takes place:

$$PP_n + [^{32}P]\text{orthophosphate} \xrightarrow{M^{2+}} PP_{n-1} + [^{32}P]\text{pyrophosphate}$$

The optimum pH for this non-enzymatic reaction lies in the alkaline region (pH 8–9). It can also occur, however, at a lower rate, in 0.6 $HClO_4$, i.e. under the same conditions as those in which the samples of Ipata and Felicioli were maintained following blocking of the enzyme reaction. Such a reaction may proceed within the cell, but its occurrence has not yet been reliably demonstrated.

Furthermore, Baltchevskii[25,31–33] working with chromatophores of the purple bacterium *Rhodospirillum rubrum*, and Mansurova and co-workers,[487,491] working with animal mitochondria, have clearly demonstrated yet another pyrophosphate-forming reaction which appears to proceed actively in living organisms. These investigations showed that significant amounts of pyrophosphate, together with ATP, may be synthesized by linkage with the operation of the electron transport chain. Consequently, it was established that pyrophosphate could be formed as a primary product of photosynthetic phosphorylation, and of the phosphorylation involved in the mitochondrial respiratory chain.

Thus, while high-molecular-weight polyphosphates may arise as products of glycolytic phorphorylation, pyrophosphate may evidently be formed as a product of phosphorylation linked with the operation of the electron transport chain. The significance of these reaction for an understanding of the physiological functions of polyphosphates and pyrophosphate at different stages of the evolutionary development of living organisms will be considered in detail in the final chapter.

Chapter 8

Pathways for the Utilization and Degradation of High-Molecular-Weight Polyphosphates

The metabolic utilization of high-molecular-weight polyphosphates and the processes by which they are degraded involve several pathways. A number of enzymes which participate in these processes have been identified, but it is extremely difficult in any given instance to identify the route by which the degradation of inorganic polyphosphate, or its consumption in the 'metabolic boiler', takes place.

We shall consider briefly each of the enzymes which are currently known to mediate the utilization and degradation of polyphosphates.

THE POSSIBILITY OF THE INVOLVEMENT OF POLYPHOSPHATE KINASE (POLYPHOSPHATE–ADP PHOSPHOTRANSFERASE) IN THE UTILIZATION OF HIGH-MOLECULAR-WEIGHT POLYPHOSPHATES

In considering the enzymatic reactions which result in the formation of condensed inorganic phosphates in living cells, it has already been pointed out that polyphosphate kinase preparations have been isolated from many micro-organisms which catalyse both the formation and the utilization of inorganic polyphosphates. Thus, high-molecular-weight polyphosphate synthesized by means of Kornberg's enzyme can, under the influence of this enzyme, again transfer its terminal phosphate residues to ADP with the formation of ATP.[140, 245,271,331,818,836] A careful investigation of the properties of the polyphosphate kinase isolated from *Azotobacter* by Zaitseva and Belozersky[836] has shown this enzyme to possess very high substrate specificity. According to their results, this enzyme effects the transfer of high-energy phosphate from high-molecular-weight polyphosphate (Graham's salt) only, and to ADP alone.

The polyphosphate kinase from *E. coli* has been shown to possess very high affinity for ADP ($K_m = 4.7 \times 10^{-5}$ M), and a much lower affinity for ATP ($K_m = 1.4 \times 10^{-3}$ M). The same ratio of the values of K_m with respect to ADP and ATP was found by Stahl and Felter[683] for the polyphosphate kinase from yeast. On the basis of these findings, Stahl and Felter considered that, *in vivo*, this enzyme functioned mainly in the direction of the utilization of polyphosphate for the synthesis of ATP. On the other hand, it has already been mentioned that Nesmeyanova et al.[540,542] have shown that during the growth of *E. coli* several maxima of activity of polyphosphate kinase occurred (cf. Fig. 24). One of these, which was comparatively small, coincided with the period of active synthesis of polyphosphate, whereas the greatest increase in activity was recorded during the period of rapid utilization of these high-energy compounds. This enzyme is apparently able to function in either direction at different stages in the development of *E. coli*, yeasts, and perhaps other microorganisms, depending on the ATP to ADP ratio in the cell. It may, therefore, be of great significance for the control and regulation of the ATP to ADP ratio within the cell. Further, the presence of this enzyme in the cells of microorganisms makes it possible for high-molecular-weight polyphosphates to be utilized via ATP in a range of synthetic reactions. Apparently Winder and Denneny[818] had in hand such an enzyme when, using a cell-free extract from *Mycobacterium smegmatis* in the presence of inorganic polyphosphate and traces of ATP, they effected the synthesis of glycerpophosphate according to the equation

$$PP_n + ADP \longrightarrow ATP + PP_{n-1} \tag{1}$$

$$\frac{ATP + glycerol \rightleftharpoons glycerophosphate + ADP}{PP_n + glycerol \rightleftharpoons PP_{n-1} + glycerophosphate} \tag{2}$$

the actual presence of polyphosphate kinase in *Mycobacterium smegmatis* was demonstrated by Suzuki et al.[706] They showed, using a partially purified enzyme, that it had a greater affinity for ADP ($K_m = 1.1 \times 10^{-3}$ M) than for ATP ($K_m = 2.5 \times 10^{-3}$ M).

Reactions which result in the utilization of phosphate from polyphosphates via ATP in the synthesis of other organophosphorus compounds (in particular, glucose phosphate from glucose) have also been demonstrated by other workers.[331] Interestingly, McElroy[505] has also demonstrated the possible utilization of polyphosphates via ATP in bioluminescent reactions.

The involvement of polyphosphate kinase in the synthesis of ATP from polyphosphates, and its reversibility of action in many cases, is therefore beyond doubt. Nevertheless, other examples have been described in the literature in which the biosynthesis and breakdown of high-molecular-weight polyphosphate is catalysed by highly specific enzymes which differ from one another mainly by their mode of action. In particular, the biosynthesis of high-molecular-weight polyphosphates in *Aerobacter aerogenes*[221,222] and *Corynebacterium xerosis*[257,530,531] is mediated by polyphosphate kinase, and their utilization and degradation by polyphosphatase.

The existence of different biosynthetic and degradative pathways for intracellular compounds, and consequently the existence of various enzymes involved in these processes, permits finer control over the rates at which these occur within the cell and enables them to be coordinated precisely with the rates of other, associated processes. This does not exclude the possibility that the polyphosphate kinase present in the cell normally preferentially catalyses either the biosynthesis of polyphosphates from ATP, or the formation of ATP from polyphosphates, although it is of course very likely that in certain situations, especially when the appropriate ATP to ADP ratios are present, this enzyme is capable of functioning in either direction in the cell.

THE UTILIZATION OF HIGH-MOLECULAR-WEIGHT POLYPHOSPHATES BY POLYPHOSPHATE–AMP PHOSPHOTRANSFERASE

The first reports of the occurrence of an enzyme which catalyses the phosphorylation of AMP to ADP by polyphosphate according to the equation.

$$PP_n + AMP \longleftrightarrow ADP + PP_{n-1} \tag{3}$$

appeared in 1957, in a paper by Winder and Denneny.[821] This reaction was observed by these workers using as an enzyme preparation a cell-free extract from *Mycobacterium smegmatis*, and a little later by Szymona,[713] using an extract from *Mycobacterium phlei*.

The enzyme was obtained in a relatively pure state only in 1964, however, by Dirheimer and Ebel.[106,112] They succeeded in isolating polyphosphate–AMP phosphotransferase from *Corynebacterium xerosis*, and in purifying it approximately 60-fold. The enzyme required Mg^{2+} for its operation, and was specific for AMP and polyphosphates of fairly high molecular weight. However, the physiological function of this enzyme in the phosphorus metabolism of the bacteria concerned is not yet fully understood, in view of its very low affinity for AMP. For example, in the case of the enzyme from *C. xerosis*, the K_m for AMP was 2×10^{-2} M.[106,112]

Nevertheless, the discovery of this enzyme lent support at the time to the hypothesis that both the biosynthesis and utilization of high-molecular-weight polyphosphate took place only by way of their involvement in adenyl nucleotide metabolism.

It was, however, also significant that the utilization of polyphosphates, by polyphosphate–AMP phosphotransferase in particular, did not require the participation of the ADP–ATP system, as was previously believed. It is also important to note that this phosphotransferase is apparently able to function in the cell only under specific conditions, such as in the presence of very high concentrations of AMP. This may limit to a substantial extent the physiological significance of this pathway for the utilization of polyphosphates.

THE DETECTION OF POLYPHOSPHATE GLUCOKINASE (POLYPHOSPHATE–D–GLUCOSE 6-PHOSPHOTRANSFERASE) IN MICROORGANISMS

In 1957, Szymona observed in *Mycobacterium phlei* a new enzyme, polyphosphate glucokinase, which was involved in the utilization of polyphosphates.[710, 715] The discovery of this enzyme was of the greatest significance for an understanding of the physiological functions of polyphosphates. This enzyme catalysed the direct phosphorylation of glucose by polyphosphate without the involvement of ATP, in accordance with the equation

$$PP_n + glucose \longrightarrow PP_{n-1} + glucose\text{-}6\text{-}phosphate \qquad (4)$$

In subsequent investigations, Szymona and co-workers[182,710,712,714,717,719,720] studied in detail the properties of their enzyme, and investigated the optimum conditions for the polyphosphate glucokinase reaction. They found polyphosphate glucokinase in many bacteria,[720] notably in *M. phlei*, *M. smegmatis*, *M. tuberculosis*, *M. friburgensis*, and *M. scotochromogenus*, and also in *Corynebacterium diphtheriae*.[720] They were, however, unable to detect it in *Aspergillus niger*, *E. coli*, or *Aerobacter aerogenes*. The enzyme from *M. phlei* phosphorylated glucosamine in addition to glucose, but it was without effect on mannose and fructose. It was purified 65-fold,[712] and partially separated from an associated enzyme, ATP hexokinase.[717] Szymona and Ostrowski[717] obtained convincing evidence for the direct phosphorylation of glucose specifically by high-molecular-weight polyphosphate, without the involvement of ATP. In a specially undertaken investigation,[182] it was established that this enzyme cleaves inorganic polyphosphates to tripolyphosphate, which constitutes the end product of the reaction. The rate of phosphorylation of glucose by this enzyme decreased as the molecular weight of the polyphosphate increased, indicating that the reaction rate was primarily dependent on the concentration of terminal phosphate groups, and that the utilization of polyphosphate for the phosphorylation of glucose proceeds stepwise from the end of the chain. The optimum effect of polyphosphate glucokinase from *Mycobacter* required a specific ionic strength in the incubation mixture (0.3 M KCl), and the presence of Mg^{2+} ions.[712,717] The enzyme possessed a high affinity for polyphosphate ($K_m = 1.75 \times 10^{-4}$ M),[717] indicating that comparatively low concentrations of polyphosphate in the mycobacterial cell can be utilized for the phosphorylation of glucose.

Shortly after the observation by Szymona that polyphosphate glucokinase was present in *Mycobacteria*, a similar enzyme was isolated from *Corynebacterium xerosis* by Dirheimer and Ebel,[106,109] and purified 25–30-fold. In this case also, the enzyme showed high specificity, phosphorylating glucose and glucosamine only, and being inactive with respect to other hexoses, pentoses, and nucleosides. Tripolyphosphate and pyrophosphate, like trimetaphosphate and tetrametaphosphate, were incapable of functioning as phosphate donors in this reaction, but the enzyme from *C. xerosis* also possessed high affinity for inorganic polyphosphate ($K_m = 0.7 \times 10^{-4}$ M). Dirheimer and Ebel[106,109] also adduced substantial evidence that the enzyme phosphorylates glucose directly by means of high-molecular-weight polyphosphate, without the intervention of ATP.

Following these investigations, it became important to determine the range of organisms which were able to utilize inorganic polyphosphate for the phosphorylation of glucose without the intervention of ATP. This work was carried out by Uryson and co-workers.[721,764-767] Since, previous to our work, polyphosphate glucokinase had been found in *Mycobacteria* and *Corynebacteria* only, it was most important to investigate with particular care and in detail the distribution of this enzyme in organisms closely related to these bacteria. According to the classification of Krasil'nikov,[347] such organisms comprise representatives of the *Actinomycetales*, namely the *Actinomycetes* themselves, the *Proactinomycetes*, the propionic and lactic bacteria, and various cocci. From these we selected for examination for the presence of polyphosphate glucokinase representatives which were closely related to the true *Actinomycetes* (*Streptomyces*; 13 species), *Proactinomyces* (*Nocardia*; 10 species), propionic and lactic bacteria (two species each), and a variety of cocci.

For purposes of comparison, we also examined some true bacteria (*Eubacteria*), fungi (*Fungi*), green algae (*Chlorophyta*), and blue–green algae (*Cyanophyta*). In all of these organisms, enzyme activity was measured at two successive stages of purification, namely, in the initial cell-free extract (0.05 M Tris–HCl buffer, pH 7.4), and after salting out the enzyme with 40–80% saturated ammonium sulphate solution (fraction II). In many cases, fraction II was further purified by heating for 1 min at 60 °C to give a fraction (fraction III) of high specific polyphosphate glucokinase activity. In one subject (*Nocardia minima*), more exhaustive purification was carried out using calcium phosphate gel, which afforded a twentyfold purification of the polyphosphate glucokinase. Even in this case, however, the purified enzyme preparation exhibited both polyphosphate glucokinase and ATP glucokinase activity. In order to prove that in *Nocardia minima*, as in *Mycobacterium phlei*, a specific enzyme was present which utilized high-molecular-weight polyphosphate only for the phosphorylation of glucose, we fractionated the partially purified fraction II on Sephadex G-100. It can be seen from Fig. 33 that, using this method, we were able to obtain several fractions which possessed either polyphosphate glucokinase activity alone, or ATP glucokinase activity alone. This clearly demonstrates that a specific polyphosphate glucokinase is present in *Nocardia minima*.

Having established this fact, we were able to approach the main task, i.e. to investigate the distribution of polyphosphate glucokinase in the various organisms. In order to obtain comparable data, the specific activity was determined in each case for the partially purified preparation (fraction II). ATP glucokinase activity was observed in all of the species examined, once more confirming the universal distribution of this enzyme in living organisms, and its paramount importance in the metabolism of all living cells. Polyphosphate

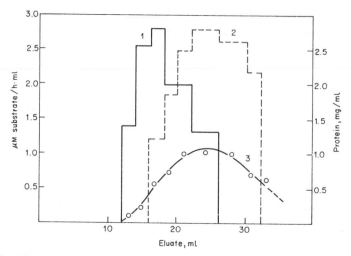

Figure 33. Separation of glucokinase from *Nocardia minima* on Sephadex G-100. Fraction III applied to the column, and eluted with 0.025 M Tris–HCl buffer at pH 7.4, in the cold; 2-ml fractions.[721] 1, Polyphosphate glucokinase; 2, ATP glucokinase; 3, protein content

glucokinase activity, on the other hand, was by no means universally present in these microorganisms. Fig. 34 shows that we were able to detect this enzyme only in those organisms assigned in the classification of Krasil'nikov to the *Actinomycetales*. In no case was this enzyme detected in representatives of the true bacteria, fungi, green or blue-green algae. Polyphosphate glucokinase was thus found to be a constitutive enzyme. In those organisms in which it occurred, its activity was detected under all growth conditions, and in cultures of all ages, whereas in organisms not assigned by Krasil'nikov to the *Actinomycetales*, it could not be found even in widely differing media, and cultures at all stages of growth. This is shown clearly by the results obtained with the fungus *Neurospora crassa*, in which the enzyme was not detected in either of two strains. It was likewise not found in an adenine-deficient mutant, even when the biosynthesis of ATP was strongly inhibited by the presence of 8-azaadenine.

Figure 34 and Table 26 show that polyphosphate glucokinase was not, however, detected in certain species regarded by Krasil'nikov as belonging to the *Actinomycetales*, but the number of these species was small (six) in comparison with the number of organisms belonging to the *Actinomycetales* in which this enzyme was detected (46 species).

These findings provide convincing evidence that the presence or absence of polyphosphate glucokinase in microorganisms is certainly not a chance occurrence. The presence of the enzyme in any given species may apparently be regarded as an additional taxonomic feature of the closely related organisms classed by Krasil'nikov as *Actinomycetales*. If this is the case, then the six species usually assigned to this class, in which polyphosphate glucokinase was not found (*Nocardia paragvensis*, *Staphylococcus aureus*, *Streptococcus faecalis*,

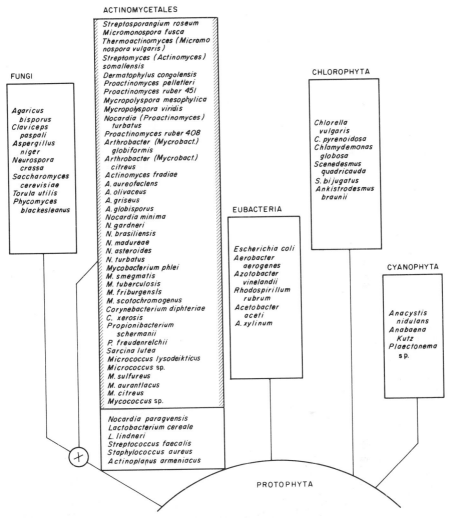

ACTINOMYCETALES

Streptosporangium roseum
Micromonospora fusca
Thermoactinomyces (Micromonospora vulgaris)
Streptomyces (Actinomyces) somallensis
Dermatophylus congolensis
Proactinomyces pelletleri
Proactinomyces ruber 451
Mycropolyspora mesophylica
Mycropolyspora viridis
Nocardia (Proactinomyces) turbatus
Proactinomyces ruber 408
Arthrobacter (Mycrobact.) globiformis
Arthrobacter (Mycrobact.) citreus
Actinomyces fradiae
A. aureofaciens
A. olivaceus
A. griseus
A. globisporus
Nocardia minima
N. gardneri
N. brasiliensis
N. madureae
N. asteroides
N. turbatus
Mycobacterium phlei
M. smegmatis
M. tuberculosis
M. friburgensis
M. scotochromogenus
Corynebacterium diphteriae
C. xerosis
Propionibacterium schermanii
P. freudenrelchii
Sarcina lutea
Micrococcus lysodeikticus
Micrococcus sp.
M. sulfureus
M. aurantiacus
M. citreus
Mycococcus sp.

Nocardia paragvensis
Lactobacterium cereale
L. lindneri
Streptococcus faecalis
Staphylococcus aureus
Actinoplanus armeniacus

FUNGI

Agaricus bisporus
Claviceps paspali
Aspergillus niger
Neurospora crassa
Saccharomyces cerevisiae
Torula utilis
Phycomyces blackesleanus

CHLOROPHYTA

Chlorella vulgaris
C. pyrenoidosa
Chlamydemonas globosa
Scenedesmus quadricauda
S. bijugatus
Ankistrodesmus braunii

EUBACTERIA

Escherichia coli
Aerobacter aerogenes
Azotobacter vinelandii
Rhodospirillum rubrum
Acetobacter aceti
A. xylinum

CYANOPHYTA

Anacystis nidulans
Anabaena Kutz
Plaectonema sp.

PROTOPHYTA

Figure 34. The occurrence of polyphosphate glucokinase in microorganisms in relation to their origins, according to the scheme of Krasil'nikov. Species in which the enzyme has been detected are indicated by hatching[374]

Lactobacterium lindneri, L. cereale, and *Actinoplanes armeniacus*), are perhaps not so closely related to the remaining representatives of the *Actinomycetales* which were studied. In several cases (for example, *Nocardia paragvensis* and *Actinoplanes armeniacus*), however, this conclusion is at variance with the opinion of microbiologists who have made a special study of the classification of the *Actinomycetales*, and it must therefore be carefully checked and compared with other, more objective features.

Further, it is of great significance to us that an enzyme which utilizes polyphosphate to phosphorylate glucose without the participation of ATP is

Table 26. Activity of PP glucokinase and ATP glucokinase in organisms assigned by Krasil'nikov to the *Actinomycetales*. Results are given in mE/mg protein

Organism	PP glucokinase	ATP glucokinase	PP glucokinase/ ATP glucokinase	Reference
Micrococcus lysodeikticus	34	7	4.9	764
Micrococcus citreus	43	15	2.9	767
Micrococcus aurantiacus	23	9	2.6	767
Micrococcus sulfureus	12	4	3.0	767
Micrococcus sp. strain 115N	97	23	4.2	767
133N	146	35	4.2	767
220N	124	39	3.2	767
Mycococcus sp. strain 36F	125	54	2.3	767
61N	14	8	1.8	767
196N	94	50	1.9	767
357N	199	97	2.0	767
12N	125	45	2.8	767
114N	66	19	3.5	767
338N	54	19	2.9	767
Sarcina lutea	34	8	4.2	764
Staphylococcus aureus	0	63	—	764
Streptococcus faecalis	0	97	—	764
Propionibacterium shermanii	100	10	10	764
Propionibacterium friedenreichii	37	Trace	1	764
Lactobacterium lindneri	0	110	—	764
Lactobacterium cereale	0	157	—	764
Mycobacterium tuberculosis	87	10	8.7	720
Mycobacterium scotochromogenus	237	118	2	720
Mycobacterium phlei	169	130	1.3	720
Mycobacterium smegmatis	263	233	1.2	720
Mycobacterium friburgensis	177	155	1.1	720
Arthrobacterium (Mycobacterium) citreus	173	97	1.8	767
Arthrobacterium (Mycobacterium) globiformis	23	11	2.1	767
Corynebacterium xerosis	125	62	2.0	109
Proactinomyces ruber strain 451	247	26	9.5	766
Nocardia (Proactinomyces) madurea	150	22	7.0	721
Nocardia (Proactinomyces) turbatus	43	15	2.9	766
Micropolyspora viridis	233	107	2.2	766
Micropolyspora mesophylica	92	51	1.8	766
Nocardia (Proactinomyces) gardneri	67	57	1.2	721
Nocardia (Proactinomyces) minima	172	158	1.1	721
Nocardia (Proactinomyces) brasiliensis	83	110	0.8	721
Nocardia (Proactinomyces) asteroides	78	187	0.4	721
Proactinomyces ruber strain 408	30	273	0.1	766
Nocardia (Proactinomyces) paragvensis	0	112	—	721
Proactinomyces pelletieri	Trace	71	—	766
Streptomyces (Actinomyces) globisporus	80	122	0.6	721
Streptomyces (Actinomyces) olivaceus	60	127	0.5	721

Table 26 (continued)

Organism	PP glucokinase	ATP glucokinase	PP glucokinase/ ATP glucokinase	Reference
Streptomyces (Actinomyces) fradiae	30	267	0.1	721
Streptomyces (Actinomyces) aureofaciens	38	252	0.1	721
Streptomyces (Actinomyces) griseus	38	312	0.1	721
Dermatophylus congolensis	3	23	0.1	766
Streptomyces (Actinomyces) somaliensis	9	174	0.05	766
Thermoactinomyces (Micromonospora) vulgaris	Trace	12	—	766
Micromonospora fusca	Trace	95	—	766
Stretosporangium roseum	Trace	252	—	766
Actinoplanus armeniacus	0	154	—	766

present only in extremely primitive organisms, apparently of very ancient origin.[315,686] In more highly developed microorganisms such as fungi and algae, on the other hand, this enzyme is absent.

The absence of polyphosphate glucokinase from such ancient organisms as the blue–green algae once more demonstrates that the presence of this enzyme would appear to be an attribute of a limited group of organisms which include the *Actinomycetes, Proactinomycetes, Mycobacteria, Corynebacteria, Propionibacteria, Micrococci, Tetracocci, Mycococci,* and other related species. The presence of this enzyme indicates the existence of certain common metabolic features in these species, and provides, as mentioned above, further evidence of their phylogenetic relationship.

Taken together with such features as the structure of the nucleic acids,[55,56] the components of the cellular envelope, and phospholipid composition,[830] polyphosphate glucokinase activity may be of assistance in developing a new and more objective classification of microorganisms.

Since the facts presented in this section have, in our view, a direct bearing on the question of the role of inorganic polyphosphates in the evolution of phosphorus metabolism, they will be discussed in greater detail in the final chapter.

DETECTION OF POLYPHOSPHATE FRUCTOKINASE (POLYPHOSPHATE–D-FRUCTOSE 6-PHOSPHO-TRANSFERASE) AND OTHER SIMILAR ENZYMES

Another enzyme which was capable of utilizing inorganic polyphosphate as phosphate donor was discovered by Szymona and Szumilo.[718] This was polyphosphate fructokinase, which catalysed the reaction

$$PP_n + \text{fructose} \longrightarrow PP_{n-1} + \text{fructose-6-phosphate} \qquad (5)$$

This enzyme was found in *Mycobacterium phlei*, when this organism was grown on a medium containing fructose. It was absent from the cells of *Mycobacteria* growing on a glucose-containing medium, and was therefore an adaptive enzyme. In addition to polyphosphate fructokinase activity, these workers[718] also observed ATP–fructokinase activity, leading to the formation of fructose-6-phosphate with ATP functioning as the phosphate donor. These activities were partially separable by gel filtration on Sephadex G-100, and they were therefore attributed to different enzymes. Two other adaptive enzymes have been discovered by Szymona *et al.*[716] namely polyphosphate mannokinase, which is synthesized in the cells of *Mycobacterium phlei* when this organism is grown on a medium containing mannose, and polyphosphate gluconatokinase, which is found in the same organism when it is cultured in the presence of gluconate.

These enzymes effect the phosphorylation of hexoses by means of polyphosphate phosphorus, as follows:

$$PP_n + \text{mannose} \longrightarrow PP_{n-1} + \text{mannose-6-phosphate}$$
$$PP_n + \text{gluconate} \longrightarrow PP_{n-1} + \text{gluconate-6-phosphate}$$

Both types of enzyme activity were shown to require the presence of Mg^{2+} ions and a high concentration of salts.

In view of the fact that many organisms contain enzymes capable of utilizing high-molecular-weight polyphosphate instead of ATP for the phosphorylation of sugars and nucleotides, it is noteworthy that it has been shown repeatedly (for review, see ref. 255) that in addition to ATP, many other high-energy compounds are able to function as direct donors of phosphate in a variety of phosphorylation reactions. This has been demonstrated for example, in the case of pyrophosphate, as has already been mentioned, in several investigations.[170,564,689]

POLYPHOSPHATES AND THEIR ROLE IN THE METABOLISM OF HIGH-MOLECULAR-WEIGHT

In addition to the enzymes mentioned above, which mediate the transfer and further utilization of the phosphate residues of polyphosphates in various biochemical processes, there occur in many organisms polyphosphatases which hydrolyse these compounds to fragments of lower molecular weight.[5,38,165,166,198, 222, 249,257,264–266,283,284,310,311,327,348–350,355,358,377,385, 396,398,399,457,481,482,498–501, 506,531,540–542,547,760, 761,781] In view of the large number of investigations devoted to this topic, it is not possible to deal with them fully here. Detailed accounts of these investigations have been given in various reviews.[220,243,257,264,363,443,501,506,635,810] In this book, we shall discuss only the more fundamental and recent work in this area.

On the basis of the work of Ingelman and Malmgren,[264] Mattenheimer,[501] and Hughes and co-workers,[257,531] the enzymes capable of cleaving condensed inorganic phosphates which are known to date can be divided into the following groups:

1. Polyphosphatases which cleave inorganic polyphosphates *in vitro* directly to orthophosphate (polyphosphate phosphohydrolases),

2. Polyphosphatases which depolymerize inorganic polyphosphates to less highly polymerized fragments (polyphosphate depolymerases, or polyphosphate polyphosphohydrolases),

3. Oligopolyphosphatases, which cleave oligomeric polyphosphates (n = 4–2) *in vitro* to form orthophosphate and even less highly polymerized polyphosphates (tetrapolyphosphate phosphohydrolase, tripolyphosphate phosphohydrolase, and pyrophosphate phosphohydrolase).

4. Metaphosphatases, which cleave cyclic metaphosphates with fission of the ring to form the corresponding linear polyphosphates (tetrametaphosphate phosphohydrolase and trimetaphosphate phosphohydrolase).

We shall consider some of these groups of enzymes, in particular those which have been investigated in some detail.

Polyphosphate phosphohydrolases

Polyphosphate phosphohydrolases, or more simply polyphosphatases, have been investigated in the greatest detail in the five microorganisms *Corynebacterium xerosis*,[257,531] *Aerobacter aerogenes*,[222] *Saccharomyces cerevisiae*,[165,166] *Endomyces mangnusii*,[1-3,5,385,404] and *Neurospora crassa*.[327,377,399,401,404,760-762] They mediate the following reaction:

$$PP_n + H_2O \longrightarrow PP_{n-1} + \text{orthophosphate}$$

This enzyme was first investigated in detail in *Corynebacterium xerosis*.[257,531]

Muhammed[531] purified the enzyme from *C. xerosis* approximately 100-fold. The purified preparation hydrolysed high-molecular-weight polyphosphates of the Graham's salt type very rapidly to orthophosphate, and had a fairly high affinity for the former ($K_m = 7.7 \times 10^{-4}$ M), but was completely inert towards oligomeric polyphosphates and metaphosphates, Pyrophosphate, tri- and tetraphosphate, tri- and tetrametaphosphates, and ATP were not hydrolysed by it to orthophosphate. Short-chain fragments were never found amongst the products of hydrolysis of high-polymeric polyphosphates, indicating that this enzyme hydrolysed polyphosphates terminally.

A very similar enzyme was isolated by Harold and Harold[222] from *Aerobacter aerogenes*. It differed from the enzyme from *Corynebacterium xerosis* in requiring Mg^{2+} ions for the expression of its activity, whereas the enzyme from *C. xerosis* was inhibited by Mg^{2+} ions. It is interesting that Harold and Harold succeeded in separating a polyphosphatase, isolated from *A. aerogenes* and actived on high-polymeric polyphosphates, from tripolyphosphatase.

A polyphosphatase capable of hydrolysing inorganic polyphosphate to orthophosphate was also isolated from baker's yeast by Felter et al.[165] This enzyme was purified approximately 20-fold. Like the enzyme from *Aerobacter aerogenes*, it was strongly activated by Mg^{2+} ions. The optimum concentration of magnesium in the incubation medium was 1 μg/ml, and the optimum ratio of the concentration of magnesium to that of the substrate undergoing cleavage was 1:1. An activating effect was also displayed by Co^{2+} ions (even stronger than that of Mg^{2+}) and Ni^{2+} ions. Ca^{2+}, Mn^{2+}, and Cu^{2+} ions were virtually without effect on the activity of the yeast polyphosphatase. The optimum pH for this enzyme was 7.5, and the optimum temperature for its operation was 45–50 °C.

Unlike Harold and Harold, who worked with a similar enzyme from *A. aerogenes*,[222] the French workers were unable to isolate from yeast a poly-phosphatase preparation which was active towards high-molecular-weight polyphosphate only, and was inactive towards tripolyphosphate. On the contrary, at all stages of purification the yeast enzyme preparation retained very nearly the same ratio of tripolyphosphatase activity to that towards Graham's salt ($\bar{n} = 30$). At all stages of the purification, hydrolysis of tripolyphosphate mediated by the yeast enzyme preparation proceeded at three times the rate of cleavage of high-molecular-weight polyphosphate, although it is true that following purification on Sephadex G-100 the maximum of tripolyphosphatase activity did not correspond with that of the hydrolysis of high-molecular-weight polyphosphates. Nevertheless, these workers concluded that yeast polyphosphatase was capable of terminal hydrolysis of the whole range of inorganic polyphosphates, with the exception of pyrophosphate. They were, therefore, unable to confirm the work of Mattenheimer,[499] who had reported the presence in yeast of a specific enzyme, tripolyphosphatase, which hydro-lysed tripolyphosphate alone. It should be pointed out that, according to the French workers, the affinity of their enzyme preparation for tripolyphos-phate was high ($K_m = 3.45 \times 10^{-4}$ M).

At present, therefore, the question of the presence in yeast of a specific enzyme capable of hydrolysing high-molecular-weight polyphosphates to orthophosphate on the one hand, and of a specific tripolyphosphatase on the other, remains unresolved, although the writer considers it more likely that yeast contains two specific enzymes, polyphosphatase and tripolyphosphatase, which Felter et al.[165] were unable to separate in consequence of the close similarity of their physicochemical properties. The writer inclines to this view, since similar enzymes were isolated by Harold and Harold[222] from bacteria, and by ourselves[327,328,399,401,760–762] from mycelia *Neurospora crassa*.

The investigation of polyphosphatase from *N. crassa*[327,344,376,399,401,404,761,762] was carried out by the writer in collaboration with Konoshenko, Mansurova, Umnov, and Egorov.

In the first stage of this work, carried out with a cell-free extract, it was shown that the optimum pH for polyphosphatase was 7.2, and that the enzyme re-quired for its activity the presence of Mg^{2+} ions (1 μg/ml). An investigation was carried out into the polyphosphatase activity of cell-free extracts of *N.*

Table 27. Investigation of the polyphosphate activity of a cell-free extract of *Neurospora crassa*.[399] Results expressed as E/mg protein

Conditions for determination of activity	Length of polyphosphate chain						
	290	180	72	47	40	30	3
Tris–HCl (pH 7.2), polyphosphate, cell-free extract	4	5	5	3	2	10	25
+ MgCl$_2$, (1μM/ml)	11	11	15	17	20	29	90
+ MgCl$_2$ (1μM/ml) + KCl (200 μM/ml)	30	34	39	29	35	40	74

crassa, using as substrates polyphosphates of various chain lengths. It was found (Table 27) that cleavage of polyphosphates by the unpurified preparation from *N. crassa* was virtually independent of chain length.

From the data presented in Table 27, it can be seen that the polyphosphatase activity of a cell-free extract of *N. crassa* towards high-molecular-weight polyphosphates increases significantly in the presence of K$^+$ ions (200 μM/ml). The activating effect of K$^+$ on polyphosphatase activity had been observed earlier by Harold and Harold,[222] working with an enzyme from *Aerobacter aerogenes*. It is interesting that the tripolyphosphatase activity of the cell-free extract is not only not increased, but is even slightly inhibited by the presence of K$^+$. This fact, taken in conjunction with inhibition analysis, makes it necessary to assume that in *N. crassa* we are probably dealing with two different enzymes, polyphosphatase (which hydrolyses inorganic polyphosphates from the chain ends), and tripolyphosphatase (which cleaves tripolyphosphate only).

Subsequent investigations in our laboratory confirmed this hypothesis by the analytical[328,760] and preparative[761] preparation of protein fractions which possessed either polyphosphatase or tripolyphosphatase activity alone. A diagrammatic representation of the electrophoretograms obtained is shown in Fig. 35. It was also established that tripolyphosphatase clearly differed from

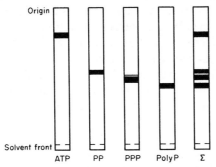

Figure 35. Distribution of phosphohydrolase activity in a cell-free extract from *N. crassa*, following electrophoresis in polyacrylamide gel[760]

pyrophosphatase and ATPase in several respects,[399] especially in its behaviour on electrophoresis in polyacrylamide gel (PAG), and it therefore constituted a strictly specific enzyme which hydrolysed tripolyphosphate exclusively.

Results obtained by Egorov and Kulaev[156] have shown that purified preparations of this enzyme cleave tripolyphosphate to pyrophosphate and orthophosphate. Further investigations[156] have shown that a homogeneous preparation of tripolyphosphatase from *N. crassa* consists of two identical subunits each of molecular weight 40 000. It was found that the presence of divalent metal ions was required for the expression of activity in this enzyme. The best activators were Co^{2+}, Mo^{2+}, and Mn^{2+} in concentrations of 10^{-3} M. The optimum pH for this enzyme was 7.0, the optimum temperature was 450 °C, and its isoelectric point was 4.75. The K_m with respect to magnesium tripolyphosphate was $5.9.10^{-4}$ M. The presence in *N. crassa* of an enzyme which hydrolysed inorganic polyphosphates terminally to orthophosphate was confirmed by Umnov *et al.*[761] They succeeded in purifying the enzyme more than 1000-fold (Table 28). This highly purified preparation was, to judge from electrophoresis in PAG, virtually homogeneous, i.e. it consisted of a single protein which was very slightly contaminated by a second, associated component. The molecular weight of the polyphosphatase from *N. crassa*, as determined by the method of Kingsbury and Masters,[307] was 49 000–52 000 daltons. It exhibited a high affinity for polyphosphate, having a K_m value of 6.8×10^{-4} M, i.e. almost identical with that found by Muhammed *et al.*[531] for the polyphosphatase from *Corynebacterium xerosis*.

In all of the properties examined, the purified polyphosphatase from *N. crassa* was similar to the unpurified enzyme investigated earlier by Konoshenko.

The effects of various divalent metal ions on the activity of a partially purified polyphosphatase preparation from *N. crassa* (degree of purification 65-fold) was examined, in concentrations ranging from 5×10^{-4} to 5×10^{-3} M. It was found that the cations Mg^{2+}, Co^{2+}, Mn^{2+}, and Fe^{2+}, in concentrations of 0.0005–0.0015 M, stimulated the activity of the polyphosphatase. A further increase in concentration resulted in inhibition of the enzyme. The cations Ca^{2+}, Sr^{2+}, Ba^{2+}, Zn^{2+}, Ni^{2+}, and Cu^{2+} inhibited *N. crassa* polyphosphatase in all concentrations, even in the presence of the optimum concentration of magnesium. In this respect (i.e. the action of different divalent cations on its activity), the polyphosphatase from *N. crassa* differed substantially from similar enzymes isolated from other biological sources. For example, the polyphosphatase from *Corynebacterium xerosis* was inhibited by any metal cation at any concentration.[531] As has already been pointed out, a wide range of divalent cations have different effects on yeast polyphosphatase.[165] For example, Ni^{2+} ions at a certain concentration activate yeast polyphosphatase, but strongly inhibit this enzyme from *N. crassa*.

The effects of divalent metal ions on polyphosphatase activity were investigated in some detail by Afanas'eva and co-workers,[1–3,5,385] using the enzyme from the fungus *Endomyces magnusii*. In a cell-free extract from *E. magnusii* cells,

Table 28. Purification of polyphosphatase from mycelia of *Neurospora crassa*[761]

Stage of purification	Volume (ml)	Activity (E/ml)	Total activity (E)	Protein (mg)	Specific activity (E/mg)	Yield (%)	Degree of purification
Homogenate	1740	0.127	202	7200	0.031	100	1
Centrifugation, 150 000 g	1740	0.126	200	4400	0.046	99	1.48
Precipitation with streptomycin sulphate*	1760	0.125	200	—	—	99	—
Precipitation with $(NH_4)_2SO_4$, 0.42–0.65 saturated	25	3.59	99.7	475	0.189	49.5	6.1
Fractionated on Bio-Gel P-60	60	1.51	100.3	470	0.191	50.3	6.15
Fractionated on cellulose DE-32	52	1.44	82.2	30	2.52	41.1	82.3
Concentrated on Amicon ultrafilter	2	36.80	82.0	30	2.52	41.0	82.2
Electrophoresis in PAG	21	2.11	42.0	1	42.10	20.8	1360
Concentrated on Amicon ultrafilter	2	21.50	41.8	1	42.10	20.7	1360

* Protein was not determined in the presence of streptomycin sulphate.

they found a polyphosphatase having an optimum pH of 7.1. The enzyme hydrolysed high-molecular-weight polyphosphates of differing chain lengths, the specific activity of the enzyme varying inversely with the chain length of the molecules serving as the substrated for this reaction. The enzyme exhibited a fairly high affinity for its substrate. Thus, the value of K_m for polyphosphate of chain lenght $\bar{n} = 290$ was 3.5×10^{-4} M. This figure is close to the corresponding value found for polyphosphatase from other biological sources.

It was found that the polyphosphatase from *E. magnusii* was a metalloenzyme, requiring for its functioning the presence in the incubation medium of Mn^{2+} and Co^{2+} (see Fig. 36). The ions Mg^{2+}, Zn^{2+}, Fe^{2+}, and Ni^{2+} were less effective, while Cu^{2+}, Ca^{2+}, Al^{3+}, Ag^+, Hg^{2+}, Cd^{2+}, Pb^{2+}, and $(UO_2)^{2+}$ inhibited this enzyme. The effects of these ions on the activity of *E. magnusii* polyphos-

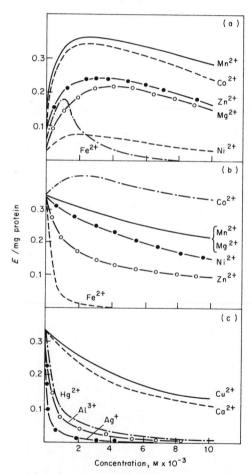

Figure 36. Effect of metal ions on polyphosphate phosphohydrolase activity from a cell-free extract of the fungus *E. magnusii*.[385] (a) Activating effect; (b) and (c) effect in the presence in the reaction mixture of 2×10^{-3} M Mn^{2+}

phatase are shown in Fig. 37. It was found that the hydrolysis of high-molecular-weight polyphosphate by this enzyme was also inhibited by p-chloromercuribenzoate and by KCN.

The kinetics of this polyphosphatase reaction do not follow the classical Michaelis–Menten law. The dependence of the rate of reaction on equimolar concentrations of substrate and activator (for instance, Mn^{2+}) is described by a sigmoid curve which passes through a maximum. There is no direct relationship between the rate of the reaction and the concentration of the enzyme. On the basis of her results, Afanas'eva has suggested that the role of the divalent metal ions participating in the polyphosphate phosphohydrolytic reaction consists in, firstly, an activating effect on the enzyme molecule, and secondly, the formation of a true substrate reaction complex of the cations with the polyphosphate molecules. This in turn may bring about a specific change in the secondary structure of the polyphosphates, rendering the P–O–P bond more susceptible to attack.

However, it remains unclear why different polyphosphatases require different divalent cations for the expression of optimum activity. It is reasonable to suppose that the overall picture we obtain from a study of the effects on polyphosphatase activity must be considerably complicated by the fact that the cations influence both enzyme and substrate. Since the enzymes from different organisms may differ somewhat from each other (particularly in their affinity for any given effector), and a very wide range of substrates can be and are being used by various workers in the investigation of polyphosphatase activity, the overall inhibitory effect of any given ion in the incubation medium may differ substantially from one case to another. Afanas'eva et al.[6] have recently detected a multiplicity of forms of polyphosphatase in *Endomyces magnusii*.

In the case of the polyphosphatase from *E. magnusii*, it was also shown that

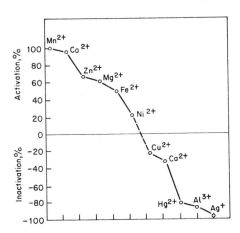

Figure 37. Effect of cations on the activity of polyphosphate phosphohydrolase from *E. magnusii*[385]

one of the products of the enzymatic hydrolysis of polyphosphate, namely orthophosphate, can function as a regulator of the activity of this enzyme in the cell. All the properties of polyphosphatase mentioned above were, however, investigated *in vitro*, and it was therefore very tempting to investigate as fully as possible its functioning *in vivo*. One approach to this problem was made by Konoshenko *et al.* who sought to establish the intracellular localization, and the nature of the linkage between *N. crassa* polyphosphatase and various subcellular structures. Even before any systematic work of this nature had been carried out, reports by Weimberg and Orton,[796–798] Souzu,[674,675] and Kulaev *et al.*[404] had appeared which suggested that polyphosphate was localized superficially in fungi, in the immediate vicinity of the more highly polymerized polyphosphates. The results of a careful investigation of this problem, using *N. crassa* polyphosphatase, carried out by Kulaev and co-workers,[401,762] are shown in Table 29, in which are summarized the results of several independent experiments. Table 29 shows that the protoplasts from *N. crassa* possess approximately half the polyphosphatase activity of the intact cells, and that this activity was absent from the nuclei and mitochondria. It was present in the membranes, but in comparatively small amounts. After further fractionation, practically the whole of the polyphosphate remaining in the protoplasts was found in the hyaloplasmic fraction. These results led to the conclusion that at least part of the polyphosphatase in *N. crassa* is located between the cell wall and the plasmic membrane, the remainder being very loosely linked with the protoplast. In a subsequent series of experiments, the effect on polyphosphatase activity of brief washing of the protoplasts (without causing structural damage) with an extraction medium containing a small amount (0.02%) of a detergent was examined. It was found that washing the protoplasts with desoxycholate solution had no effect on their polyphosphatase activity, but washing with a solution of Triton X-100 in the same concentration resulted in a marked reduction in activity.

In order to establish whether the Triton X-100 inactivated the polyphosphatase, or whether it removed it from the surface of the protoplast, Mansurova

Table 29. Specific activity of polyphosphatase in cells of *Neurospora crassa*[401]

Structure	Polyphosphatase (mE/mg protein)
Cells	50
Protoplasts	30
Nuclei	0
Mitochondria	0
Cytoplasmic membranes	5
Microsomes	2
Hyoloplasma	40

Incubation medium: Tris–HCl, pH 7.2 (50 μM); $MgCl_2$ (0.5 μM); KCl (100 μM); polyphosphate ($\bar{n}$ = 290) (0.1–0.15 mg); protein (100–200 mg); total volume, 0.5 ml; time of incubation at 37 °C, 20 min.

and Velichko carried out an experiment, the results of which are shown in Table 30. Protoplasts of *N. crassa* were incubated for 15 min in an extraction medium, either without an additive or with the addition of 0.02% or 0.1% of Triton X-100, followed by removal from the incubation medium during the course of the next 10 min by centrifugation. Following this treatment, samples of both the incubation medium and the protoplasts were tested for polyphosphatase activity. The results showed that when the incubation was carried out in the presence of 0.02% Triton X-100 solution, virtually all of the enzyme was removed from the protoplasts, in contrast to the controls.

Bearing in mind that such incubation did not destroy the integrity of the protoplasts, these results suggest that in *N. crassa* the polyphosphatase is located exclusively in the cell at the external surface of the plasmatic membrane, and apparently loosely bound to it.

The results of this experiment (Table 30) show that inactivation of the enzyme does in fact occur in the presence of Triton X-100, but that it becomes substantial only at concentrations of 0.1% and above. It is important to note that, from our preliminary data, removal of Triton X-100 from a solution containing polyphosphatase inactivated by the detergent results in reactivation of the enzyme. One of the most significant results of the work which has been carried out on the localization of polyphosphatase in *N. crassa* mycelia was the demonstration that the enzyme and tripolyphosphatase differ not only in their nature and properties, but also their localization in the cells of this organism. It was shown (Table 31) that the whole of the tripolyphosphatase activity of *N. crassa* cells was retained in the protoplasts following lysis of the cell walls with snail enzyme. It is of interest that, apart from the hyaloplasma, the activity of this enzyme was found only in the mitochondria, out of the varied subcellular structures isolated from the protoplasts. It can be seen from the data presented in Table 32 that, expressed as per gram of cells (from the cytochrome a content), from 70 to 100% of the tripolyphosphatase activity is attributable to the mitochondria. Even on careful purification of the mitochondrial preparation in a sucrose concentration gradient (Fig. 38), it was not possible to separate the tripolyphosphatase activity from the mitochondrial protein peak.

It can be concluded from these results that the enzyme which specifically

Table 30. Effect of Triton X-100 on the isolation and activity of polyphosphatase from protoplasts of *Neurospora crassa*[344]

Additive	Total enzyme activity (ME)			
	In protoplasts	In incubation medium	Total	Lose of activity (%)
Control (no detergent)	210	20	230	—
+ 0.02% Triton X-100	10	160	170	26
+ 0.1% Triton X-100	0	40	40	83

Table 31. Specific activity of tripolyphosphatase in cells and subcellular structures of *Neurospora crassa*[401]

Structure	Tripolyphosphatase (mE/mg protein)
Cells	150
Protoplasts	60
Nuclei	0
Mitochondria	200
Plasmatic membranes	6
Microsomes	20
Hyoloplasma	160

Incubation medium: Tris–HCl, pH 7.2 (60 μM); MgCl$_2$ (0.5 μM); tripolyphosphate (0.1–0.15 mg); protein (100–200 mg); total volume, 0.5 ml; time of incubation at 37 °C, 20 min.

Table 32. Tripolyphosphatase activity of mitochondria from *Neurospora crassa*[401]

Structure	Total enzyme activity	
	mE/g of cells taken	% of activity of cells taken
Cells	3400–3800	100
Protoplasts	3400–4400	100–117
Mitochondria	2300–3800	70–100

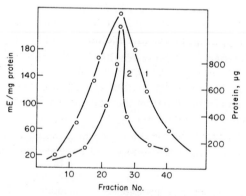

Figure 38. Distribution of tripolyphosphatase activity in *N. crassa* mitochondria following centrifugation in a sucrose concentration gradient.[401] 1, Protein; 2, enzyme activity

hydrolysed tripolyphosphate is localized in *N. crassa* either exclusively or mainly in the mitochondria, and it apparently plays a special part in the phosphorus and energy metabolism of this organism. For instance, it is possible that tripolyphosphatase, like ATPase and pyrophosphatase,[668] may play a leading role in ion transport across the mitochondrial membrane, and in other energy-dependent processes.[668] It is important to note that, according to the results of Egorov and Kulaev,[156] tripolyphosphatase is present in mitochondria isolated from different biological materials.

It should be mentioned that up to the present time *N. crassa* tripolyphosphatase is perhaps the only oligopolyphosphatase whose properties and cellular location have been investigated in detail.

Among investigations into the oligopolyphosphatases which have not so far been mentioned, that of Mattenheimer[499-501] is worthy of attention. He found certain oligophosphatases in yeast, as follows:

(a) tetrapolyphosphatase, which cleaves tetrapolyphosphate according to the equation

$$\text{tetrapolyphosphate} \longrightarrow \text{tripolyphosphate} + \text{orthophosphate}$$

(b) tripolyphosphatase, which cleaves tripolyphosphate according to the equation

$$\text{tripolyphosphate} \longrightarrow \text{pyrophosphate} + \text{orthophosphate}$$

(c) pyrophosphatase, which cleaves pyrophosphate to orthophosphate according to the equation

$$\text{pyrophosphate} \longrightarrow \text{two molecules of orthophosphate}$$

It is interesting that Mattenheimer found yeast tripolyphosphatase to be a completely specific enzyme which does not cleave other polyphosphates, including more highly polymerized polyphosphates. This finding is not in accordance with later work by Felter and Stahl.[166]

Oligopolyphosphatases have also been found in animal[198] and plant tissues[79, 283,284] The oligopolyphosphatases also include various metaphosphatases which mediate the hydrolysis of cyclic metaphosphates, the ring being cleaved to form the corresponding polyphosphates. Those so far discovered are (a) a tetrametaphosphatase, and (b) a trimetaphosphatase (see, for example, refs. 499-501). Although metaphosphatases have been found in a wide variety of organisms (yeasts,[499-501] higher plants,[596,597,617] and animals,[64,149,198,502]) the presence of their substrates, the metaphosphates, has not (as already pointed out) been established beyond doubt in any of these groups of organisms.[106]

Polyphosphate polyphosphohydrolases

Polyphosphate polyphosphohydrolases (or polyphosphate depolymerases) have so far been found in only a very small number of organisms.[264-266,355,356,358,]

[376,402,457,481,482,562,846] They have been investigated in the greatest detail in the fungi *Aspergillus niger*[264-266,481,482,562] and *Neurospora crassa*.[355,356,358,376,377,402]

We shall now consider in more detail the properties of the polyphosphate depolymerases found in *N. crassa* Chernysheva and Kritsky. They used a viscometric method to determine the activity of these enzymes, enabling them to follow changes in the molecular weight of the substrate by the changes in the specific viscosity (η_{sp}) of the incubation medium. A method of calculation was developed[358] which enabled them to determine the number of P–O–P bonds in the substrate attacked by the enzyme, and to express the activity of the enzyme in terms of standard units of activity (E). The number of micro-equivalents of bonds cleaved in the polyphosphate molecule (A) are calculated from the equation

$$ A = \left(\frac{\bar{n}_0}{\bar{n}_t} - 1 \right) \left(\frac{C_{PP}}{\bar{n}_0} \right) $$

where $\bar{n}_0$ is the mean chain length of the substrate and n_t that of the reaction product, and C_{PP} is the concentration of polyphosphate in micromoles of labile phosphorus introduced into the reaction mixture.

An analysis of the polyphosphate depolymerase reaction products by paper chromatography did not reveal any significant amounts of ortho- or pyrophosphate, or polyphosphates of low molecular weight ($n = 3–5$). This led to the conclusion that the reaction involves the enzymatic hydrolysis of P–O–P bonds situated at some distance from the ends of the polyphosphate chains.

In an investigation into the more important properties of the polyphosphate depolymerases, Kritsky et al.[358] followed the effects of changes in the concentration of the enzyme on the rate of the reaction. It was found that the rate of enzymatic depolymerization of inorganic polyphosphate ($\bar{n} = 290$) was proportional to the amount of enzymatic protein.

It is important to note that the optimum pH for the enzymatic degradation of inorganic polyphosphates lies in the acidic region, at pH 3.2–3.4, in common with the polyphosphate depolymerase reaction observed previously in the cells of certain bacteria and fungi.[266,482,562] It is stressed that during the course of the depolymerase reaction the terminal phosphate groups of the substrate are not affected, although they are attacked by the group of enzymes mentioned above, the phosphohydrolases, which mediate the hydrolysis of polyphosphate which operate, in contrast to the polyphosphate depolymerases, in the neutral pH region.[399] It is interesting that a similar difference in the concentrations of hydrogen ions which effect the hydrolysis of intermediate or terminal P–O–P bonds in the high-polymeric polyphosphate molecule is also observed in the non-enzymatic hydrolysis of polyphosphates.[504]

An examination of the effects of the ions of divalent metals on the polyphosphate depolymerase activity in *N. crassa* showed that Zn^{2+} ions had the greatest activating effect on the enzyme, increasing the activity of the dialysed cell-free extract 6.5-fold in a concentration of 1×10^{-3} M. In concentrations of 1×10^{-4} and 1×10^{-5} M, Mg^{2+}, Ba^{2+}, and Mn^{2+} ions had almost the same activating

117

effects, whereas Ca^{2+} ions, in low concentrations, had a somewhat inhibitory effect. Such an activating effect of divalent ions on the enzymic degradation of polyphosphate is also typical of the enzymes from other mould fungi.[482,562]

In an investigation into the effect of the substrate ($\bar{n} = 290$) concentration on the rate of the polyphosphate depolymerase reaction, the rate of enzymatic hydrolysis was found to increase as the polyphosphate content of the sample increased. The relationship between reaction rate and substrate concentration departed from hyperbolic, as was made apparent by plotting the reciprocal of the substrate concentration against reaction rate. The resulting curve, which deviated from a straight line, was typical of enzymatic reactions which do not follow classical Michaelis–Menten kinetics. Similar results were also obtained using a substrate with $\bar{n} = 180$.

Since the concentration of orthophosphate in the medium and in the cell is the key factor in the control of the phosphorus-metabolizing enzymes, this investigation[358] included an examination of the effects of different concentrations of orthophosphate on the activity of *N. crassa* polyphosphate depolymerase. The presence of inorganic orthophosphate in the reaction medium, at the polyphosphate concentration of 0.85×10^{-2} M normally used in the experiments, reduced the specific activity of this enzyme, the inhibitory effect of the orthophosphate increasing substantially when its concentration in the sample was increased. At a concentration of 5×10^{-2} M, orthophosphate completely suppressed polyphosphate depolymerase activity (Fig. 39). However, further detailed study has shown that the effect of orthophosphate on the enzyme is of a more complex nature.

An examination of the effect of a constant concentration (1.3×10^{-2} M) of orthophosphate on depolymerase activity at differing substrate concentra-

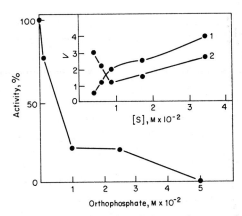

Figure 39. Effect of orthophosphate on the rate of enzymatic depolymerization[358] Substrate PP ($\bar{n} = 290$); substrate concentration 0.85×10^{-2} M. The inset shows the dependence of the reaction rate on substrate concentration: (1) without orthophosphate; (2) in the presence of 1.3×10^{-2} M orthophosphate

tions has revealed the following behaviour. At low substrate concentrations (of high-molecular-weight polyphosphate, $\bar{n} = 290$), orthophosphate displayed a substantial stimulating effect on the activity of the enzyme (Fig. 39, inset). Thus, at a polyphosphate concentration of 0.43×10^{-2} M, the specific activity of the polyphosphate depolymerase was increased more than four-fold by the addition of 1.3×10^{-2} M orthophosphate. Increasing the concentration of polyphosphate in the incubation medium to 0.85×10^{-2} M resulted in a sharp decrease in, followed by suppression of polyphosphate depolymerase activity by the orthophosphate. Hence, the effect of orthophosphate on polyphosphate depolymerase activity is of a very complex nature–depending on the substrate concentration, orthophosphate at any given concentration may either inhibit or accelerate the degradation of polyphosphates, i.e. it is possible that the ratio of the concentration of the substrate to that of orthophosphate is the controlling factor in the fine regulation of the activity of the enzyme at its site of action. It may be that in the case of other polyphosphate-metabolizing enzymes also, the ratio of the concentrations of polyphosphate and orthophosphate at the reaction sites is a factor of great importance in regulating their activity.

One more feature of N. crassa polyphosphate depolymerase which remains to be mentioned is that it is almost completely inhibited by fluoride. Thus, when this inhibitor is present in the incubation mixture in a concentration of 5×10^{-3} M, the specific activity of the polyphosphate depolymerase falls to as little as 6% of its original value in the cell-free extract.

The specificity of action of different polyphosphate depolymerases in the metabolism of the various polyphosphate fractions is an important question. Examination of the results obtained by various workers before our own work was initiated, using a wide variety of microbiological sources, led to the conclusion that the cells of microorganisms contain several polyphosphate depolymerases, viz.: (a) depolymerase, which cleaves very high molecular weight polyphosphates having $\bar{n} = 1000$ and below;[264–266,348] (b) depolymerase, which cleaves polyphosphates having $\bar{n} = 30$ and above to form fragments with $\bar{n} = 10$ together with a small amount of orthophosphate;[499–501] and (c) depolymerase, which cleaves polyphosphates with $\bar{n} = 10$ to form polyphosphates with $\bar{n} = 4–9$, together with a small amount of orthophosphate.[499–501]

Such enzymes were found, for instance, in yeasts[499–501] and in Aspergillus niger,[264,348,562] but no thorough experimental investigations had been carried out into the possibility of the presence in any given organism of specific polyphosphate depolymerases acting only on inorganic polyphosphates of a certain chain length.

In collaboration with Chernysheva and Kritsky, the author made the first attempt to discriminate between these activities, working with N. crassa polyphosphate depolymerase, but none of our results, with the exception of those relating to the behaviour of polyphosphate depolymerase activity during the development of N. crassa mycelia, provided an answer to the problem. It appeared to us that the only substantial indication of the possible presence in N. crassa cells of a range of polyphosphate depolymerases specifically

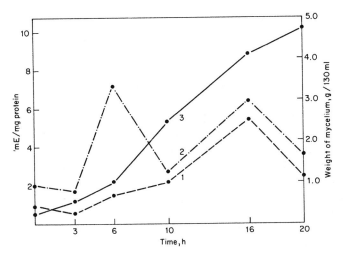

Figure 40. Changes in polyphosphate depolymerase activity during the development of fungal mycelia.[358] 1, For substrate PP ($\bar{n}$ = 290); 2, for substrate PP ($\bar{n}$ = 180), concentration of the substrates 0.85 × 10^{-2} M; 3, growth curve of the culture

'tuned' to the various polyphosphate fractions differing in their degree of polymerization was the differing nature of the changes in polyphosphate depolymerase activity towards two substrates having different $\bar{n}$ values (180 and 290), during ontogenesis (Fig. 40).

It was of special interest to determine the intracellular localization of the polyphosphate depolymerases, since a knowledge of their localization in the cell could provide a key to the understanding of the physiological functions of these important enzymes in the metabolism of polyphosphates in fungi.

In order to determine the intracellular location of the polyphosphate depolymerases present in mycelia and conidia of *N. crassa*, Kulaev *et al.*[402,376] carried out an investigation of the following cellular structures: protoplasts, nuclei, mitochondria, microsomes, and hyaloplasma.

Table 33 shows that removal of the cell wall by treatment with an enzyme couples from the digestive juice of the vine snail, *Helix ponate*, resulted in almost complete loss of enzyme activity. The specific activity of the polyphosphate deplymerases of the intact cells was 8.3 mE when PP ($\bar{n}$ = 290) was used *crassa* (mE/ mg protein)[402]

Table 33. Intracellular localization of polyphosphate depolymerase activity in *Neurospora crassa* (ME/ mg protein)[402]

Substrate	Intact cells	Proto-plasts	Nuclei	Mito-chondria	Micro-somes	Hyalo-plasma
Polyphosphate ($\bar{n}$ = 290)	8.3	0.6	0.15	0.0	2.2	0.0
Polyphosphate ($\bar{n}$ = 180)	9.4	1.3	0.20	0.0	—	0.0

Incubation medium: Sodium acetate buffer, pH 3.2–3.4 (0.3 μM); polyphosphate (25 μM, calculated as orthophosphate); total volume, 3.0 ml; time of incubation at 30°C, 30 min.

as substrate, whereas the specific activity of the protoplasts did not exceed 0.6 mE.

In no experiment was polyphosphate depolymerase activity detected in the mitochondria and hyaloplasma. This absence of enzymatic activity in the mitochondria again shows that the metabolism of inorganic polyphosphates is not linked directly to oxidative phosphorylation in the respiratory chain. Slight enzymatic activity was found in other structural elements of the cell, namely the nuclei and microsomes, but purification of the nuclear preparation in a sucrose concentration gradient resulted in the total loss of this enzymatic activity. The presence of small amounts of activity in the microsomal fraction may in all probability be explained as being due to contamination of the microsomes by fragments of the external cell membrane.

Although the absence of polyphosphate depolymerase from the nuclei and microsomes strictly cannot be regarded as having been established conclusively, the weight of the evidence given above tends to show that polyphosphate depolymerases are completely, or almost completely, localized at the periphery of the N. crassa cell. These enzymes are most probably sited in the space between the cell wall and the outer cytoplasmic membrane, i.e. in the same localization as their substrates, the inorganic polyphosphates.[381,405]

However, it is possible that polyphosphate depolymerases are present within the cell (in particular, in the nucleus and volutin granules) which cleave polyphosphates of fairly low molecular weight ($\bar{n} = 40$–10), i.e. the same as those found in the intracellular structures of microorganisms. Unfortunately, the viscometric method which we employed to determine polyphosphate depolymerase activity did not permit measurement of the depolymerization of such relatively low molecular weight polyphosphates.

In considering the possible physiological functions of polyphosphate depolymerase in the metabolism of the polyphosphates, it is necessary to consider the views of Kritsky and Chernysheva.[356] They suggest, on the basis of research into the metabolism of high-molecular-weight polyphosphates in Neurospora crassa, and in particular the findings on the localization of polyphosphate depolymerase in this organism, that these enzymes may, for example, participate in the penetration of newly synthesized polyphosphate molecules through the membrane into the cell. Bearing in mind the variety of chain lengths of the various polyphosphate fractions found by these workers in the mycelia of N. crassa, it might well be thought that the energy liberated by the fission of a single P–O–P bond (enzymatic depolymerization) would not be dissipated, and that the fission of the P–O–P bond in the middle of a chain is a justifiable operation, since the energy thus liberated is more readily utilized for the transport of the two fragments formed from the original polyphosphate molecule through the membrane into the cell. Since those polyphosphate fractions with the lowest molecular weights (PP_1 and PP_2) occur within the cell, whereas those with higher molecular weights are found outside it, it is proposed that the migration of polyphosphates in different regions of the cell, and their involvement in metabolism, occur with the participation of depolymerases, which in effect fulfil the function of translocases. It is also possible that movement of poly-phosphates both inside and outside the cell involves an enzyme–polyphosphate complex.

To summarize, in concluding this section on phosphatases which cleave condensed phosphates, it is noteworthy that, in living cells, these enzymes function both as phosphatases and as phosphotransferases. Such behaviour has been recorded in the literature on phosphatases which hydrolyse other high-energy phosphorus compounds. For example, the phosphatase which hydro-lyses acetyl phosphate to orthophosphate was capable of transferring phosphate from acetyl phosphate to glycerol[192] and other compounds.[18] The phosphatase which cleaves phosphoenol pyruvate was able to transfer phosphate from this high-energy compound not only to water, but also to glycerol and other organic compounds.[564] The enzyme which hydrolyses creatine phosphate to orthophos-phate may also transfer the phosphate residue from creatine phosphate to glucose and glycerol.[255,513] The phosphatase which cleaves simple phospho-amides to free orthophosphate is able to phosphorylate glucose, using phos-phoamides as phosphate group donors.[255] In this connection, interesting work carried out by Fisher and Stetten[170,689] and Nordlie and Arion[564] has provided substantial evidence that rat liver and kidney pyrophosphatase also possess *in vivo* pyrophosphate–glucose transferase and glucose-6-phosphatase activity.

Thus, it is possible that both the polyphosphatases and metaphosphatases found in cell-free extracts from many organisms also fulfil other functions within the cell. In particular, polyphosphatases may function as transferases, while the metaphosphatases may perhaps utilize substrates other than meta-phosphate within the cell.

In concluding this section on the degradation and utilization of inorganic polyphosphates, it remains only to point out that the particular polyphosphate-cleaving enzymes which occur in a given organism determine which polyphos-phate fractions predominate in it. The differences in the particular enzymes present are apparently responsible for the fact that, in many organisms, there occur only acid-insoluble[257,559,768,771] or predominantly acid-soluble,[162,391,709,745] or the whole range of condensed inorganic phosphates of different molecular weights, from high-molecular to tripolyphosphate inclusive.[442,444,447]

Chapter 9

The Accumulation and Utilization of Polyphosphates in Living Organisms, and the Control of their Metabolism

It has become clear in recent years that the metabolism of polyphosphates in different organisms, although partly involving similar pathways, nevertheless possesses certain characteristic features. In this section we shall consider the principal features of the more typical aspects of polyphosphate metabolism in those organisms which have been subjected to the closest and most thorough examination, and for which it is at least in outline possible to depict the general features of their metabolism. Special attention will be paid to the particularly complex and as yet little investigated problem of the intracellular control of polyphosphate metabolism in various groups of organisms. In this part of the review, therefore, we shall consider some specialized aspects of the biochemistry of inorganic polyphosphates.

We begin our discussion with a description of the results obtained with one of the more ancient groups of microorganisms, assigned by Krasil'nikov to the *Actinomycetales*.

ACTINOMYCETALES

Mycobacteria and corynebacteria

The metabolism of polyphosphates in these two genera of bacteria has many features in common, and it is best to discuss them together. Both types of organism, under normal growth conditions, accumulate substantial amounts of volutin granules,[257] and possess the widest range of polyphosphate-cleaving enzymes so far known.[106,714]

1. General features of the metabolism of high-molecular-weight polyphosphates

The most detailed investigations into the polyphosphate metabolism of the *Mycobacteria* have been carried out by Drews,[123] Mudd and co-workers,[529,834] Winder and Denneny,[821] and Szymona.[714] These workers have shown that rapidly dividing cells of *Mycobacteria*, in the logarithmic growth phase, accumulate minimal amounts of polyphosphate. When the cells enter the stationary growth phase, the more rapid biosynthetic processes cease, the polyphosphate content increases rapidly. Then the development of the cultures is inhibited by nitrogen starvation,[625] antimetabolites (azaserine[529,834] or ethionine[133]), or a deficiency of Zn^{2+} in the medium, polyphosphate accumulates rapidly. When normal conditions of growth are restored, the polyphosphates are actively utilized for the biosynthesis of nucleic acids and phospholipids, for which they serve as the source of phosphorus.[529,821,834] Polyphosphate metabolism under various growth conditions and at different stages of development has also been investigated in the *Corynebacteria*.[106,257,625]

Thus, Hughes and Muhammed[257] showed that the polyphosphate content of *Corynebacterium xerosis* was a function of the phase of growth. When cells of this bacterium were placed in fresh medium, these workers observed an accumulation of polyphosphate during the latent period which precedes the phase of rapid growth. On entering the logarithmic growth phase, the polyphosphates were actively utilized, and in the stationary phase they again accumulated. As in *Mycobacteria*, rapid accumulation of polyphosphate occurred in this organism when the cells were starved of nitrogen. Polyphosphate metabolism in these bacteria was shown to be very closely linked with that of the nucleic acid.

It was established in separate experiments using ^{32}P that, in *Mycobacteria*,[529] polyphosphate phosphorus was incorporated into nucleic acids, and was subsequently involved directly in the biosynthesis of these most important biopolymers.

Further, Sall et al.,[625] in a separate investigation using a synchronous culture of *Corynebacterium diphtheriae*, showed that polyphosphates are involved in cell division in these bacteria. Thus, it was shown that acid-insoluble polyphosphates accumulate in the greatest amounts at the stage preceding cell division, and are consumed very rapidly during this process. The accumulation of acid-insoluble polyphosphates before the onset of cell division takes place in parallel with that of RNA.

Relatively little is known about polyphosphate biosynthesis in these groups of bacteria. It has been shown only that the normal course of this process requires a constant input of energy, and that the formation of polyphosphates in these organisms involves ATP and polyphosphate kinase.[257,530,819] The ways in which polyphosphates are utilized in these organisms may differ widely, however.

2. Enzymes of metabolism of high-molecular-weight polyphosphates and the regulation of their operation

The work of Hughes and Muhammed[257,531] has shown that there are present in *C. xerosis* polyphosphatases which hydrolyse inorganic polyphosphates to orthophosphate. It can be seen from Fig. 41 that the polyphosphatase activity changes during the growth of a culture, in parallel with the accumulation of polyphosphates and polyphosphate-containing granules in the cells. This demonstrates the great importance of this enzyme in the metabolism of polyphosphates in *Corynebacteria*. It may be that the level of polyphosphate in the cells of *C. xerosis* controls in some manner the activity (and perhaps also the biosynthesis) of this enzyme. Further, the results obtained lend further support to the possible involvement *in vivo* of polyphosphatase, both in the utilization and the biosynthesis of polyphosphates.

Of fundamental importance for an understanding of many of the problems of polyphosphate metabolism is the discovery in *Mycobacteria* and *Corynebacteria* of polyphosphate glucokinase, an enzyme which utilizes polyphosphates without the intervention of ATP, for the phosphorylation of glucose.[106,109,710,714,715,720]

According to Dirheimer and Ebel,[109] the affinity of this enzyme for polyphosphate in *C. xerosis* is greater than that of polyphosphatase. This may indicate that the utilization of polyphosphates in these organisms involves polyphosphate glucokinase to a greater extent than polyphosphatase. It is interesting that the specific activity of *C. xerosis* polyphosphate glucokinase changes during ontogenetic development, increasing with the requirement for polyphosphates during rapid growth and cell multiplication.[106]

As has been mentioned in the preceding chapter, it has been shown in the

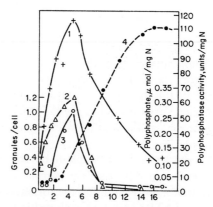

Figure 41. Changes in the amounts of polyphosphate, granules, and polyphosphatase activity in *Corynebacterium xerosis*.[257] 1, Polyphosphatase activity; 2, polyphosphates; 3, granules; 4, increase in biomass

laboratory of Szymona[716,718] that three adaptive enzymes are present in *Mycobacteria* which are involved in polyphosphate utilization. The investigations of Szymona and co-workers have shown that if *M. phlei* is grown on a medium containing fructose, mannose, or gluconate, the formation of the appropriate adaptive enzymes is induced to utilize polyphosphate, namely, polyphosphate fructokinase, polyphosphate mannokinase, or polyphosphate gluconatekinase. Each of these enzymes phosphorylates its own, and no other, specific substrate (fructose, mannose, or gluconate) to form fructose-6-phosphate, mannose-6-phosphate, or gluconate-6-phosphate, respectively. It is interesting that when this bacterium is grown on a medium containing glucose, these enzymes are not found. Therefore, the conditions of growth, and in particular the composition of the nutrient medium, control very closely the metabolism of polyphosphates in these organisms. This example also shows clearly how strongly the conditions of growth influence polyphosphate metabolism. Finally, it should be pointed out that another enzyme has been found in *Corynebacteria* and *Mycobacteria*, namely AMP phosphotransferase, which phosphorylates AMP to ADP by means of inorganic polyphosphates, similarly without the participation of ATP.[106,110,112,714,821]

It therefore appears that there exist no less than four pathways for the utilization of polyphosphates in these groups of organisms. At the present time, however, virtually nothing is known concerning the actual mechanism by which the functioning of these enzymes is controlled within the cell. This control may be effected both genetically and metabolically, and by specific compartmentalization of the polyphosphate-metabolizing enzymes within the cell. Polyphosphate metabolism in *Mycobacteria* and *Corynebacteria* is shown schematically in Figs. 42 and 43. Although very similar in principle, it is reason-

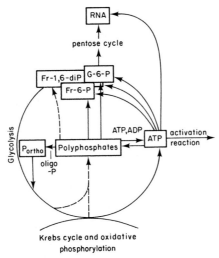

Figure 42. Possible pathways of polyphosphate metabolism in *Mycobacteria*[714]

126

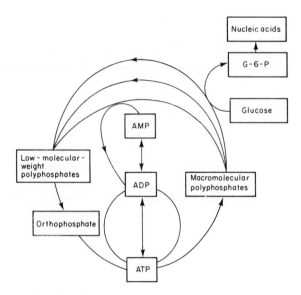

Figure 43. Possible pathways of polyphosphate metabolism in *Corynebacteria* [106]

able to present both schemes, since they emphasize different aspects of the metabolism of these very closely related organisms.

In Fig. 42, from a paper by Szymona,[714] which summarizes the literature data on possible pathways for the metabolism of polyphosphates in *Mycobacteria*, possible interconnections between polyphosphate and nucleic acid metabolism can clearly be seen. On the other hand, the close connection between the energy-liberating glycolysis reaction and the di- and tricarboxylic acid cycle and polyphosphate metabolism is graphically shown. Of great interest in this scheme are the proposed pathways for the synthesis of polyphosphates (by glycolytic phosphorylation without the participation of ATP), and their utilization in the phosphofructokinase reaction.

In the scheme (Fig. 43) put forward by Dirheimer,[106,110] the possible interconnections of the metabolism of polyphosphates with that of the adenyl nucleotides are clearly shown. Apart from this, it shows graphically that the pathways for the synthesis and utilization of polyphosphates in *Corynebacteria* do not appear to be identical, and are mediated by different enzymes.

Propionibacterium shermanii

1. General features of the metabolism of high-molecular-weight polyphosphates

Konovalova and Vorob'eva[326] have examined the polyphosphate content of *P. shermanii*. A study of the results revealed some significant facts. First by, at all growth stages of cultures of the organism (using lactic acid as the source of carbon), 70–80% of the total polyphosphates were found in the fraction extracted from the cells only by our modification[409] of the method of Langen and Liss, using hot perchloric acid (i.e. a very highly polymerized fraction). Secondly, the remaining 20–30% of the polyphosphates were distributed fairly evenly between the salt-soluble and alkali-soluble fractions. Finally, the virtually complete absence of low-molecular-weight, acid-soluble polyphosphates in the propionic bacteria at all growth stages was demonstrated. As is shown in Fig. 44, the total amount of polyphosphate increases rapidly during the growth of the culture. Thus, under normal growth conditions, between the first and third days of growth the amount of polyphosphate (per 100 ml of medium) increased nearly 6-fold, whereas during the stationary growth phase (between the third and fifth hours), the accumulation of polyphosphate remained at approximately the same level. It is interesting that this mode of accumulation of polyphosphates is mainly determined by the behaviour of

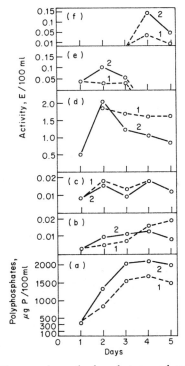

Figure 44. Changes in polyphosphates and polyphosphate-metabolising enzymes during the development of a culture of *Propionibacterium shermanii* under normal conditions, and in the presence of polymyxin M.[437] (a) Total polyphosphates: 1, polymyxin M; 2, normal conditions. (b) – (e): 1, Normal conditions; 2, polymyxin M. (b) 1,3-Diphosphoglycerate–polyphosphate phosphotransferase; (c) ATP–polyphosphate phosphotransferase; (d) polyphosphate–D–glucose 6-phosphotransferase; (e) tripolyphosphate phosphohydrolase; (f) polyphosphate phosphohydrolase ($\bar{n} = 290$)

the fraction which is soluble in hot perchloric acid, and the salt-soluble fraction. The curve for the accumulation of alkali-soluble polyphosphates differs, in that the amount continues to increase in the stationary growth phase also.

The abovementioned investigation[326] also showed that the accumulation of all the polyphosphate fractions was strongly inhibited by adding to the medium 50 μg/ml of the antibiotic polymyxin M. Figure 44 shows that the presence of this antibiotic in the culture medium substantially retards (by approximately 30%) the accumulation of these compounds, while having no effect on the overall appearance of the curve for their accumulation during the development of the propionic bacteria. This observation is of great interest in view of the fact that the site of attack of this antibiotic in the bacterial cell is the cytoplasmic membrane. This is evidence of the close link between polyphosphate metabolism and the functioning of the cytoplasmic membrane in *P. shermanii*. The retardation of the accumulation of polyphosphates in the presence of polymyxin M could be due either to inhibition of polyphosphate-synthesizing enzymes, or conversely to activation of polyphosphate-cleaving enzymes. In order to resolve this question, an attempt was made in our laboratory to determine the activities of most of the enzymes currently known to be involved in the synthesis or utilization of polyphosphates in microorganisms.[437]

2. Enzymes of metabolism of high-molecular-weight polyphosphates

In our investigation[437] into the behaviour of polyphosphate-metabolizing enzymes in *P. shermanii* under normal culture conditions and in the presence of polymyxin M (50 μg/ml of medium), firstly the activities of the polyphosphate-synthesizing enzymes polyphosphate kinase and 1,3-DPGA–polyphosphate phosphotransferase were determined.

The specific activities of these two enzymes in *P. shermanii* at all growth stages examined (from the beginning of the logarithmic deep into the stationary phase), in the presence and absence of polymyxin M, are shown in Table 34. It can be seen that polyphosphate kinase, under normal growth conditions, is the more active in the 24-h culture, and 1,3-DPGA–polyphosphate phosphotransferase in the 120-h culture. In this connection, although in the young, actively dividing cells the synthesis of polyphosphate is apparently mediated for the most part by the former enzyme, by the end of the stationary growth phase (120 h growth) the contribution of the latter enzyme has become predominant. Polymyxin M has no effect on polyphosphate kinase activity, but a slight, but definite, effect on 1,3-DPGA–polyphosphate phosphotransferase activity. These results suggest that the functioning of the latter enzyme is associated with the membrane apparatus of these bacteria to a greater extent than is that of polyphosphate kinase.

Figure 44 presents data on the activity of the enzymes examined, calculated on the biomass of cells harvested from 100 ml of medium. Consideration of the effects of polymyxin M on the accumulation of polyphosphate, and the total

Table 34. Specific activity of polyphosphate-metabolizing enzymes when *Propionibacterium shermanii* is grown in the presence and absence of polymixin M (results expressed as mE/mg protein in the cell-free extract)[437]

Enzyme	Growth conditions	Time of growth (h)				
		24	48	72	96	120
ATP–polyphosphate phosphotransferase	No addition	1.15	0.90	0.82	0.70	0.45
	+ 50 μg/ml of polymixin M	—	1.05	0.78	0.65	0.60
1,3-DPGA–polyphosphate phosphotransferase	No addition	0.18	0.25	0.38	0.47	0.60
	+ 50 μg/ml of polymixin M	—	0.55	0.74	0.40	0.25
Polyphosphate–D-glucose 6-phosphotransferase [fraction precipitated by $(NH_4) SO_4$]	No addition	210	270	240	190	210
	+ 50 μg/ml offpolymixin M	—	260	230	200	170
Polyphosphate phosphohydrolase ($\bar{n}$ = 290)	No addition	0.0	0.0	0.0	5.8	1.1
	+ 50 μg/ml of polymixin M	—	0.0	0.0	15.7	7.6
Tripolyphosphate phosphohydrolase (n = 3)	No addition	10.3	3.7	4.0	0.0	0.0
	+ 50 μg/ml of polymixin M	—	13.2	8.2	0.0	0.0

activity of the polyphosphate-synthesizing enzymes presented, shows it to be very unlikely that this antibiotic inhibits the accumulation of polyphosphate by inhibiting the activity of these enzymes.

Polyphosphate glucokinase activity was next examined in *P. shermanii*, and the effect thereon of the presence of polymyxin M in the growth medium. Inspection of Table 34 shows that during the whole period of growth of this organism, cellular polyphosphate glucokinase activity remained at a high level, two to three orders of magnitude greater than that of both the polyphosphate-synthesizing enzymes studied by us. It is true that the data in Table 34 relate to a partially purified polyphosphate glucokinase preparation, but the corresponding values for the specific activity of this enzyme found for the cell-free extract differ from those shown in Table 34 by a factor of only two or three.

The total polyphosphate glucokinase activity per 100 ml of culture (cf. Fig. 44) is also 100–200 times greater than that of the polyphosphate-synthesizing enzymes. As can be seen from Table 34 and Fig. 44, polymyxin M has scarcely any effect on polyphosphate glucokinase activity in *P. shermanii*. The results suggest the absence of any significant connection between this enzyme and the functioning of the cytoplasmic membane, and show that the substantial suppression of polyphosphate accumulation caused by polymyxin M is not a result of the activation of polyphosphate glucokinase.

Polyphosphatase activity was then determined at all growth stages in *P. shermanii*, both in a normal medium and in the presence of polymyxin M, and the results are shown in Table 34 and Fig. 44. They show that two different polyphosphatases are present in the propionic bacteria, one of which hydrolyses polyphosphates terminally to orthophosphate, and the other is a tripolyphosphatase. Tripolyphosphatase activity was found only during the first 3 days' growth of the culture, i.e. in young cells, whereas the polyphosphatase which cleaves higher polyphosphates was observed in the cells only during the stationary growth phase, i.e. when one type of activity was observed in *P. shermanii* cells, the other was absent, and *vice versa*.

The results shown in Table 34 and Fig. 44 also demonstrate that polymyxin M, when added to the growth medium, significantly increases (by a factor of 2–7) the activity of both polyphosphatases. This is an indication that the functioning of these enzymes is closely bound (as in other microorganisms[401]) with the cytoplasmic membrane, which is the site of action of polymyxin M. It has already been pointed out that tripolyphosphatase is localized in fungi (specifically, in *N. crassa*) almost exclusively in the mitochondria. In the propionic bacteria, in which mitochondria are not present, this enzyme is evidently bound to some parts of the cytoplasmic membrane, and perhaps also to mesosomal elements. However, as in other microorganisms, it is at present difficult to say what the function of this enzyme in the propionic bacteria may be.

The polyphosphatase which hydrolyses high-molecular-weight polyphosphates to orthophosphate evidently plays an important part in the degradation and utilization of these compounds in *P. shermanii*. As soon as this enzyme activity begins to appear in normally growing cells, the accumulation of polyphosphate ceases completely. In the presence of polymyxin M, when polyphosphatase activity ($\bar{n} = 290$) is higher still, a corresponding reduction in the level of accumulation of polyphosphate in the cell is observed.

It may therefore be concluded that the suppressive effect of polymyxin M on the accumulation of high-molecular-weight polyphosphates is in some way connected with its specific activating effect on polyphosphatase. To summarize the situation regarding the polyphosphate-metabolizing enzymes present in *P. shermanii*, the synthesis of polyphosphates in these bacteria apparently involves two enzymes, polyphosphate kinase and 1,3-DPGA-polyphosphate phosphotransferase.

The degradation, and possibly the utilization, of polyphosphates in an old culture is mediated by polyphosphatase, but the function of polyphosphate glucokinase remains obscure. Most perplexing is the fact that, as has been mentioned, in *P. shermanii* this enzyme possesses extremely high *in vitro* activity which is in no way comparable to the activity of the other polyphosphate-metabolizing enzymes which have been investigated in this organism. Either this enzyme displays a different type of activity and specificity *in vitro* from that shown in *in vivo* experiments, or there is present in the cells of these organisms an as yet undetected polyphosphate-synthesizing system which is comparable in activity to that observed for polyphosphate glucokinase.

In any event, it is clear from these observations that it is by no means always possible to obtain reliable information on the functioning of these enzymes *in vivo* from determinations of their *in vitro* activity. Nevertheless, in many cases, for example in establishing the role of polyphosphatase in the metabolism of high-molecular-weight polyphosphates in the propionic bacteria, this approach is obviously fully justifiable.

Actinomyces (Streptomyces) aureofaciens

1. *General features of the metabolism of high-molecular-weight polyphosphates*

Both the propionic bacteria, and the *Myco-* and *Corynebacteria* considered above, were assigned by Krasil'nikov, as has been pointed out, to the class Actinomycetes. They apparently constitute ancient representatives of this class of microorganisms. It was of interest to compare the general features of polyphosphate metabolism, and of those enzymes known to be involved in the metabolism of these macromolecules, in the less highly organized groups of this class with those of the more highly developed representatives. Following the work on the metabolism of polyphosphates in the propionic bacteria, a similar investigation was therefore carried out in our laboratory using a younger group of microorganisms of this class, namely the true Actinomycetes, or Streptomycetes. The experimental subject chosen was *Actinomyces (Strep-*

tomyces) aureofaciens, a chlortetracycline producer. The main results of this investigation are shown in Figs. 45 and 46.[396] Figure 45 shows the results obtained with a strain of *Actinomyces aureofaciens* which was a poor producer of chlortetracycline, and Fig. 46 shows the results obtained using a highly productive strain (a mutant of the first strain). It can be seen from Fig. 45 that the low-producing strain, under normal conditions of growth (nutrient carbon source, glucose), accumulated large amounts of polyphosphate which, as in the propionic bacteria, was largely (60–80%) extractable from the cells by hot perchloric acid only. The remaining 20–40% was found for the most part in the salt-soluble fraction, and to a very small extent as low-polymeric, acid-soluble polyphosphates. On average, the highly productive strain accumulated one

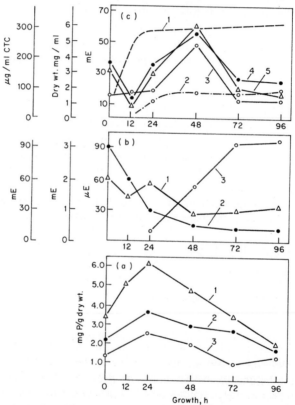

Figure 45. Changes in biomass and production of chlortetracycline, inorganic polyphosphate, and polyphosphate-metabolizing enzymes during the growth of a culture of the low-producing strain *Actinomyces aureofaciens* 2209.[396] (a) 1, Total acid-insoluble polyphosphate; 2, polyphosphates extracted with hot perchloric acid; 3, salt-soluble polyphosphates. (b) 1, ATP–polyphosphate phosphotransferase (centre scale); 2, 1,3-diphosphoglycerate–polyphosphate phosphotransferase (right-hand scale); 3, polyphosphate–D–glucose 6-phosphotransferase (left-hand scale). (c) 1, Biomass; 2, chlortetracycline; 3, polyphosphate phosphohydrolase ($\bar{n} = 290$); 4, pyrophosphatase; 5, tripolyphosphatase

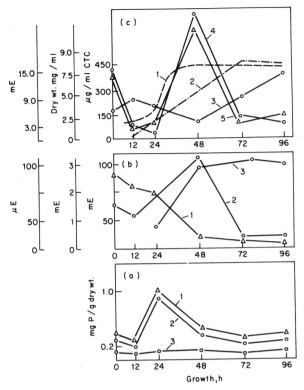

Figure 46. Changes in biomass and production of chlortetracycline, inorganic polyphosphate, and polyphosphate-metabolizing enzymes during the growth of the high-producing strain *Actinomyces aureofaciens* 8425.[396] (a) 1, Total acid-insoluble polyphosphate; 2, polyphosphates extracted by hot perchloric acid; 3, salt-soluble polyphosphates. (b) 1, ATP–polyphosphate phosphotransferase (centre scale); 2, 1,3-diphosphoglycerate–polyphosphate phosphotransferase (left-hand scale); 3, polyphosphate–D–glucose 6-phosphotransferase (right-hand scale). (c) 1, Biomass; 2, chlortetracycline; 3, polyphosphate phosphohydrolase ($\bar{n} = 290$); 4, pyrophosphatase; 5, tripolyphosphatase

tenth as much polyphosphate. It is interesting that in both strains, the maximum accumulation of polyphosphate occurred during the second half of the logarithmic and at the beginning of the stationary growth phases. This indicates that the accumulation of polyphosphates is accompanied by extremely rapid biosynthetic processes in the Actinomycetes, resulting in the synthesis of the principal cell components.

The fact that cells of Actinomycetes which are actively producing the antibiotic chlortetracycline accumulate so much smaller amounts of polyphosphate than the parent low-producing strain may indicate the possible utilization of these high-energy phosphorus compounds in the production of metabolites formed as a result of so-called 'secondary metabolism'. However, the reduction in the accumulation of polyphosphates in the highly productive strain could also be due to 'damage' to its mechanism of synthesis of these compounds.

In order to settle this question, it was necessary to investigate the behaviour of the principal polyphosphate-metabolizing enzymes in each strain.

2. Enzymes of metabolism of high-molecular-weight polyphosphates

The following enzymes were found in the cells of both strains of *Actinomyces aureofaciens* during the growth of cultures: polyphosphate kinase, 1,3-DPGA–polyphosphate phosphotransferase, polyphosphate glucokinase, the polyphosphatase which hydrolyses high-molecular-weight polyphosphates to orthophosphate, tripolyphosphatase, and pyrophosphatase.[396]

It is interesting that in the Actinomycetes, as in the propionic bacteria, the least activity is displayed by the polyphosphate-synthesizing enzymes. The polyphosphatases are much more active, but polyphosphate glucokinase remains the most active of all. These findings again emphasize the very important role of the last enzyme in all the organisms regarded by Krasil'nikov as belonging to the Actinomycetes.

We shall first consider the polyphosphate-metabolizing enzymes in the low-producing strain. In young cells, which are rapidly accumulating polyphosphates, both the polyphosphate-synthesizing enzymes (ATP–polyphosphate phosphotransferase and 1,3-DPGA–polyphosphate phosphotransferase) display relatively high activity. From the time when the culture enters the stationary phase and the accumulation of polyphosphate is at a maximum, the activity of these enzymes falls substantially, and remains at a fairly constant level for the whole of the remaining period of growth. Concurrently, at the time when the accumulation of cellular polyphosphate reaches a maximum, the activity of the enzymes involved in the utilization of polyphosphates (polyphosphatase and polyphosphate glucokinase) begins to rise sharply. As a result, the cellular polyphosphate levels fall approximately 3-fold. It therefore appears that both of these enzymes are actually involved in the utilization of high-molecular-weight polyphosphates in *A. aureofaciens*. It is interesting that, in an old culture (after 72 h), polyphosphates are utilized mainly by means of polyphosphate glucokinase, since polyphosphatase activity has decreased substantially by this time. These findings are in good agreement with those obtained for the propionic bacteria, and suggest that polyphosphate glucokinase plays a particularly important part in the stationary growth phase of cultures of organisms belonging to the Actinomycetes.

Further, these results, as with the propionic bacteria, suggest that we do not yet know any powerful polyphosphate-synthesizing partner of polyphosphate glucokinase. Likewise, we have not yet succeeded in detecting in *A. aureofaciens* an enzyme which is active enough to synthesize the amount of polyphosphate which, despite the very high activity of polyphosphate glucokinase, is present in the cells. We have in fact found significant amounts of polyphosphates even in the stationary growth phase of the low-producing strain of *A. aureofaciens*.

We now compare the behaviour of the polyphosphate-metabolizing enzymes in the low- and high-producing strains of *A. aureofaciens* (Figs. 45 and 46).

It should be noted firstly that, despite the very low accumulation of polyphosphate in the high-producing strain (Fig. 46), the activity of the polyphosphate-synthesizing enzymes in the young culture of this strain is approximately the same as that of cells of the low-producing strain of the same age.

This is taken to indicate that at least those polyphosphate-synthesizing enzymes which we have investigated function normally in young cells of the high-producing strain.

The behaviour of polyphosphate kinase in the high-producing strain at all stages of growth differs little from that in the low-producing strain. This, however, is certainly not true of 1,3-DPGA–polyphosphate phosphotransferase. This enzyme doubles its activity at the beginning of the stationary growth phase of the high-producing strain, i.e. at the point when active production of antibiotic begins. At the end of the process of formation of chlortetracycline, the activity of the 1,3-DPGA–polyphosphate phosphotransferase falls rapidly. These results may be explained as follows.

The tetracyclines are formed by condensation of large amounts of malonyl-CoA, formed from acetyl-CoA. It would appear that the reduction in the cellular concentration of acetyl-CoA consequent upon its utilization in the synthesis of tetracycline stimulates glycolysis. The resulting large amounts of 1,3-DPGA must undergo more rapid dephosphorylation, requiring more rapid formation of 1,3-DPGA–polyphosphate phosphotransferase. It is interesting that, in parallel with the increase in the activity of this enzyme and the formation of tetracycline, the polyphosphate glucokinase activity also increases, more rapidly in the high- than in the low-producing strain, reaching a plateau in the 48-h culture. The acceleration of glucose phosphorylation by polyphosphate when tetracycline production increases occurs, it seems, for the same reason as in the case of 1,3-DPGA–polyphosphate phosphotransferase, i.e. in consequence of an acceleration in glycolysis. Of no less interest is the reduction in polyphosphatase activity which is observed in the high-producing strain when the two polyphosphate-metabolizing enzymes in these two strains are compared. The polyphosphatase which hydrolyses high-molecular-weight polyphosphate behaves in general in completely different ways in the high-producing and low-producing strains. Unlike tripolyphosphatase and pyrophosphatase, it does not display its maximum activity at the beginning of the stationary growth phase in the high-producing strain, i.e. when tetracycline production is proceeding with particular rapidity. It is only at the end of the stationary phase, when antibiotic production has ceased, that polyphosphatase activity increases somewhat. These results are particularly significant when it is recalled that in those organisms in which it has been possible to localize polyphosphatase (mould fungi and yeasts), it has been found to occur mainly in the vicinity of the cytoplasmic membrane. The results obtained with the propionic bacteria (see above) also provide circumstantial evidence that in these organisms, too, which are evidently closely related to the Actinomycetes, polyphosphatase functions in association with the cytoplasmic membrane. It might therefore be concluded that in Actinomycetes polyphosphatase is a peripherally localized enzyme. If this is indeed the case, then we have an interesting situation.

When high-molecular-weight polyphosphates are utilized for the biosynthesis of tetracycline, polyphosphatase plays no significant part (cf. Fig. 46), and it may therefore be located peripherally, whereas the biosynthesis of tetracycline takes place within the cell, apparently in the cytosole, and salt-soluble polyphosphates are consumed in its formation. It is also possible that in the high-producing strain polyphosphate depolymerizing activity is enhanced, resulting in the degradation and movement from the periphery to the interior of the cell of the more highly polymerized polyphosphate fractions which are extractable by hot perchloric acid. It might therefore be thought that the metabolism of polyphosphates becomes, as it were, intracellular when the biosynthesis of chlortetracycline increases. Their utilization, and to some extent their biosynthesis, are closely bound up with the actively functioning glycolytic process, which is known to take place in the cytosole. However, the polyphosphatase which hydrolyses high-molecular-weight polyphosphate to orthophosphate, and which is apparently primarily responsible for the degradation of these compounds at the cell surface, functions very weakly under these conditions.

It only remains to draw attention, in considering Figs. 45 and 46, to the strict synchronization of the three phosphohydrolases [polyphosphatase ($\bar{n} = 290$), tripolyphosphatase, and pyrophosphatase] in the low-producing strain, and the absence of such synchronization in the high-producing strain. Perhaps in *A. aureofaciens*, as in *Aerobacter aerogenes* and *Escherichia coli* (see below), these enzymes are contained in a single regulon. Under conditions of rapid biosynthesis and excretion of tetracycline, however, coordinated control of phosphohydrolase synthesis is disturbed. Figure 46 shows that in this case the highest polyphosphatase activity is not displayed at the stage at which maximum tripolyphosphatase and pyrophosphatase activity is observed.

The physiological function of pyrophosphatase, and especially of tripolyphosphatase, in the metabolism of microorganisms remains obscure. However, the fact that these enzymes display their highest activity at a time when high-molecular-weight polyphosphates (perhaps together with other high-energy phosphorus compounds) are being consumed rapidly in biochemical processes provides indirect evidence of the important function of these enzymes in the bioenergetics of microorganisms.

EUBACTERIA

Aerobacter aerogenes

1. General features of the metabolism of high-molecular-weight polyphosphates

Aerobacter aerogenes is one of a group of bacteria in which polyphosphates do not normally accumulate (cf. Table 2 ref. 257). It has, however, been known for some time that under unfavourable conditions of growth these bacteria begin to accumulate condensed phosphates in their cells.[671] Detailed investigations have been carried out by Wilkinson and Duguid[815] into the effects of different conditions, and phases, of growth on the polyphosphate content of these organisms. These investigations showed in particular that a very substantial accumulation of polyphosphate occurred when the organisms were cultured in a liquid medium at a low pH (4.5), when the medium was deficient in sulphur, and following addition of phosphate to a culture which had been subjected to phosphorus starvation. It was shown clearly that in order for polyphosphate accumulation to occur in *A. aerogenes*, the presence of K^+ and Mg^{2+} ions in the culture medium was essential, in addition to phosphate and an energy source. The work of Wilkinson and Duguid[815] showed that polyphosphate and nucleic acid metabolism in this bacterium are closely linked. When growth and nucleic acid synthesis ceased, the total amount of cellular polyphosphate increased. When growth was resumed, the previously accumulated polyphosphates (present in the form of volutin granules) disappeared from the cell.

2. Enzymes of metabolism of high-molecular-weight polyphosphates and the regulation of their biosynthesis and activity

The most fundamental work on the metabolism of polyphosphates in *A. aerogenes* has been carried out by Harold and co-workers.[216–222,227] These investigations have provided an explanation at the molecular level of the mechanisms underlying the dependence of polyphosphate metabolism in this bacterium on conditions of growth.

The work of Harold's group is of fundamental importance in elucidating the regulatory mechanisms which control the metabolism of polyphosphates in all those organisms in which they are present. It was shown that the effect of growth conditions, especially the composition of the nutrient medium, on polyphosphate metabolism is mediated by two entirely different mechanisms.

The first of these comes into operation when nucleic acid synthesis ceases as a result of the absence of an essential metabolite from the nutrient medium. Such a component, which is required for the biosynthesis of nucleic acids in *A. aerogenes*, is sulphate ion (SO_4^{2-}). Figure 47(a) shows that during the period of sulphur deprivation, a gradual accumulation of polyphosphate occurs, while growth and nucleic acid biosynthesis are totally suppressed. The polyphosphate kinase and polyphosphatase activity found in this organism by Harold[216,218] hardly changes during this period. A basically similar situation was observed by Harold[216] during an examination of polyphosphate metabolism in auxotrophic mutants of *A. aerogenes*, which required the presence of uracil or methionine for normal growth and nucleic acid synthesis, when these components were not added to the medium. When the conditions necessary for normal growth were restored by adding the necessary components to the medium, accumulation of

138

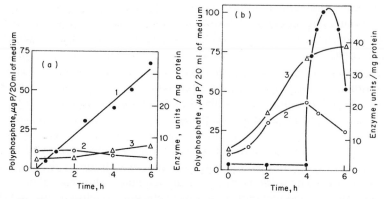

Figure 47. Accumulation of polyphosphates in *Aerobacter aerogenes.*[220] (a) Sulphur deprivation, cells placed at O h in a sulphur-free medium; (b) hypercompensation, cells placed at O h in a phosphorus-free medium, and orthophosphate added after 4 h. 1, Polyphosphate; 2, polyphosphate kinase; 3, polyphosphatase

polyphosphate ceased, and the phosphorus of the polyphosphate was utilized quantitatively for the biosynthesis of nucleic acids.

It was shown that polyphosphate is not required for the biosynthesis of proteins under these conditions. For instance, addition to the culture medium of chloramphenicol, a specific inhibitor of protein biosynthesis, had no effect on the rate of utilization of the accumulated polyphosphate for nucleic acid biosynthesis. In a convincing series of experiments, it has been established that a two-fold relationship exists between nucleic acid and polyphosphate metabolism, i.e. resumption of nucleic acid synthesis resulted in both a reduction in the rate of polyphosphate synthesis, and in an increase in their rate of breakdown.[216,219,222] The mechanism by which sulphur deprivation increases the rate of synthesis of polyphosphates in *A. aerogenes* appears to be very complex. From the work of Harold[219,227] it appears that the intracellular compound normally responsible for the inhibition of polyphosphate biosynthesis in *A. aerogenes* when sulphate was present in the medium was oxidized glutathione, or some derivative thereof. In this case, therefore, as in other cases in which polyphosphate accumulation is enhanced when a component essential for growth is absent from the medium, the functioning of the enzymes involved in polyphosphate metabolism is controlled metabolically. We point out in passing that the inhibition of polyphosphate accumulation by oxidized glutathione has also been shown in the bacterium *E. coli*, which similarly does not normally accumulate polyphosphate.[600]

A fundamentally different regulatory mechanism was observed by Harold and Harold[218,219,221] in an examination of polyphosphate metabolism under conditions of phosphorus deprivation followed by growth on a phosphate-containing medium. An attempt was made to explain the phenomenon known as hypercompensation (Überkompensation, or 'polyphosphate overplus'), first discovered by Jeener and Brachet[276] in yeasts, and investigated in detail in these organisms by Liss and Langen.[466] This is the name given to the phenomenon in which, following impoverishment in polyphosphates resulting from growth in a phosphorus-free medium, the cells of a microorganism, when grown in a medium containing phosphorus, accumulate amounts of these compounds far in excess of their initial levels. The change in polyphosphate content, together with that of polyphosphate kinase and polyphosphatase, when hypercompensation occurs in *A. aerogenes*, is shown in Fig. 47(b). It can be seen that in this case the rapid accumulation of polyphosphate is associated not with an increase in the activity of the pre-existing enzymes polyphosphate kinase and polyphosphatase, but rather with increased biosynthesis of these enzymes. The absence of phosphate from the medium derepressed polyphosphate kinase and polyphosphatase synthesis, resulting in a substantial accumulation of these enzymes. In this case, this facilitated the rapid synthesis of unusually large amounts of polyphosphate. It was concluded from these findings that hypercompensation involved regulation of polyphosphate metabolism at the gene level. This hypothesis was indeed confirmed by the convincing work of Harold and Harold,[218,221,222] who obtained mutants of *A. aerogenes* in which polyphosphate metabolism was profoundly modified. Table 35 shows that Harold was able to obtain and isolate four such mutants.

In the mutant P_{n-2}, polyphosphate accumulation was blocked completely, irrespective of the conditions of growth. This mutant was devoid of polyphosphate kinase.[218] The mutant P_{n-4}, on the other hand, synthesized and accumulated polyphosphates, but their degradation was completely inhibited. This mutant was found to be devoid of polyphosphatase.[222] These findings lead to some very important conclusions.

Firstly, if mutants are capable of existence which lack the ability to synthesize and accumulate polyphosphates, then these compounds are not strictly obligatory cellular components in contemporary organisms, and these organisms can function in their absence.

Secondly, since those mutants which did not produce polyphosphates likewise did not contain volutin granules, this provides unambiguous proof of the polyphosphate composition of these granules, at least in *A. aerogenes*.

Thirdly, these results lead to the conclusion that nucleic acid synthesis in *A. aerogenes* (and apparently in other organisms also) can take place without the participation of polyphosphates, since mutants lacking polyphosphatase were unable to utilize polyphosphates for this purpose, whereas nucleic acid synthesis proceeded normally.

Finally, these results show that the biosynthesis of polyphosphates in *A. aerogenes* is normally mediated by polyphosphate kinase only, and their degradation and utilization by polyphosphatase.

In addition to the *A. aerogenes* mutants mentioned above, Harold and Harold[218,222] were able to isolate two further mutants which possessed defects in the regulation of their phosphorus metabolism (Table 35). In the strain P_n-1, all of the enzymes which participate in polyphosphate metabolism were present in the cell, but it was not possible to derepress their biosynthesis by phosphorus

Table 35. Mutants of *Aerobacter aerogenes* possessing modified polyphosphate metabolism[219]

Mutant	Physiological characteristics	Enzymes			Functional characteristics of polyphosphate-metabolizing enzymes
		Polyphosphate kinase	Polyphosphatase	Alkaline phosphatase	
Wild Strain	Polyphosphates not detected in growing cells, but accumulate in deficient media during hypercompensation	+	+	+	All enzymes derepressed by phosphate deprivation
Mutant P_n–1	Lack of ability to hypercompensate, but maintenance of the ability to accumulate polyphosphate in deficient media	+	+	+	Enzymes present, but not derepressed by phosphate deprivation
Mutant P_n–2	Lack of ability to accumulate polyphosphate under all growth conditions	–	+	+	Polyphosphate kinase absent, and the other enzymes derepressed by phosphate deprivation
Mutant P_n–3	Continuous accumulation of polyphosphate in growing cells	+	+	+	All enzymes constitutively increased in activity
Mutant P_n–4	Blockage of polyphosphate degradation	+	–	+	Polyphosphatase absent, and the other enzymes derepressed by phosphate deprivation

deprivation, and this mutant did not display the hypercompensation effect. It was, however, capable of accumulating polyphosphates when nucleic acid synthesis was suppressed by nutrient deficiencies. That is to say, in this mutant the genetic regulation of polyphosphate biosynthesis was disturbed, but metabolic control functioned normally. In mutant P_n-3, all of the polyphosphate-metabolizing enzymes were constitutively derepressed, as was apparent from the uninterrupted, uncontrolled synthesis and accumulation of polyphosphates during the whole of the exponential growth phase.[222] As has been pointed out, normally this is never observed in *A. aerogenes*. It is interesting that these mutations also affect alkaline phosphatase in a very similar manner to polyphosphate kinase and polyphosphatase. These observations undoubtedly show that all three enzymes are controlled by a common regulator gene. However, a special investigation carried out by Harold[219] showed that the rates of synthesis, repression, and derepression of these three enzymes were not the same, showing that the genes controlling their synthesis do not constitute a single operon. The results of Harold are summarized in the scheme for the metabolism of polyphosphates and its genetic control in *Aerobacter aerogenes*, shown in Fig. 48.[220,222]

In conclusion, it is necessary to draw attention to one further conclusion drawn by Harold[219] from his experiments with this strain and mutants of *A. aerogenes*. He showed clearly that polyphosphate kinase activity in *A. aerogenes* was controlled metabolically, not only by oxidized glutathione but also by the concentration of cellular ATP.

In addition, Harold considers that the inhibition of polyphosphate synthesis during the period of rapid biosynthesis of nucleic acids is due to a large extent to the competition of these two processes for ATP.

Very little is so far known about the metabolic regulation of polyphosphate utilization in *A. aerogenes*, but it is clear that by some mechanism polyphosphatase activity is strongly stimulated during the biosynthesis of nucleic acids.

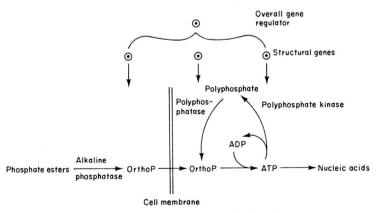

Figure 48. Genetic control of polyphosphate metabolism in *Aerobacter aerogenes* [220]

The nature of the stimulant and its mode of action at the molecular level, however, remain unknown.

Escherichia coli

1. General features of the metabolism of high-molecular-weight polyphosphates

In recent years, Nesmeyanova and co-workers[539-546] have carried out a broad investigation into the metabolism of high-molecular-weight polyphosphates, and its control, in *Escherichia coli*. Contrary to earlier work (see, in particular, Pine,[599,600] which reported that these compounds were absent from *E. coli*, Nesmeyanova and co-workers were able to detect them in this organism also. They carried out a fractionation of the phosphorus compounds from *E. coli* cells, showing the absence of any significant amounts of polyphosphate in the acid-soluble fraction. Polyphosphates were found, however, in the salt and alkaline fractions in addition to the fraction extracted from the cells by hot perchloric acid. The high-molecular-weight polyphosphates in these fractions were identified by the formation of trimetaphosphate, which is a specific product of their partial hydrolysis by acid.

In investigating the polyphosphate content at various stages of growth of the culture, fractionation was not carried out, but the total polyphosphates in the acid-insoluble fraction were determined by extracting them all simultaneously with hot perchloric acid. The results showed (Fig. 24) that the polyphosphate content was dependent on the age of the culture. The greatest amount of polyphosphate, amounting to 0.2–0.4% of the dry weight of the bacteria, accumulated at the end of the latent and the beginning of the logarithmic phases, at the stage preceding the rapid synthesis of biomass. When the cells entered the exponential growth phase, the level of intracellular polyphosphate fell sharply (by a factor of 5 or 10).

Thus, accumulation of polyphosphate in *E. coli* precedes rapid synthesis of biomass, but from the onset of active growth in the culture utilization of these high-energy compounds occurs. Similar behaviour had previously been noted in the polyphosphate metabolism of several organisms. Thus, accumulation of polyphosphates in the latent phase, and their active utilization on entering the exponential growth phase, was observed in the fungi *Aspergillus niger*[58] and *Penicillium chrysogenum*,[391] and in the bacteria *Corynebacterium xerosis*,[257] *C. diphtheriae*,[625] and *Azotobacter vinelandii*.[835a] In parallel with polyphosphate, the levels of orthophosphate in the cell and culture medium were determined during the growth of the culture. It was shown (Fig. 24) that the accumulation of polyphosphate in this organism proceeded in step with the accumulation of intracellular orthophosphate, accompanied by a fall in the level in the surrounding medium. It is important to note that at all growth stages of the experimental culture, the amount of polyphosphate exceeded to a greater or lesser extent the free orthophosphate content. As can be seen from Fig. 24(a), the level of intracellular orthophosphate in general changed very little at any growth stage of the culture.

It must be emphasized that, according to the findings of Nesmeyanova and co-workers, *E. coli* cells are able to accumulate polyphosphates on a complete medium only at the end of the latent growth phase of the culture. In those cases in which high-molecular-weight polyphosphates were not found in *E. coli*,[599,600] the investigators were apparently working with cells which were at the growth stage at which the polyphosphate content is at a minimum.

2. Enzymes of metabolism of high-molecular-weight polyphosphate and the regulation of their biosynthesis and activity

Nesmeyanova and co-workers[540–543] have investigated the changes in activity during the growth of *E. coli* of the two enzymes currently known to be involved in polyphosphate synthesis, namely ATP–polyphosphate phosphotransferase and 1,3-diphosphoglycerate–polyphosphate phosphotransferase. Both of these enzymes had previously been observed in *E. coli*.[331,333,395] The changes in their activity in *E. coli* had, however, been investigated only in a much older culture.[395] In the same investigations, and in others,[539,544–546,655] in addition to the enzymes which bring about polyphosphate biosynthesis, changes in those enzymes which were assumed to be involved in polyphosphate utilization during the growth of cultures, and under various growth conditions were also examined. These were the polyphosphatase which hydrolyses high-molecular-weight polyphosphates to orthophosphate, and tripolyphosphatase, which cleaves tripolyphosphate to orthophosphate. Neither of these enzymes has yet been detected in *E. coli*. In parallel with these enzymes, changes were also examined in the activity of alkaline phosphatase, an enzyme of the greatest importance in phosphorus metabolism, which has been investigated thoroughly in this organism.

Fig. 24 shows that all of the enzymes which were studied are present in *E. coli* cells at all stages of growth, but that their activity varies considerably with the age of the culture. It was found that all of the enzymes shared a common peak of activity at the end of the latent growth period, coinciding with the maximum polyphosphate and intracellular orthophosphate content. The maximum polyphosphate accumulation must be accompanied either by the activation and induction of the enzymes responsible for their synthesis, or by repression and inhibition of the enzymes responsible for their utilization, or both together. The coincidence of the activity maximum of the polyphosphate-synthesizing enzymes with the maximum accumulation of polyphosphates thus has a logical basis, and evidently reflects the involvement of these enzymes in the synthesis of polyphosphates. The similar levels of activity of both the polyphosphate-synthesizing enzymes at this time is an indication that each enzyme makes a similar contribution to polyphosphate synthesis in *E. coli* at this growth stage. It can be seen, however, from Fig. 24 that, as has already been mentioned, polyphosphate kinase activity oscillates during the growth of the culture, exhibiting two additional maxima. The activity of this enzyme increases with particular rapidity at the end of the exponential growth stage, and it is very interesting that this increase in polyphosphate kinase activity coincides with a fall in the concentration of intracellular polyphosphate. It is possible that at this stage of development this enzyme may function to a larger extent in the utilization of polyphosphates than in their synthesis.[167,331]

In addition to the case described above, it is also surprizing that the activity of enzymes which participate in polyphosphate utilization corresponds directly not to the decrease, but to the increase in intracellular polyphosphate. Thus, polyphosphatase exhibits a maximum of activity which coincides with the maximum polyphosphate and intracellular orthophosphate content.

It is possible that under these metabolic conditions, when the level of intracellular orthophosphate is increasing and the exogenous orthophosphate remains at a constant, high level, the polyphosphatase may mediate the biosynthesis rather than the utilization of polyphosphates. In this case, the polyphosphatase itself would make the main contribution to polyphosphate biosynthesis, since its activity is several orders of magnitude greater than that of the polyphosphate-synthesizing enzymes mentioned above.

It is also possible, however, that the correlation between polyphosphatase activity and polyphosphate content has a different significance. For example, it may be that polyphosphatase is activated by the increasing amounts of its substrate, high-molecular-weight polyphosphate, but that it does not begin to break them down until, for instance, their chain length or intracellular concentration reach a certain level. This period could coincide with the beginning of rapid cell growth and the onset of high demand for orthophosphate.

It may therefore be suggested that the factor which regulates the activity and even the direction of operation of polyphosphatase in the cell is the ratio of the concentration of orthophosphate to that of polyphosphate. The changes in the activities of tripolyphosphatase and alkaline phosphatase are similar to those of polyphosphatase.

It is scarcely feasible, however, to suggest the possibility that the intracellular level of the high-polymeric polyphosphate affects the activity of these highly specific phosphohydrolases. The increased activity of alkaline phosphatase and tripolyphosphatase at the stage preceding rapid cell growth is perhaps related to their participation in some phosphotransferase reactions.

The investigations of Harold and co-workers[216,218–222,227] on *Aerobacter aerogenes* mentioned above have shown conclusively that regulation of the biosynthesis of the polyphosphate-metabolizing enzymes by orthophosphate present in the medium is possible. Since orthophosphate is the end product of the reactions mediated by phosphohydrolases, and since its presence is necessary for the functioning of the polyphosphate-synthesizing enzymes, it must be a very important regulator of polyphosphate-metabolizing enzymes in *E. coli*.

In order to be certain of this, Nemeyanova and co-workers[539,540,542–545,655] investigated the effect of phosphorus deprivation on the intracellular high-molecular-weight polyphosphate levels in *E. coli*, and on the activities of alkaline phosphatase, polyphosphatases, and the polyphosphate-synthesizing enzymes.

Figure 49 shows some typical results from this series of experiments. It can be seen that the growth of the culture during phosphorus deprivation continues, but less rapidly than in a medium containing orthophosphate, and that in a medium without orthophosphate there is greater consumption of intracellular orthophosphate and polyphosphate. Rapid growth of the culture on a complete medium containing orthophosphate is also accompanied by a fall in the level of these phosphorus compounds in the cells. In the final analysis, however, the level remains higher (sometimes by a factor of two) than in the deprived culture.

An investigation of the changes in enzyme activity during phosphorus deprivation showed that the polyphosphate-synthesizing enzymes (ATP–polyphosphate phosphotransferase and 1,3-DPGA–polyphosphate phosphotransferase) react to only a slight extent in the absence of orthophosphate from the medium. Figure 50 shows that polyphosphate kinase activity both in the presence and absence of orthophosphate in the medium varies around a mean which is close to the original value. A deficiency of orthophosphate does not give rise in *E. coli* to an increase in the activity of this enzyme such as was observed by Harold[218] in *Aerobacter aerogenes*. The activity of 1,3-DPGA–polyphosphate phosphotransferase even decreased to a slight extent during phosphorus deprivation, as compared with a culture growing in a medium containing orthophosphate.

The effect of phosphorus deprivation on the three remaining enzymes was

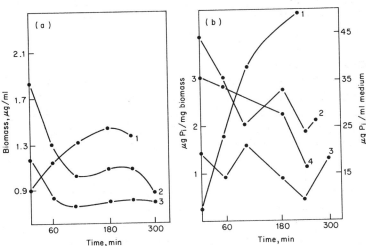

Figure 49. Effect of exogenous orthophosphate on the levels of intracellular ortho- and polyphosphates in *E. coli*. (a) Deficiency of exogenous orthophosphate; (b) excess of exogenous orthophosphate. 1, Growth of biomass; 2, polyphosphates; 3, intracellular orthophosphate; 4, exogenous orthophosphate[542]

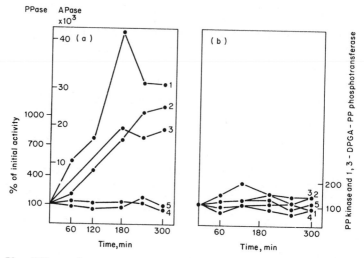

Figure 50. Effect of exogenous orthophosphate on polyphosphate-metabolizing enzymes in *E. coli*. (a) Deficiency of exogenous orthophosphate; (b) excess of exogenous orthophosphate. 1, Alkaline phosphatase; 2, tripolyphosphatase; 3, polyphosphatase; 4, ATP–polyphosphate phosphotransferase; 5, 1,3-DPGA-polyphosphate phosphotransferase[542]

more marked. As would be expected, phosphorus deprivation resulted in a substantial increase in alkaline phosphatase activity in comparison with the controls. Likewise, there was a substantial increase in the specific activity and the activity per unit dry weight of bacteria in the two polyphosphate phosphohydrolases. The extent of the increase was however different, the alkaline phosphatase activity increasing by a factor of 70–400 (depending on the duration of starvation and the age of the starved culture), whereas the activity of the polyphosphate phosphohydrolases increased by a factor of 2–10 only. The fact that, of the three phosphohydrolases, polyphosphatase alone hydrolyses inorganic polyphosphates to orthophosphate suggests that it is the latter enzyme which plays the leading part in the utilization of polyphosphate during phosphate starvation.

In a subsequent series of experiments, Nesmeyanova and co-workers[540,542] investigated the effects of adding orthophosphate to a culture previously starved of phosphate. The addition of orthophosphate to the starved culture (Fig. 51) resulted in a rapid restoration of the accumulation of biomass, whereas when starvation was continued this did not occur to any significant extent. Despite this rapid synthesis of biomass, however, the cellular orthophosphate level rose from the outset, followed by a rise in polyphosphate, reaching the level typical of cells growing rapidly on a complete medium containing orthophosphate.

From these results, it seems likely that exogenous orthophosphate regulates the polyphosphate level, and the accumulated polyphosphates in turn regulate the level of intracellular orthophosphate.

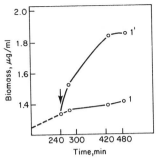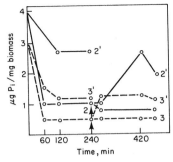

Figure 51. Effect of the addition of orthophosphate to a previously starved culture on the levels of intracellular polyphosphates and orthophosphate in *E. coli*. 1, Biomass; 2, polyphosphates; 3, intracellular orthophosphate (in orthophosphate-deficient medium); 1′, 2′, and 3′, the same in a medium containing excess of orthophosphate. The time of addition of orthophosphate to the previously starved medium is shown by an arrow[542]

It is known[219,220] that in many microorganisms the addition of orthophosphate to a culture previously deprived of phosphorus results in a rapid accumulation of polyphosphate to an extent many times exceeding the original level (hypercompensation). This phenomenon had not previously been investigated in *E. coli*, and our investigations failed to detect its occurrence.[542] All that was observed was replenishment of the reserves of polyphosphate to a level typical of that in cells growing on a complete medium. In order for excessive accumulation to occur, certain other conditions appear to be required in addition to an excess of orthophosphate in the medium.

What, then, happens to the polyphosphate-metabolizing enzymes following the transfer of a culture of *E. coli*, previously starved of phosphate, to a medium containing phosphate? The addition of orthophosphate to a previously deprived culture results in some increase in the activity of both of the polyphosphate-synthesizing enzymes, whereas the activity in the control culture varies about its initial level (Fig. 52). The maximum of activity of the polyphosphate-synthesizing enzymes coincides with, or precedes, polyphosphate accumulation. These enzymes doubtless participate in the biosynthesis of polyphosphate when orthophosphate is added to a previously deprived culture. The increased activity of 1,3-DPGA–polyphosphate phosphotransferase following addition of orthophosphate had been observed previously,[426] but the behaviour of polyphosphate kinase under the conditions described, as in the absence of phosphorus, is different from that in *Aerobacter aerogenes* under similar conditions.[219,220] In *A. aerogenes*, polyphosphate kinase activity falls when orthophosphate is added.

As far as the phosphohydrolases are concerned, addition of orthophosphate to the previously starved culture stops its increase in activity (Fig. 52). The increase in polyphosphatase and alkaline phosphatase activity during phosphorus starvation could be due either to activation of pre-existing enzymes, or to an increase in the amounts present.

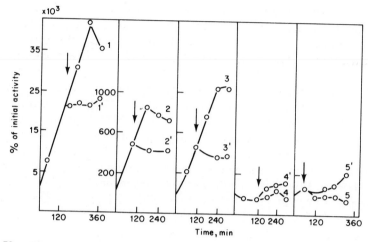

Figure 52. Effect of addition of orthophosphate to a previously starved culture on the activity of phosphorus-metabolizing enzymes in *E. coli*. 1, Alkaline phosphatase; 2, polyphosphatase; 3, tripolyphosphatase; 4, 1,3-DPGA-polyphosphate phosphotransferase; 5, ATP–polyphosphate phosphotransferase (all in an orthophosphate-free medium); 1′, 2′, 3′, 4′, and 5′, with the addition of orthophosphate. The time of addition of the orthophosphate is shown by an arrow[542]

In order to decide between these alternatives, Nesmeyanova *et al.*[540] subjected *E. coli* cells to phosphorus starvation in the presence of chloramphenicol, which is a specific inhibitor of protein synthesis, at concentrations of 50–100 μg/ml. In this case (Fig. 53), no increase in the activity of any of the three

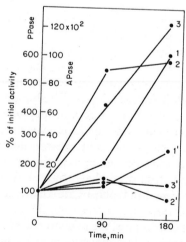

Figure 53. Effect of chloramphenicol on phosphohydrolase synthesis in *E. coli*. 1, Polyphosphatase; 2, tripolyphosphatase; 3, alkaline phosphatase (deficiency of phosphorus); 1′, 2′, and 3′, the same, in the presence of chloramphenicol[540]

enzymes was observed. It follows that the biosynthesis of these enzymes is controlled by orthophosphate, and the increase in their activity during phosphorus starvation results from derepression of synthesis rather than activation of preexisting enzymes. Calculation of the total amount of enzyme synthesized per millilitre of medium in the presence and absence of orthophosphate showed that this derepression was only an enhancement of the basic capacity of the organism. The extent of derepression depended on the stage of growth and the number of cells subjected to starvation. It is suggested that other extra- and intracellular factors, in addition to orthophosphate, may have an effect on the derepressed biosynthesis of these enzymes. The differing extents of derepression of alkaline phosphatase and polyphosphatase, in the opinion of Nesmeyanova, shows that control of the repression of alkaline phosphatase may conserve protein synthesis. In the case of the polyphosphatases, repression serves especially to control the formation of the end-product, conservation of protein synthesis not being of prime significance.

Thus, the phosphohydrolases are regulated by exogenous orthophosphate, being derepressed by a deficiency, and repressed by an excess of phosphorus. The similarity of the response of the enzymes to the presence of orthophosphate in the medium provides confirmation of the generality of their functions in regulating the level of orthophosphate in the cell.

It is important to note that Nesmeyanova and co-workers[545,546,655] have shown that in *E. coli* polyphosphatase, like alkaline phosphatase, is located at the cell periphery, on the external surface of the cytoplasmic membrane. The polyphosphatase in *E. coli* is tightly bound to the cytoplasmic membrane, and can be removed completely only by the use of detergents (in particular, Triton X-100).[546,655] In view of the different effects of membranolytic agents (detergents, proteases, etc.) on polyphosphatase and alkaline phosphatase activities, it is suggested that these phosphohydrolases are located at different sites in the cytoplasmic membrane of *E. coli*.[546]

In order to confirm the determination of the interrelated regulation of phosphohydrolases by genetic means, Nesmeyanova *et al.*[540] studied mutants with respect to one of the regulatory genes for alkaline phosphatase. The regulation of alkaline phosphatase is known[155] to be controlled by two regulator genes, R_1 and R_2. One of these (R_1) is responsible for induction, since it is only mutation in this region of the chromosome which renders the enzyme non-inducible under conditions of phosphorus starvation. Investigation of the effect of orthophosphate on the synthesis of the three phosphohydrolases in the non-inducible mutant showed that none of them was derepressed by phosphorus starvation (Fig. 54). It may be that the R_1 gene, which is responsible for the induction of alkaline phosphatase, is common to all three phosphohydrolases. Figure 55 shows a possible scheme for the regulation of the biosynthesis of the three phosphohydrolases of *E. coli* which we have examined.[540]

The total of the results obtained by Nesmeyanova and co-workers on the regulation of the phosphohydrolases lead to the conclusion that these enzymes, which are not components of a single metabolic pathway but possess similar

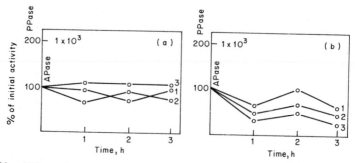

Figure 54. Effect of exogenous orthophosphate on the production of phosphohydro-lases in the non-inducible mutant of *E. coli* C_{85} (R_1^-, R_2^+ + P). (a) Deficiency of exo-genous orthophosphate; (b) excess of exogenous orthophosphate. 1, Polyphosphatase; 2, tripolyphosphatase; 3, alkaline phosphatase[540]

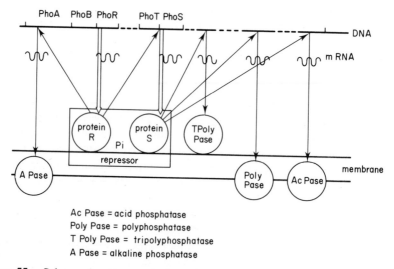

Ac Pase = acid phosphatase
Poly Pase = polyphosphatase
T Poly Pase = tripolyphosphatase
A Pase = alkaline phosphatase

Figure 55. Scheme for the regulation of phosphohydrolases in *E. coli*.[539,544] S = structural gene; R = regulatory gene

functions, are regulated in *E. coli* by a single metabolite, possess a common regulatory gene, and constitute a single regulon.

Azotobacter

1. *General features of the metabolism of high-molecular-weight polyphosphates*

Polyphosphate metabolism in several species of *Azotobacter* has been investiga-ted in some detail by Zaitseva and co-workers.[835–842] These investigations were initially concerned with changes in polyphosphate content during the onto-genetic development of cultures of the various species of *Azotobacter*. It was shown that, as in *E. coli*, acid-insoluble polyphosphates accumulated during the latent phase of development, and were subsequently utilized actively during the. exponential growth of the culture. In several cases (*Azotobacter agile*), they disappeared completely from the cells at this growth stage, reappearing only in the stationary phase. It is interesting that acid-insoluble polyphosphates accumulated in the early part of the stationary phase, followed at a later stage in the development of the culture by the accumulation of acid-soluble polyphosphates.[835,837,838] Zaitseva *et al.*[843] also studied the behaviour of polyphosphates in a synchronized *Azotobacter* culture. It was found that acid-insoluble polyphosphates attained their highest levels in synchronously growing *Azotobacter* cells immediately prior to the onset of cell division, and that during cell division they were degraded to acid-soluble polyphosphates of lower molecular weight, and to orthophosphate. It is interesting that the increase in acid-insoluble polyphosphates before the onset of cell division is clearly correlated with increased RNA polymerase activity. These findings provide further confirmation of the close relationship between the metabolism of polyphosphate and that of RNA, and constitute evidence for the validity of the hypothetical scheme which we have presented above for the biosynthesis of inorganic polyphosphates from pyrophosphate produced during the functioning of RNA polymerase.

These investigations by Zaitseva and co-workers also demonstrate the dependence of polyphosphate metabolism in *Azotobacter* on the composition of the nutrient medium. In particular, in a medium containing ammonium salts, polyphosphates accumulated in larger amounts than under nitrogen fixation conditions in the absence of a nitrogen source in the medium. Following Esposito and Wilson,[160] Zaitseva *et al.*[839] established conclusively the inhibitory effect of Ca^{2+} ions on the biosynthesis of polyphosphates, and simultaneously on nitrogen fixation. However, Zaitseva *et al.* showed that these two phenomena were not directly related, and that polyphosphates do not provide the source of phosphorus and energy in nitrogen fixation, as suggested by Esposito and Wilson. It was found that Ca^{2+} has an inhibitory effect on glycolytic phosphorylation, the common energy source for both processes, which is in turn (in *Azotobacter*) associated both with nitrogen fixation and polyphosphate synthesis. It is also possible that the inhibitory effect of Ca^{2+} in *Azotobacter* is due to its ability to form macromolecular complexes with the phosphate ions[568,570] which are essential for the normal course of glycolytic phosphorylation.

2. *Enzymes of metabolism of high-molecular-weight polyphosphates*

In a series of investigations, Zaitseva and co-workers have shown that the biosynthesis of polyphosphates in *Azotobacter*, and also to a large extent their utilization, are brought about by the reversible effect of polyphosphate kinase, mediated by ATP. This was shown both *in vivo*[835,840] and *in vitro*.[836] This enzyme was isolated by Zaitseva and Belozersky[836] from *Azotobacter vinelandii*, and purified to a considerable extent. As has already been pointed out, it is highly specific with respect to its substrates, i.e. adenyl nucleotides and high-molecular-weight polyphosphates.

The most important and distinctive feature of the metabolism of high-molecular-weight polyphosphates in *Azotobacter* is their apparent ability to utilize the phosphorus and energy which they contain for the direct synthesis of ATP mediated by polyphosphate kinase. This ability, as we have already said, is certainly not found in all microorganisms, being absent for example from *Mycobacteria*, *Corynebacteria*, and *Aerobacter aerogenes*. It is possible that the range of enzymes participating in the metabolism of high-molecular-weight polyphosphates in *Azotobacter* is not restricted to the single enzyme polyphosphate kinase.

For example, it is likely that a depolymerizing polyphosphatase is also present which brings about the rapid hydrolysis of polyphosphates to fragments of low molecular weight during cell division in *Azotobacter*.[843]

Photosynthesizing bacteria

1. *General features of the metabolism of high-molecular-weight polyphosphates*

In the photosynthesizing bacteria, polyphosphate metabolism was first investigated in any detail in the green sulphur bacterium *Chlorobium thiosulfatophillium*.[89,162,163,256,659] Fedorov[162,163] and Fedorov and Shaposhnikov[659] have followed the accumulation of polyphosphate during the ontogenetic development of a culture of this organism. They showed that the maximum accumulation of polyphosphate occurred in the stationary growth phase, when the basic synthetic processes were retarded. During rapid growth of cultures of these bacteria, the proportion of acid-soluble to acid-insoluble polyphosphates shifted towards the latter. It is interesting that Fedorov was able to show that large amounts of acid-soluble polyphosphates accumulate in these bacteria. When the cells were illuminated in the absence of CO_2, polyphosphate synthesis proceeded, the larger proportion of the ^{32}P from the medium taken up during this time finding its way into the acid-soluble fraction. It was suggested that the polyphosphates were synthesized first, in the absence of CO_2 fixation (and consequently in the absence of biosynthetic activity), in acid-insoluble form, and then immediately stored in the form of lower molecular weight, acid-soluble polyphosphates. These findings could, however, be interpreted differently, in the light of more recent work, using another phototropic bacterium, *Rhodospirillum rubrum*, carried out by Baltchevskii and co-workers[25,32,33,772] and in our laboratory.[425]

Baltschevskii and co-workers[25,32,33,772] showed that when *Rh. rubrum* is grown in the light, photosynthetic phosphorylation gives rise to substantial amounts of pyrophosphate. A similar observation was made in our laboratory by A. Shadi and co-workers using the same microorganism.[425,657]

These results give reason to suppose that, in Fedorov's experiments on the incorporation of ^{32}P in *Chl. thiosulfatophillium* in the absence of CO_2, the label was initially incorporated into the pyrophosphate which was formed. At that time, reliable analytical methods for the separate determination of pyrophosphate and higher molecular weight polyphosphates in the acid-soluble fraction were not yet in existence, and it was impossible to say into which of these compounds the ^{32}P was incorporated. It appears very likely that in cells of *Chl. thiosulfatophillium*, in the absence of CO_2, in the first instance large amounts of pyrophosphate may accumulate. Subsequently, according to a suggestion put forward by Baltschevskii,[25,32,33] the pyrophosphate produced by photosynthetic phosphorylation could be used for the synthesis of higher molecular weight polyphosphates. In an investigation carried out by Kulaev *et al.*,[425] it was shown *inter alia* that when the photosynthesizing bacterium *Rh. rubrum* was grown anaerobically in the light, its chromophores accumulated salt-soluble polyphosphates in addition to pyrophosphate. In the corresponding membrane equivalents of the chromatophores of a dark culture of this bacterium, neither polyphosphate nor pyrophosphate was present. These workers later showed[657] that high-molecular-weight polyphosphate biosynthesis is brought about in chromatophores of *Rh. rubrum* from ATP only, and not from pyrophosphate. It should be pointed out, however, that in *Rh. rubrum*, according to results obtained in our laboratory,[425] both in the dark and in the light, accumulation takes place not only of salt-soluble, but also of more highly polymerized polyphosphate fractions (alkali-soluble, and extractable with hot perchloric acid), the total amounts of polyphosphate in these bacteria differing little whether growth took place in the light or in the dark. Under conditions of phototrophic nutrition, however, the ratio of salt-soluble polyphosphates to total polyphosphate in cells of *Rh. rubrum* at all stages of growth was much greater than in the control cells grown in the dark.

In the main, our findings[425] are in good agreement with previous investigations into polyphosphate metabolism in the purple bacteria (*Rh. rubrum* and *Rhodopseudomonas spheroides*) carried out by Weber.[794] However, the two species of purple bacteria which he examined differed substantially in the distribution of polyphosphate in the various fractions. In *Rh. rubrum*, the greater part of these compounds were found by Weber to be present in the more highly polymerized acid-insoluble fractions, whereas in *Rh. spheroides* the acid-soluble polyphosphates constituted a much greater part of the total polyphosphate. It appears that in the latter case a greater amount of pyrophosphate accumulated, resulting in substantial accumulation of acid-soluble polyphosphates in this organism. Polyphosphate metabolism in *Rh. spheroides* has also been studied by other workers,[81] but this work was reported so briefly that it affords no additional information. In addition, inorganic polyphosphates have been identified in certain other photosynthesizing bacteria, namely

Chromatium sp.,[163] *Chromatium okenii*,[632] and *Rhodopseudomonas palustris*,[163] but their metabolism has not been studied in any detail.

2. *Enzymes of metabolism of high-molecular-weight polyphosphates and the regulation of their activity*

The mechanism of the biosynthesis and utilization of polyphosphates in phototrophic bacteria has been studied in *Chlorobium thiosulfatophillium* and *Rhodospirillum rubrum*[657] only. These investigations were carried out by Hughes and co-workers[89,256] and in our laboratory.[657] Examination of the results shows that the biosynthesis and utilization of polyphosphates in this phototrophic bacterium, as in *Azotobacter*, may be mediated by polyphosphate kinase with the participation of ATP. Hughes and co-workers[89,256] found that degradation of polyphosphates in *Chl. thiosulfatophillium* occurred in the absence of poly-phosphatase, by means of the following two coupled reactions:

$$PP_n + ADP \longrightarrow PP_{n-1} + ATP$$
$$\underline{ATP \longrightarrow ADP + orthophosphate}$$
$$PP_n \longrightarrow PP_{n-1} + orthophosphate$$

The ATP formed in the polyphosphate kinase reaction is hydrolysed by ATPase to ADP and orthophosphate. In this case, degradation of polyphosphate is not mediated by polyphosphatase, but by a more complex route, which was elucidated as follows. Firstly, Cole and Hughes[89] observed polyphosphate kinase and ATPase activity in cell-free extracts from this bacterium, but they were unable to detect polyphosphatase activity. Secondly, they showed that the breakdown of polyphosphate to orthophosphate by means of a cell-free extract of *Chl. thiosulfatophillium* occurred only in the presence of ADP.

It is most important to note that Hughes *et al.*[256] established that the ATP-dependent synthesis of polyphosphate in this organism bore no direct relationship to photophosphorylation. The cleavage of polyphosphates to orthophosphate was not connected with photosynthetic light reactions. The possibility that high-molecular-weight polyphosphates could have been formed from pyrophosphate, probably formed during phosphorylation[25–37] was not considered by Hughes and co-workers.

As far as the regulation of polyphosphate metabolism in this organism is concerned, likewise almost nothing is known at the present time. Hughes *et al.*[256] have adduced several arguments in support of the view that polyphosphate metabolism in *Chl. thiosulfatophillium* is controlled by the level of phosphate in the medium.

Similar results were obtained by Weber[794] in an investigation into the metabolism of high-molecular-weight polyphosphates in *Rhodospirillum rubrum* and *Rhodopseudomonas spheroides*. Weber showed clearly that the metabolism of these compounds in these phototrophic bacteria is controlled by the concentration of phosphorus in the surrounding culture medium.

Other bacteria

Polyphosphate metabolism has also been investigated to a certain extent in other bacteria, but the information available is clearly insufficient to reveal even the general features of the metabolism of these compounds, much less its regulation.

The work of Kaltwasser and Schlegel on polyphosphate metabolism in the hydrogen-oxidizing[288,289,631,634] and nitrifying bacteria[290] is undoubtedly of great interest. It is of importance primarily because it demonstrates the possibility of polyphosphate synthesis by the oxidation of inorganic materials, but nevertheless no clear picture of the metabolism of polyphosphates in these bacteria under normal growth conditions is to be found in the literature. All of the work carried out in this laboratory on polyphosphate metabolism utilized phosphorus-starved cells, no control comparison being made with their behaviour under normal growth conditions. From the work of Kaltwasser and Schlegel with *Hydrogenomonas*[288,289] and *Micrococcus dentrificans*,[290] however, it is clear that these bacteria display the phenomenon of polyphosphate hypercompensation following phosphorus starvation. Eigener and Bock[159] described the biosynthesis and breakdown of polyphosphates in *Nitrobacter vinogradskii*.

Some investigations of polyphosphate metabolism in other bacteria have a certain interest. In particular, it was shown that in *Bacillus cereus, Bacillus megaterium*, and *Agrobacterium radiobacter*,[726] accumulation of high-molecular-weight polyphosphates does not occur under normal conditions, but only when the growth of these organisms is restricted in some way.

Thus, these investigations and those of other workers have shown that the accumulation of significant amounts of inorganic polyphosphates in bacterial cells is not essential.

FUNGI

1. *General features of the metabolism of high-molecular-weight polyphosphates*

The metabolism of polyphosphates is very similar in all of the fungi which have been examined, and it is therefore convenient to present the available data in a single section. The metabolism of polyphosphates during ontogenetic development has been investigated in mould fungi[20,58,214,279,351,357,365,367,383,386,391,392,421, 510,562,563,573] and yeasts,[50,51,53,73–75,85,166,197,244,246,251,286,287,297,335,407,432,447,450,468,469,702,754,781, 796,808] as well as in mucor fungi,[272] ergot,[430] mushrooms,[360,361,390,409] and other Basidiomycetes.[352,354] It has been found that rapid synthesis of acid-insoluble high-molecular-weight polyphosphates takes place during the germination of fungal spores. By using ^{32}P, it has been shown that the formation of highly polymerized condensed inorganic phosphates at this stage of development involves utilization of less highly polymerized, acid-soluble polyphosphates, resulting in the complete conversion of the latter into the acid-insoluble form.[280,386]

In the light of our present ideas, the acid-soluble polyphosphates which are present in fungal cells in the early stages of spore germination are mainly composed of pyrophosphate. It was, in fact, pyrophosphate which appeared to be used for the synthesis of higher polyphosphates at the end of the latent phase and the beginning of the logarithmic growth phase.

It is interesting that, during this period of development, polyphosphate depolymerase (which cleaves the higher polymeric polyphosphates to lower polymeric fragments) occurs in an inactive state in mould fungi.[355,356,358,457] This appears to facilitate to a great extent the accumulation of large amounts of inorganic polyphosphate in the cells of these organisms. On the other hand, at this time the utilization of polyphosphate for the synthesis of a variety of compounds has already begun. Nishi[562,563] has shown that even during the first few hours of spore germination in *Aspergillus niger*, polyphosphates have already begun to be utilized for the synthesis of nucleotides and sugar phosphates, as well as in the formation of RNA.

From the onset of rapid growth and fission of the fungal cells, acid-insoluble high-molecular-weight polyphosphates are utilized with increasing rapidity in the synthetic processes which are proceeding at a high rate at this stage, especially in the biosynthesis of RNA.

The involvement of high-molecular-weight polyphosphate in RNA biosynthesis has been demonstrated both in yeasts[50,51,53,73–75,85,335,407,678,685,808] and other fungi.[58,213,215,352,354,357,365,367,368,386,390,391,563]

In one of our early investigations in which ^{32}P was employed, it was shown that polyphosphates functioned as phosphate donors in the biosynthesis of RNA to the same extent for all the nucleotides present in RNA.[386] This was later confirmed by Stahl *et al.*[681,685] It was shown also that polyphosphates can be utilized for the synthesis of all the known types of RNA, and in particular ^{32}P was incorporated equally in all the nucleotides of both transport and ribosomal RNA.

Kulaev *et al.*[407] have attempted to elucidate the relationship between the metabolism of the various polyphosphate fractions, RNA, and DNA in synchronous cultures of the yeast *Schizosaccharomyces pombe* (synchronicity index 0.7–0.8). As is shown in Fig. 56, in the interval between the two episodes of division, i.e. during the growth of the dividing cells, substantial changes take place in the amounts of a wide variety of phosphorus compounds present. The total phosphorus content of the yeast cells increases exponentially immediately following the completion of cell division, reaching a maximum approximately midway through this period.

Consideration of the data presented in Fig. 56 leads to the conclusion that the shape of the curve for the accumulation of total phosphorus during this period is determined to a large extent by the accumulation of RNA in the cells. The RNA content, like the total phosphorus, increases rapidly during the first half of this period, but it decreases substantially at the time of cell division.

The DNA content of yeast cells, on the other hand, remains constant during the first two thirds of the latent period. The amount of DNA doubles over a

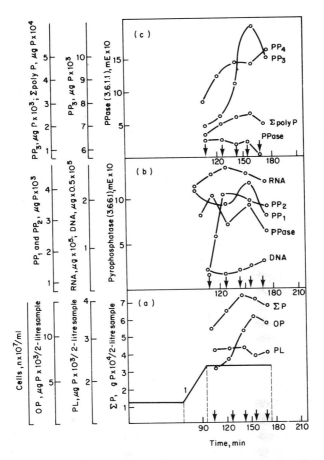

Figure 56. Changes in nucleic acids, phospholipids, orthophosphate, polyphosphates, polyphosphate phosphohydrolase (3.6.1.11), and pyrophosphatase (3.6.1.1) during synchronous growth of a culture of *Schizosaccharomyces pombe.*[407] (a) 1, Number of cells; PL, phospholipids; OP, orthophosphate; Σ P, total phosphate. (b) PP_1 and PP_2, acid-soluble and salt-soluble polyphosphates. (c) PP_3, PP_4 and Σ poly P, alkali- and hot perchloric acid-soluble polyphosphates and total polyphosphate

short time interval (approximately 15 min), and reaches a maximum at the beginning of the next episode of cell division (after 5–10 min). The ratio of total RNA to DNA during this period varies from 10:1 to 20:1, depending on the growth phase of the yeast.

A parallel investigation was carried out into the behaviour of various polyphosphate fractions and some polyphosphate-metabolizing enzymes. Figure 56 shows that during the first two thirds of the period of synchronous growth of the yeast under examination, the total polyphosphate increases. In the period immediately preceding cell division, on the other hand, a slight fall in the accumulation of these phosphorus compounds is observed. The slight increase which we observed in the total amount of polyphosphate in the early stages of growth of the yeast correlates well with the reduction in its rate of utilization by polyphosphatase (PPase) (see Fig. 56). Further, it is undoubtedly connected with the rapid increase in the biosynthesis of the more highly polymerized polyphosphate fractions (PP_4 and PP_3). It can be seen from Fig. 56 that both of these fractions accumulate rapidly at the beginning of the period, accumulation of the more highly polymerized PP_4 fraction taking place first, followed, perhaps in consequence of its degradation, by a substantial increase in fraction PP_3. Shortly before the beginning of the new episode of cell division, degradation of PP_4 ceases, resulting in its accumulation, accompanied by a parallel reduction in the level of PP_3.

It has already been noted above that in another yeast (*Saccharomyces carlsbergensis*) it has been found that the biosynthesis of these polyphosphate fractions is very closely linked to the biosynthesis of the cell wall polysaccharides mannan and glucan.[432,433] These results, taken in conjunction with those obtained by us (see above), leads to the conclusion that, during this period of growth of a synchronous culture of *Schiz. pombe*, active synthesis of cell wall components is taking place as would be expected, and this process apparently also leads to the accumulation of the more highly polymerized polyphosphate fractions.

Of no less interest was the behaviour of fractions PP_2 and PP_1. As can be seen from Fig. 56, the PP_2 fraction is the only polyphosphate fraction whose accumulation in the cells of synchronous cultures of the yeast is directly correlated with RNA biosynthesis. The shape of the curve for the accumulation of this fraction follows exactly that for total RNA accumulation. It now appears very likely that the formation of fraction PP_2 and nucleic acid biosynthesis are linked as described in the scheme shown in Fig. 28.

It is interesting that the curve for DNA accumulation does not show the same correlation with fraction PP_2 as it does in the case of RNA. This, however, is not surprizing. As has already been mentioned, DNA is present in *Schiz. pombe* cells at all stages of growth of the synchronous culture, in amounts of one tenth to one twentieth that of RNA. It is therefore difficult to detect changes in the content of fraction PP_2 associated with the accumulation of what are essentially very small amounts of DNA during the period when it is being replicated. At the same time, it is during this period, when uniform but antidirectional reduction in the amount of RNA and increase in DNA is observed, that the level of fraction PP_2 remains constant.

In connection with our proposed scheme for the relationship of PP_2 biosynthesis to that of nucleic acids, it is of interest to consider the behaviour of pyrophosphatase activity during this period of synchronous growth of yeast cells. As Fig. 56 shows, the curve for the change in the specific activity of this enzyme displays two maxima during the period in question. One of these occurs at the time of rapid accumulation of RNA, and the other corresponds to the period of DNA replication. These findings are in good agreement with our hypothetical mechanism for the relationship of the biosynthesis of polyphosphates of fraction PP_2 to that of nucleic acids.

In conclusion, it is interesting to follow the behaviour of the lowest molecular weight fraction (PP_1), which is localized in fungi (especially in yeasts) in volutin granules.[381] As Fig. 56 shows, the shape of the curve for the changes of PP_1 during the whole period of synchronous growth of *Schiz. pombe* is a mirror image of curve PP_2, the more highly polymerized fraction. On the other hand, during almost the whole of this period the behaviour of fraction PP_1 is the converse of that of the total RNA. This polyphosphate fraction is apparently the main reserve of phosphate for the biosynthesis of RNA and DNA. It appears very likely that it is primarily this reserve of high-energy phosphate which is stored in the cell in the form of volutin-like granules, and is available at any time for the biosynthesis of nucleic acids, as well as, perhaps, for other purposes.

On the basis of these results, the following outline of the relationship between nucleic acid and polyphosphate metabolism in *Schiz. pombe* is proposed. The biosynthesis of nucleic acids, primarily RNA, and especially the formation of nucleoside triphosphates, utilizes the intracellular reserves of high-energy phosphate previously accumulated in the cell in the form of inorganic polyphosphate of fraction PP_1. On the other hand, the pyrophosphate formed in the cell nucleus during the biosynthesis of nucleic acids may, by means of the appropriate phosphotransferase (possible pyrophosphatase), lead to the formation in the nucleus, and perhaps in the vacuoles, of the more highly polymerized fraction PP_2. This fraction, by enzymatic depolymerization, is able once more to provide cellular reserves of the less polymerized fraction PP_1, which is stored in the cytoplasm in the form of volutin granules. The more highly polymerized fractions PP_4 and PP_3, being located superficially in the cells of yeasts and other fungi, are apparently not directly involved in nucleic acid biosynthesis.

Returning to the behaviour of high-molecular-weight polyphosphates in yeasts under normal, non-synchronous culture conditions, it is noteworthy that even in the logarithmic growth phase these compounds are invariably found in the cells, in varying amounts. In the stationary growth phase, their utilization for synthetic purposes more or less ceases, usually resulting in even greater accumulation.

Polyphosphate metabolism during this time varies widely, according to the conditions of culture of the fungi.[105,215,272,298,351,391,678,679] This is due mainly to the fact that the conditions of growth have a profound effect on the activity of the polyphosphate-cleaving enzymes. Thus, in *Penicillium chrysogenum*, the depolymerization of polyphosphates under deep culture conditions gives rise directly to orthophosphate bypassing the intermediate formation of low-polymeric polyphosphates, whereas under surface culture conditions oligophosphates, especially pyrophosphate, are formed as intermediates.[391] In *Mucor racemosus*, it was observed[272] that ageing mycelia of this fungus underwent autolysis, and the polyphosphates present passed without hydrolysis into the growth medium, accumulating therein in large amounts. It was shown in the case of *Saccharomyces carlsbergensis* that in the stationary growth phase the accumulation of polyphosphate was strongly dependent on polyphosphatase, which hydrolyses these polymers terminally.[754]

Under more favourable growth conditions, when growth is stimulated, the overall accumulation of polyphosphate in the cells decreases.[212,213,391,392] This may be due to the fact that under conditions in which the overall rate of metabolism is higher, polyphosphates are also utilized more rapidly for synthetic requirements. On the other hand, it may be that, under conditions of active growth, polyphosphate biosynthesis is inhibited. The metabolism of high-molecular-weight polyphosphates is highly dependent on the composition of the growth medium, and in particular on its content of phosphate, glucose, and a nitrogen source. We have dwelt in some detail on the effect of the composition of the growth medium on polyphosphate metabolism in a variety of organisms, including yeasts. In particular, it was noted that very substantial intensification of the biosynthesis of these compounds occurred when the microorganisms were grown on a phosphate-enriched medium following previous phosphorus starvation (hypercompensation effect).[157,158,215,220,383,388,389,466]

In bacteria, especially *Aerobacter aerogenes*,[220] it was shown that this effect is associated with derepression of polyphosphate kinase when growth takes place in the absence of phosphate. In yeasts, however, the existence of the hypercompensation effect is due to some other, as yet undetermined mechanism. Felter and Stahl[167,683] have shown that, in yeasts, polyphosphate kinase functions only in the direction of polyphosphate utilization. This was confirmed in our own laboratory for other fungi, in which polyphosphate-synthesizing activity has not been detected in this enzyme.[395] The non-involvement of polyphosphate kinase in the phenomenon of hypercompensation in yeasts has been still more clearly demonstrated by Felter and Stahl,[166] in experiments *in vivo*.

In considering the influence of the nutrient carbon source, interesting papers have appeared[197,450] describing an investigation of high-molecular-weight polyphosphate metabolism in hydrocarbon-oxidizing yeasts when the latter were grown on media containing glucose or hydrocarbons as the sole carbon source. These investigations showed that, in all cases, much more rapid accumulation (2–3 times) of polyphosphate occurred in the yeast cells when they were grown on a hydrocarbon-containing medium then when grown on a medium containing glucose.

It would be premature to discuss here the reasons for such marked differences in polyphosphate metabolism. We shall mention one possible reason a little later.

Several papers have dealt with the effect on polyphosphate metabolism in yeasts and other fungi of the presence in the nutrient medium of varying amounts of nitrogen source.[197,382,383,432,483] This environmental factor was found to have a considerable effect on the metabolism, and especially the distribution of polyphosphates between the various fractions.

2. Enzymes of metabolism of high-molecular-weight polyphosphates and the regulation of their activity

Polyphosphate-metabolizing enzymes have perhaps been investigated in the greatest detail in yeasts and other fungi. Moreover, nearly all of the enzymes known to participate in the biosynthesis and utilization of inorganic polyphosphates were first found in these organisms. For example, polyphosphate–ADP phosphotransferase (polyphosphate kinase) was first found in cell-free extracts of yeast.[243,245,248,831,834] This enzyme is also currently isolated from yeast in its most highly purified form.[167,168] 1,3-DPGA–polyphosphate phosphotransferase was first observed by us in the fungus Neurospora crassa,[394,496] and subsequently in other subjects.[395] It is interesting to note that the presence of this enzyme in yeasts has as yet not been proved. However, it has been shown recently by Ermakova and Mansurova that pyrophosphate, rather than high-molecular-weight polyphosphates, may be formed in yeasts of the Saccharomyces genus, in a similar system. Polyphosphate phosphohydrolase (polyphosphatase) was found in yeast by Kitasato[310,311] as long ago as 1928, and since then it has been studied intensively in this organism. This enzyme was recently obtained in a homogeneous form from N. crassa by Umnov et al.[761]

Work on polyphosphate phosphohydrolase (polyphosphate depolymerase) was initiated, and has continued with some success, in mould fungi and yeasts.[264–266,353,355,358,376,377,402,481,482,562] For instance, the very important role of polyphosphate depolymerases in polyphosphate metabolism was demonstrated using these organisms.

In in vivo experiments using fungi we and Langen and Liss showed simultaneously that during polyphosphate biosynthesis, exogenous [^{32}P]orthophosphate was found mainly in the acid-insoluble polyphosphates. These investigations also showed that acid-soluble polyphosphates of low molecular weight are only formed in fungi as secondary products, by degradation of the more highly polymerized forms.[58,351,367,368,386,388,391,409,442–444,447,464–466,468,469]

The degradation of inorganic polyphosphates has been investigated most fully in yeasts. Langen and Liss,[442–444,446,447] working with these organisms, showed that the first polyphosphate fraction to be formed from exogenous orthophosphate was the most highly polymerized PP_4 fraction ($\bar{n} = 260$). Hydrolysis of PP_4 then afforded the lower molecular weight PP_3 fraction ($\bar{n} = 55$), followed by the still lower molecular weight PP_2 fraction ($\bar{n} = 20$), and finally the lowest molecular weight PP_1 fraction ($\bar{n} = 5$), which was then hydrolysed to orthophosphate.

It was in fungi that we investigated in some detail the localization of the various polyphosphate fractions (see the earlier section on localization). It was found that the more highly polymerized fractions (PP_5, PP_4, and PP_3) were sited peripherally in the fungal cells, in the region of the plasmatic membrane. The PP_2 fraction was localized in the cell nuclei and vacuoles, and PP_1 was found in the cytosol. On the basis of these findings, the important hypothesis was formulated that the degradation of the individual polyphosphate fractions, mediated by the appropriate depolymerases, is usually (perhaps always) associated with their transport through cell membranes.[402]

A study of the localization of the enzymes participating in the metabolism of polyphosphates in fungi, and the relationship of the metabolism of individual fractions with other metabolic entities, demonstrated[376,377] the metabolic diversity of the polyphosphate fractions. In the metabolism of the peripherally-sited fractions PP_5, PP_4, and PP_3, polyphosphatase and polyphosphate depolymerase play an important part. The biosynthesis of these more highly polymerized fractions in fungi is apparently in the first instance related to cell-wall polysaccharide biosynthesis.

We propose that the linking of these two processes occurs by the utilization of the dolichol pyrophosphate arising during the biosynthesis of the cell-wall polysaccharides (see Fig. 32).

The metabolism of the PP_2 polyphosphate fraction, which is located in the fungal cell nucleus,[406,667] is apparently connected very closely with nucleic acid metabolism. One possible mechanism linking the metabolism of polyphosphates with that of nucleic acids is depicted in Fig. 28. It envisages the biosynthesis of this fraction from pyrophosphate formed during nucleic acid biosynthesis.

Finally, the PP_1 polyphosphate fraction, which is located in the fungal cytoplasm, for the most part in volutin granules, could be synthesized and utilized by means of 1,3-DPGA–polyphosphate phosphotransferase.[400] It is thus very likely that the metabolism of this fraction is most closely associated with glycolysis, and its formation may be the result of the esterification of intracellular orthophosphate during glycolytic phosphorylation. It should be pointed out that, up to the present time, the metabolism of high-molecular-weight polyphosphates has been examined in the greatest detail in *Neurospora crassa*.

Changes in the activity of the enzymes participating in polyphosphate metabolism with respect to the amounts of the various polyphosphate fractions present have been investigated in some detail in the same fungus. Some of the results have been discussed in earlier chapters, and the remainder are presented in Fig. 57. The activity of the polyphosphate depolymerase which cleaves highly polymerized polyphosphates ($\bar{n} = 290$) increases with age, reaching a maximum at the end of the logarithmic growth phase (Fig. 41), at which point a sudden fall in high-molecular-weight polyphosphate content is observed. This suggests that this enzyme begins to function with particular efficiency at the point when the main biosynthetic processes are beginning to slow down. According to findings in this laboratory,[426] 1,3-DPGA–polyphosphate phosphotransferase activity also increases gradually with the age of the *N. crassa* culture, reaching a maximum at the beginning of the stationary phase.

Also, as can be seen from Fig. 57, accumulation of polyphosphate in *N. crassa* is at its most rapid at the beginning of the logarithmic phase, i.e. when the main biosynthetic processes are occurring most actively in the cell. It is, as yet, difficult to say precisely what is the enzymatic system which is primarily res-

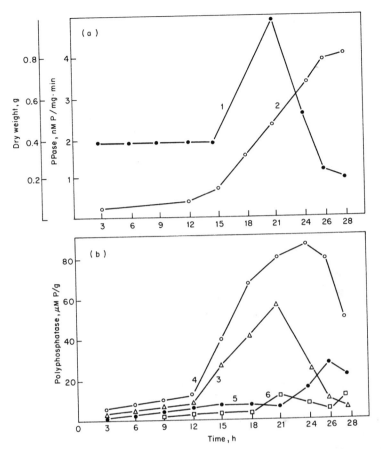

Figure 57. Polyphosphatase activity (a) and polyphosphate levels (b) during growth of *N. crassa* mycelia.[762] 1, Polyphosphatase activity; 2, dry weight of mycelium; 3, PP$_4$; 4, PP$_2$; 5, PP$_1$; 6, PP$_5$

ponsible for polyphosphate biosynthesis in *N. crassa*, but from results obtained with yeasts it is suggested that polyphosphate biosynthesis at the beginning of the logarithmic growth phase is effected by dolichol·pyrophosphate formed at this time as an end-product of cell-wall polysaccharide biosynthesis. Further, it is noteworthy that polyphosphatase activity rises strictly in parallel with the accumulation of highly polymerized polyphosphates in cells of *N. crassa* and then falls. We have already noted that polyphosphatase behaves in a similar manner in other organisms.

The correlation between the behaviour of polyphosphatase and that of the higher polymeric polyphosphates can be interpreted in either of two ways. Either polyphosphatase participates at this stage in polyphosphate biosynthesis, i.e. it behaves like ATPase, largely fulfilling in the cell the function of an

ATP synthetase,[668] or the biosynthesis of polyphosphates in the cell is activated by its own substrate, polyphosphate. In the latter case, it would be thought that the biosynthesis of polyphosphates in the cell would be mediated by some other enzyme system, and their utilization would involve polyphosphatase. It is as yet difficult to say which of these two alternatives is correct. In addition, data are now available on the functional role of polyphosphatase in fungi.

We have already pointed out that in bacteria the role of polyphosphatase is reduced virtually to the regulation of the levels of intracellular orthophosphate. The biosynthesis of this enzyme is repressed in the presence of orthophosphate, and is induced in its absence. In fungi, derepression of polyphosphatase during phosphorus starvation is not observed, as was shown in yeasts by Felter and Stahl,[166] and in *N. crassa* in our laboratory.[762] However, in an investigation into the behaviour of polyphosphatase in *N. crassa* under a variety of culture conditions, it was shown that its biosynthesis was subject to catabolic repression by glucose (the 'glucose effect'). When the medium contained large amounts of glucose, biosynthesis of the enzyme was repressed, whereas when *N. crassa* was grown on a medium containing only small amounts of glucose, or none at all, induction of polyphosphatase biosynthesis proceeded.

These results are of considerable interest in connection with the work of Neville *et al.*[548] This investigation revealed the presence in *N. crassa* cells of two glucose transport systems. One of these, involving active transport, was repressed when the cells were grown in a medium containing glucose, and was induced when the glucose was removed from the medium. The close analogy between the regulation of the biosynthesis of the active transport system in *N. crassa* and polyphosphatase biosynthesis led us to postulate the possible participation of polyphosphatase in the active transport of glucose in this fungus.

Umnov *et al.*[762] have put this hypothesis to the test. The growth conditions for conidia of *N. crassa*[548] were reproduced, whereupon the active glucose transport system came into operation. Further investigation showed (see Fig. 58) that during the first 6 h, active uptake of glucose by the conidia occurred, accompanied by an increase in dry weight. By the sixth hour of incubation, the glucose in the medium was exhausted, and mycelial growth ceased. As Fig. 58 shows, a substantial decrease in the rate of uptake of glucose occurred during the first few hours. On the other hand, at a somewhat later stage, as the glucose content of the medium decreased, the initial transport rate increased, and following the complete disappearance of glucose from the medium it increased even more rapidly, i.e. under these conditions complete derepression of the active glucose transport system occurs. A parallel investigation into the changes in polyphosphatase activity under the same conditions showed it to be analogous to the changes in the initial rate of glucose uptake. It is most important to note that, at all growth stages of *N. crassa*, the ratio of initial glucose uptake rate to polyphosphatase activity was close to unity. It is therefore suggested that, firstly, polyphosphatase is involved directly in the active transport of glucose in cells of *N. crassa*, and secondly, that the active transport

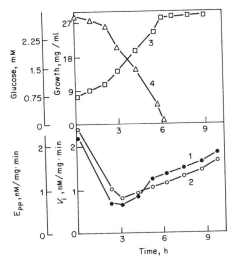

Figure 58. Repression and autoderepression of the active glucose transport system in *N. crassa* conidia during germination.[762] 1, Initial glucose transport rate; 2, polyphosphatase activity; 3, dry weight of mycelium; 4, glucose in the growth medium

of each molecule of glucose is accompanied by the fission of one phosphoric anhydride bond in polyphosphate. In order to test these suggestions, polyphosphatase activity and the initial rate of glucose uptake in *N. crassa* mycelia were measured in a medium without glucose, but containing the protein synthesis inhibitor actidione (cycloheximide). As Table 36 shows, in the presence of actidione and in the absence of glucose, derepression of the active glucose transport system did not occur, whereas in the control samples (without actidione) the activity of this system was restored to its initial level over the chosen time of incubation. The change in polyphosphatase activity under these conditions corresponded exactly with the initial rate of active glucose transport, and in nearly all of the samples the ratio of these values was close

Table 36. Effect of actidione on the derepression of the biosynthesis of the active glucose transport system and polyphosphatase biosynthesis in *Neurospora crassa*[762]

Time of incubation (h)			Initial glucose Transport rates V_i (nM/mg.min)	Polyphosphatase activity, mE (nM/mg.min)	$\dfrac{V_i}{mE}$
With glucose	Without glucose				
	Without actidione	With actidione			
0	0	0	2.50	2.58	0.970
5	0	0	0.66	1.10	0.600
5	2	0	3.50	3,72	0.940
5	4	0	2.63	2.83	0.920
5	0	2	0.53	0.57	0.910
5	0	4	0.58	0.61	0.950

to unity. The constancy of this ratio in all the experiments suggests that poly-phosphatase and the active glucose system are included in a single operon, and they may be functionally linked. These findings may also indicate that it is the high-molecular-weight polyphosphates located at the periphery of the cell (PP_5, PP_4, and PP_3), close to the site of the polyphosphatase, which function as energy donors for the active transport of glucose in *N. crassa*. Control experiments on active glucose transport and polyphosphatase behaviour in an adenine-deficient mutant of *N. crassa* (ad-6) in the presence of 8-azaadenine, i.e. under conditions which inhibit the metabolism of ATP,[429] completely excluded the possibility of the participation of ATP in this process. These re-sults provide confirmation of the existence of a close link between the func-tioning of the polyphosphate–polyphosphatase system and the active glucose transport system in *N. crassa*.

It is interesting that, judging from the results of Kotyk,[340] the active transport of sugars in fungi is not linked with phosphorylation. Polyphosphates are evidently consumed in active glucose transport, not for the activation of glu-cose itself by the formation of glucose-6-phosphate, but for the phosphoryla-tion of a carrier, or of some enzyme which mediates this process.

It is appropriate at this point to mention the work of van Steveninck,[690–692] van Steveninck and Booij,[693] and Deierkauf and Booij,[104] which resulted in the experimentally based hypothesis that polyphosphates are involved in the transport of galactose and other sugars in yeasts, phosphorylating phospha-tidylglycerol to phosphatidyl glycerophosphate (as shown in Fig. 59). In the opinion of these workers, it is the latter which is the specific donor of phos-phorus for the activation of the carrier–sugar complex and its transport across the membrane.

In the light of our results, it is possible that in fungi polyphosphatase partici-pates in the transfer of phosphoryl residues from polyphosphate to phos-phatidylglycerol located in the plasmatic membrane. It is interesting to note that in neither *N. crassa* nor yeast, nor any other fungus, has polyphosphate glucokinase (i.e. the enzyme which phosphorylates glucose itself by means of polyphosphate) yet been detected.[106,720,721,733] It appears very likely that in those organisms in which polyphosphate glucokinase has been found to be present

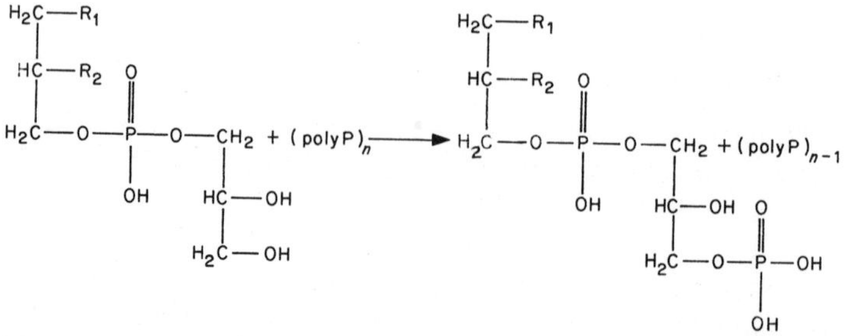

Figure 59. A possible mechanism for the polyphosphate–phosphatidylglycerol phosphotransferase reaction[104]

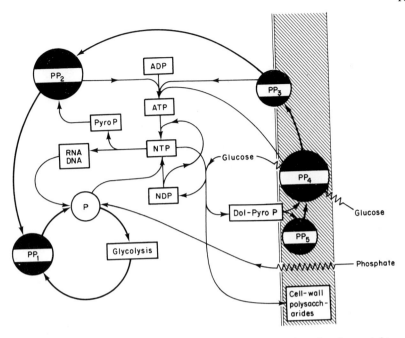

Figure 60. Hypothetical scheme for the reactions of high-molecular-weight poly-phosphates in fungi

(representatives of the Actinomycetes), this enzyme may be the one which is responsible for the transport of sugar across the cell membrane.

In concluding this section, in which we have endeavoured to shed light on the general features of the metabolism of high-molecular-weight polyphosphates in fungi, we believe it to be useful to present an integrated scheme for this metabolism. This scheme, shown in Fig. 60, incorporates both established experimental facts and certain plausible hypotheses.

ALGAE

1. *General features of the metabolism of high-molecular-weight polyphosphates*

The fact that algae are autotrophs has a profound effect on the metabolism of high-polymeric polyphosphates in these organisms.[7,22–24,42,46,81,93,97,98,115,228,240,250, 253,292,293,363,364,424,431,476,521–524,552,558,559,576,583–586,615,621,636,670,673,695,696,705,753,759,768,770,771,786,787, 825–827]

Since the metabolism of polyphosphates in those algae which have been studied has many features in common, we shall consider here all the available literature data together.

As in other microorganisms, the polyphosphate content of algal cells is highly dependent on the stage growth, being lowest during the exponential phase, and highest in older cultures.[670,771] The existence of close links between polyphosphate and RNA content during ontogenetic development has also been demonstrated in algae, in particular in *Euglena*.[670] During the growth and development of synchronous cultures of *Chlorella pyrenoidosa*, the accumulation of polyphosphate, RNA, and DNA took place in parallel, although polyphosphate biosynthesis during the first few hours of growth outpaced to some extent the synthesis of RNA and DNA.[240] From their experimental findings, Schmidt and co-workers came to the conclusion that under normal conditions of growth in the light and on a phosphate-containing medium, *Chlorella* did not utilize polyphosphate for the synthesis of nucleic acids. In Schmidt's view, during phosphorus starvation, *Chlorella* is able to utilize polyphosphate for the synthesis of nucleic acids and phospholipids.[24,636]

The results obtained in this laboratory, however, related to the total amount of polyphosphate and the overall nucleic acid content. Miyachi and co-workers have already shown convincingly[228,293,521-524] that when investigating the function of polyphosphates in algae it was necessary to examine the behaviour of the different polyphosphate fractions. These workers showed in particular that only some of the polyphosphate fractions with specific intracellular localizations were involved in RNA and DNA synthesis.[522] Their investigations together with those of Okuntsov and Grebennikov[576] showed that the metabolism of the various polyphosphate fractions in *Chlorella* proceeded differently in the dark and in the light, and in the presence and in the absence of phosphate in the medium.[293,521,522]

Before describing in more detail the work of Miyachi and co-workers, however, it is necessary to discuss briefly the general features of the action of light on the accumulation of polyphosphates in algal cells in relation to photosynthesis.

Early work by Wintermans[825,826] and subsequently by other investigators[42,43,293,476,558,559,576,585,695,696,705,753,759] showed that the formation of polyphosphates and polyphosphate-containing granules in algae proceeded much more rapidly in the light than in the dark. It is interesting that Ullrich[757] has shown that the synthesis of high-molecular-weight polyphosphates in *Ankistrodesmus braunii* is strongly stimulated as the concentration of oxygen in the medium is increased. These investigations lead to the conclusion that there is a close connection between the formation of polyphosphates in algae and photosynthesis. but it is not yet possible to come to a firm conclusion as to whether the accumulation of polyphosphates in the light in algae is directly linked to the photosynthetic process itself, or whether their formation is merely promoted by increased ATP (and perhaps pyrophosphate) synthesis by photosynthetic phosphorylation, or by the accumulation during photosynthesis of large amounts of sugars and other easily oxidisable substrates.

Kanai *et al.*[293] have shown that, although the incorporation of ^{32}P into orthophosphate proceeded more rapidly in the light and decreased substantially in the dark, nevertheless in the latter case the biosynthesis of polyphosphates did continue to some extent. It is therefore concluded that this process occurs in algae without the involvement of photosynthesis, although it is very strongly promoted by the latter process. Similar results were obtained by Kanai and Simonis[292] and Domanski-Kaden and Simonis,[115] using *Ankistrodesmus braunii*, and by Overbeck[583,584] using *Scenedesmus quadricauda*. Overbeck further showed that *Scenedesmus quadricauda* was able to accumulate excessive amounts of polyphosphate in the dark when it was grown on a phosphate-containing medium following phosphorus starvation. In other words, he was able to demonstrate that algae, like heterotrophic microorganisms, display the hypercompensation effect. That photosynthetic conditions are not obligatory for polyphosphate biosynthesis in algal cells is also shown by experiments with *Euglena*. In this alga, substantial amounts of polyphosphate were found under heterotrophic growth conditions.[670] Further, as has already been pointed out, work by Kulaev and Vagabov,[431,770] using *Scenedesmus obliquus*, showed that when this alga was grown in the dark, polyphosphates were produced by glycolytic phosphorylation.

It can therefore be concluded that part, if not all, of the polyphosphate formed in the algal cell is produced independently of photosynthesis and photosynthetic phosphorylation.

A further contribution towards an understanding of this problem was provided by the investigations of Miyachi and co-workers,[228,521–524] which showed that only one of the four polyphosphate fractions present in *Chlorella* (fraction C) was formed in the light.[293] This fraction in the author's view, was localized in *Chlorella* either in the chloroplasts, or in their immediate vicinity.[228,522] It was shown that both biosynthesis and utilization of this polyphosphate fraction occurred in the light, and that as in heterotrophs, the polyphosphate fraction extractable by cold 8% TCA (fraction A) was found in the volutin from *Chlorella*.[228] Its accumulation was to some extent dependent on photosynthesis, since it was formed by the degradation of fraction C, which is synthesized only in the light. These results, obtained with algae, confirmed those obtained with heterotrophic microorganisms, showing that acid-soluble polyphosphates were formed by degradation of acid-insoluble polyphosphates.

The biosynthesis and utilization of the other two polyphosphate fractions (D and B), observed by Miyachi and co-workers in *Chlorella*, appeared to be wholly unrelated to photosynthesis. Their metabolism depended solely on the presence of phosphate in the incubation medium. Similar results were obtained subsequently by Kanai and Simonis[292] with *Ankistrodesmus braunii*. These investigations have thus revealed a highly complex picture of the metabolism of polyphosphates in algae. Although the general behaviour and metabolic features of polyphosphate metabolism in these organisms are apparently identical with those investigated in some detail in heterotrophs, nevertheless the presence of chloroplasts in their cells which are capable of photosynthesis renders the metabolism of these compounds considerably more complex.

Miyachi and co-workers have shown that the utilization of the different polyphosphate fractions for the biosynthesis of nucleic acids in *Chlorella* proceeds differently in the light and in the dark. The reactions undergone by polyphosphates in *Chlorella* under different culture conditions, according to Miyachi *et al.*[522] are shown schematically in Fig. 61.

In the light, fraction C which is localized in the vicinity of the chloroplasts, is a phosphorus donor for the biosynthesis of chloroplast DNA. Fraction A, which is found in volutin, is involved in the synthesis of nuclear DNA. In cells of *Chlorella* under normal growth conditions in the light, RNA is not formed from polyphosphate, but rather directly from exogenous orthophosphate. On the other hand, when phosphate is absent from the medium, polyphosphates of fractions B and D are utilized for RNA biosynthesis. In the light, polyphosphates of these fractions are hydrolysed to orthophosphate, which is then utilized for the biosynthesis of RNA and other compounds. In the dark, and in the absence of phosphate, polyphosphates appear to be able in some manner to convey phosphate directly for the synthesis of RNA.

In considering the interrelationship between the metabolism of polyphosphates and nucleic acids in algae, one cannot ignore another question, namely, whether polyphosphate biosynthesis could occur in the absence of the cell nucleus, the principal organelle in which RNA and DNA biosynthesis takes place. The results of Richter[615] are of interest here. He observed in *Acetabularia* the biosynthesis of polyphosphates in parts of the cell not containing nuclei. These findings lead indirectly to the conclusion that the presence of the cell nucleus is not essential for the accumulation of at least some of the polyphosphate fractions. In general, the investigation of polyphosphate metabolism in this giant single-cell alga shows great promise. For example, Kulaev *et al.*[424] have shown that in young cells of *Acetabularia*, it is mainly the acid-soluble polyphosphates which accumulate, together with small amounts of salt-soluble polyphosphates. The more highly polymerized PP_5, PP_4, and PP_3 fractions are absent. On the other hand, in mature cells which are about to form cysts, substantial amounts of the more highly polymerized forms are found. Rubtsov *et al.*[621] has recently shown that high-molecular-weight polyphosphates are absent from the chloroplasts of this organism. They also showed that

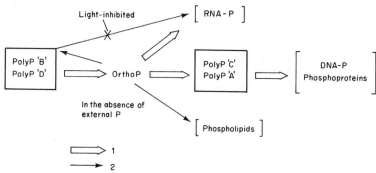

Figure 61. Polyphosphate metabolism in *Chlorella*.[522] 1, Processes induced by light; 2, dark processes

chloroplasts of *Acetabularia* are incapable of synthesizing high-molecular-weight polyphosphates from [^{32}P]orthophosphate. However, in these chloroplasts, alongside the synthesis of ATP and independently of it, the light-dependent synthesis of inorganic pyrophosphate takes place.[620] This group of investigations has therefore rigorously demonstrated on the one hand that there is no direct link between the biosynthesis of high-molecular-weight polyphosphates and photosynthesis and photosynthetic phosphorylation, and on the other hand has provided proof that it is possible for pyrophosphate to be formed and to function as an alternative product of photophosphorylation to ATP.

2. Enzymes of metabolism of high-molecular-weight polyphosphates and the regulation of their activity

An enzyme which possessed polyphosphate kinase activity, i.e. which transferred phosphate from polyphosphate to ADP to form ATP, has been observed in a cell-free extract from *Chlorella*.[271] Other than this finding, the literature has until recently contained virtually no information on polyphosphate-metabolizing enzymes in algae. All that can be hypothesized, on the basis of Miyachi and co-worker's findings, is that depolymerases are present which mediate, in particular, the degradation of polyphosphates of the acid-insoluble fraction C to the lower molecular weight acid-soluble fragments comprising fraction A.

It might be considered also that polyphosphate biosynthesis in algae should involve the participation of ATP with the assistance of polyphosphate kinase. This must be regarded as a very likely hypothesis in the light of results obtained by Schmidt and co-workers,[23,98,240] showing that the accumulation of polyphosphates during synchronous growth of cultures of *Chlorella* cells proceeded strictly in parallel with the accumulation of ATP. At the same time, the formation of high-molecular-weight polyphosphates from ATP in algae has not been conclusively proved. It is interesting that Lysek and Simonis[476] suggested, on the basis of the close relationship between the metabolism of polyphosphates and sugar phosphates in *Ankistrodesmus braunii*, that polyphosphate glucokinase was present in this organism.

However, the results of Uryson and Kulaev[765] disproved this assumption. Neither in this organism, nor in other algae, was polyphosphate glucokinase detected. Likewise, very little was known until recently concerning the control of polyphosphate metabolism in algae. Recently, however, Rubtsov and Kulaev[622] have obtained significant information concerning polyphosphate-metabolizing enzymes and their regulation in *Acetabularia mediterranea*.

The activities of two most important enzymes, ATP–polyphosphate phosphotransferase (polyphosphate kinase) and polyphosphatase, were detected in a cell-free extract from *Acetabularia*. It was shown conclusively that the activity of these enzymes is controlled by the level of phosphate in the culture medium. On the other hand, substantial evidence was obtained for the participation of polyphosphate kinase and polyphosphatase in regulating the intracellular concentrations of ATP and orthophosphate in this organism. The most significant result obtained in this investigation was the demonstration that, in algae, the linkage of the biosynthesis and accumulation of polyphosphates with photosynthesis is an indirect one. Even in those fractions in which accumulation is light-dependent, synthesis occurs via the intermediate formation of ATP, mediated by polyphosphate kinase. What is clear is that, as in heterotrophs, the intracellular concentration of ATP and orthophosphate, and the orthophosphate level in the culture medium, exert a regulatory effect on the biosynthesis and degradation of polyphosphates.[97,98,240,521,522,583,584,622,636] There is likewise no doubt that polyphosphate metabolism in algae is strongly influenced by light,[522,523] and by the concentration of oxygen in the medium.[757] However, no specific mechanisms for the regulation of polyphosphate metabolism in algae are yet known.

HIGHER PLANTS AND ANIMALS

Despite the fact that in recent years polyphosphates have been found to occur in a wide range of tissues of higher animals and plants (see the section on the distribution of inorganic polyphosphates), very little is so far known concerning their metabolism in these organisms. For example, Khomlyak and Grodzinskii[252,253] have shown that [^{32}P]orthophosphate, introduced into tomato leaves via the stem conductive system, is first incorporated into the low-molecular-weight, acid-soluble polyphosphate fraction, and subsequently into the higher molecular weight, acid-insoluble fractions. These findings indicate that in the higher plants the metabolism of polyphosphates differs substantially from that in yeasts and other lower organisms.

Polyphosphate-metabolizing enzymes, notably specific poly- and metaphosphatases, are very widely distributed in both higher plants[285,596,597,617] and animals.[64,149,198,498,502]

It is interesting to note that in higher plants, in Jungnickels's view,[285] two different specific polyphosphatases are present. One of these is a constitutive enzyme, and the other appears when the plant is grown under conditions of phosphorus deprivation, i.e. it is repressible. However, the role of polyphosphatases in the metabolism of phosphorus in higher organisms remains obscure.

Concerning the regulation of polyphosphate metabolism, from the scanty data available in the literature it can only be concluded that in the higher animals and plants the ability to accumulate significant amounts of polyphosphate in their tissues is highly dependent on the levels of exogenous and endogenous phosphate (see, for example, ref. 285). Further, there are indications that polyphosphate metabolism in higher organisms is closely linked with nucleic acid metabolism. In particular, Shiokawa and Yamana[662] have recently shown that polyphosphate is involved in the synthesis of RNA during the embryonal development of the frog foetus. The close link between the metabolism of high-molecular-weight polyphosphates and that of nucleic acids in rat liver nuclei is also indicated by results obtained by Mansurova *et al.*[492] These workers studied the biosynthesis of high-molecular-weight polyphosphates during the period of rapid synthesis of RNA in nuclei at early stages in the regeneration of rat liver. It was shown that changes in the rate of biosynthesis of high-molecular-weight polyphosphates took place in parallel with changes in the rate of RNA synthesis.

Further, Schmidt[637–639] has adduced data supporting the view that, in the higher plants also, polyphosphates are linked in some way with the metabolism of nucleic acids, or at least with the development and regulation of gene activity. However, the currently available data are totally inadequate to formulate any views concerning the metabolism of polyphosphates in higher animals and plants.

Chapter 10

The Physiological Functions of Inorganic Polyphosphates

From what has already been said, it will be apparent that much information is now available on the reactions of high-molecular-weight polyphosphates in the cells of living organisms. However, the physiological functions of these compounds remain obscure. Contemporary biochemistry is providing ever-increasing evidence for the polyfunctional nature of cellular metabolites, and it would not be unreasonable to suppose that high-molecular-weight polyphosphates are no exception in this respect. It appears to be very likely that they perform more than one function in the vital processes of those organisms in which they are present, but this consideration is frequently not taken into account in evaluating the physiological role of the polyphosphates.

Further, many facts have been presented in this book which indicate that the different polyphosphate fractions differ in more respects than their molecular weights and localization within the cell. Their formation and utilization are frequently linked with various metabolic pathways and physiological processes. Apparently the specific sets of enzymes exist participating of the formation and utilization of different volyphosphate fractions. It is therefore reasonable to suppose that, in addition to certain general functions, each polyphosphate fraction could, in any given organism, and depending on the metabolic processes involved, play its characteristic part in the metabolic processes specific to the polyphosphate fraction in question.

We shall first discuss the earlier and the current views concerning the physiological significance of high-molecular-weight polyphosphates in general, as specific components of living cells. [145,220,243,363,382,383,389,443]

THE CONCEPT OF HIGH-MOLECULAR-WEIGHT POLYPHOSPHATES AS PHOSPHAGENS OF MICROORGANISMS, AND ITS INADEQUACY

Many workers believe that polyphosphates, as high-energy phosphorus compounds, must function in microorganisms as reserves and donors of phosphorus and energy in a variety of biosynthetic processes.[243,834,838] In this sense, their metabolic function in living organisms is regarded as being comparable to that of phosphagens in animal cells such as, for instance, creatine phosphate and arginine phosphate (for a discussion of phosphagens, see review[526]). The originators of this hypothesis consider that, in a similar manner to creatine phosphate and arginine phosphate, utilization of the activated phosphate of polyphosphates for various energy-consuming processes must be mediated solely by the ATP–ADP system. According to this hypothesis, polyphosphates, like other phosphagens, cannot function as phosphate acceptors from any compound other than ATP, and are capable of transferring their phosphate residues directly to ADP only, with the formation of ATP.

This concept, first put forward by Hoffman-Ostenhof and Weigert[248] in 1952 on purely theoretical grounds, has found support following investigations by other workers. The following arguments are usually adduced in support of the identity of the functions of condensed inorganic phosphates and phosphagens.

1. The high energy of the phosphoric anhydride bonds in the condensed inorganic phosphate molecule, comparable to that of the terminal pyrophosphate bonds in ATP. It was shown by Yoshida[834] that fission of the phosphoric anhydride bonds in polyphosphates to form orthophosphate is accompanied by the liberation of substantial amounts of heat. His results are shown in Fig. 62, from which it can be seen that the curve for the liberation of heat during the enzymatic hydrolysis of polyphosphate is closely correlated with the formation of orthophosphate.

2. The existence of polyphosphate kinase enzymes which mediate *in vitro*, and perhaps also on occasion *in vivo*, the reversible transfer of phosphate re-

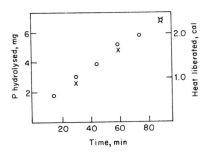

Figure 62. Evolution of heat during the hydrolysis of polyphosphates[834]

sidues from ATP to polyphosphate according to the equation

$$ATP + PP_n \xrightleftharpoons{\cdot} PP_{n+1} + ADP$$

3. The accumulation of polyphosphates in microorganisms in which the basic energy-requiring processes, which have an obligatory requirement for ATP for their operation, are inhibited, and the rapid utilization of these polyphosphates when these processes are resumed.

There is nevertheless a substantial amount of reliable information which is in conflict with this view to varying extents, and is not in accordance with the hypothesis that polyphosphates function like phosphagens. This information comprises the following facts and observations.

1. The very low energy accumulation in polyphosphates compared with the energy liberated during the oxidation of glucose (in yeast, for example, it is approximately 1% of the latter,[244,246]) and with the utilization of the energy stored in biochemical photosynthetic reactions (in *Chlorella*, a total of 0.5%).[826] These observations are in agreement with those of other workers[227,289,466] who showed that the accumulation of polyphosphates in various microorganisms corresponded to only a very small proportion of the ATP produced by the cells.

2. The absence of consumption of polyphosphates under conditions in which ATP synthesis by oxidative phosphorylation is inhibited, i.e. under conditions of limited or blocked energy production.[215,289] In these conditions, for example in *Aerobacter aerogenes*, maintenance of the concentration of ATP at the required level by consumption of inorganic polyphosphate was not observed,[215] although Langen[443] and Langen and Liss[444] found that degradation of polyphosphate occurred following suppression of glycolytic phosphorylation in yeast by poisoning with iodoacetic acid. Formation of ATP from polyphosphate did not, however, appear to take place. In addition, Shaposhnikov and Fedorov[659] observed a small amount of degradation of polyphosphate in the cells of the photosynthesizing bacterium *Chlorobium* during incubation in the dark in the presence of CO_2. This was rationalized as being due to the utilization of polyphosphate as a source of activated phosphate for the synthesis of ATP in the absence of photosynthesis, although only a very small proportion (approximately 10%) of the total polyphosphate present in the cells was consumed.

3. In contrast to the phosphagens, the production and utilization of polyphosphates are very rapid processes,[123,509] indicating that they do not constitute long-term reserves of phosphate and energy.

4. The utilization of polyphosphates in many organisms does not require the presence of polyphosphate kinase. It has, for instance, already been mentioned that in *Aerobacter aerogenes* there is conclusive proof[222] of the involvement of polyphosphatase in the utilization of polyphosphates. The presence of polyphosphate hexokinase in many microorganisms which are capable of transfering phosphate residues directly (without the participation of ATP) to glucose,[106,109,710,712,720,721,764–767] fructose,[718] and other sugars[716] is also evidence against designating polyphosphates as microbial phosphagens.

5. Polyphosphate biosynthesis in many bacteria and fungi can take place without the involvement of the ATP–ADP system, notably, with the aid of 1,3-diphosphoglyceric acid.[394–396,400,426,437,540–542]

It has thus been shown repeatedly that polyphosphate metabolism does not necessitate the participation of the ATP–ADP system.

On the one hand, polyphosphates may be formed as a direct consequence of the operation of various energy-liberating processes, and on the other hand they may be utilized as donors of active phosphate in a variety of energy-consuming processes.

Thus, the only feature common to condensed inorganic phosphates and phosphagens is that they both provide a reserve of substantial amounts of phosphate and energy. In all other respects they are totally different types of compound, and their physiological significance in the organisms in which they occur is quite dissimilar.

THE CONCEPT OF INORGANIC POLYPHOSPHATES AS A RESERVE OF PHOSPHATE IN THE CELL. LIMITATIONS OF THIS CONCEPT

Many workers have adhered to the view that inorganic polyphosphates function primarily as a reserve of phosphate, on which the cell is able to draw at any time, but more especially during periods of phosphorus starvation.[220,257,289,443,636] In the opinion of Harold,[220] polyphosphates, being polymers, constitute highly convenient compounds for the storage of large amounts of phosphate in the cell, since the accumulation of polymeric polyphosphate molecules, unlike orthophosphate and ATP, will have little effect on the osmotic pressure within the cell and on the concentration of such important metabolites as free orthophosphate and adenylic nucleotides. Since phosphorus is an element of the greatest importance, without which organisms are totally incapable of functioning normally, during the course of evolution cells have developed the ability to store surplus phosphate in the form of polyphosphate.

It has been shown repeatedly that contemporary microorganisms may on occasion accumulate sufficient quantities of polyphosphate to enable them to increase their biomass many-fold during growth on a phosphorus-free medium.[24,220,289,466]

If this is the case, what are the processes in which the phosphorus stored as polyphosphate is consumed?

Many investigators have shown that polyphosphates can function as a source of phosphorus during phosphorus starvation, for the biosynthesis of nucleic acids and phospholipids.[24,215,216,222,343,522,529,681,685,809] It is not only during phosphorus starvation, however, that polyphosphates are able to function as reserves of phosphate, but also under certain conditions of normal development. For example, our experiments[365,386] and those of Nishi[563] have shown that polyphosphates, accumulated in large amounts in the spores of *Aspergillus niger*, are actively utilized during germination to provide phosphorus for the biosynthesis of many compounds, primarily RNA and phospholipids. Similar results were obtained by Zaitseva *et al.*[837] for the latent phase of development of *Azotobacter*. It is therefore possible that the most important function of poly-phosphates in the vital processes of living organisms is the provision of specific reserves for the biosynthesis of the various nucleic acids[681] and of a variety of metabolites which participate in nucleic acid metabolism.

It is also possible that the close connection which has long been known to exist between cell division in microorganisms and polyphosphate metabolism is associated with their participation in nucleic acid biosynthesis.[298,325,388,409,458,459,559,614,625,677–679,843]

In addition to these processes, polyphosphates may serve as a reserve of phosphorus for other reactions. Thus in yeast, under normal conditions of growth and development, polyphosphates may be consumed during the uptake of exogenous sugars.[690–694] In this connection, it has been shown that they are utilized for the phosphorylation of the carrier, rather than of the sugar itself, the former effecting the active transport of the latter from the medium into the cell through the external cytoplasmic membrane. Thus, it would appear difficult to dispute the hypothesis that polyphosphates function primarily as a reserve of phosphorus within the cell.

However, the workers who advance this hypothesis[220,257,289,443] consider that polyphosphates function solely as a source of phosphorus for various biochemical reactions, and do not at the same time constitute to any significant extent a reserve of energy. These workers arrived at this conclusion on the basis of own and other workers' findings that in several organisms (in particular *Aerobacter aerogenes*), only one pathway for the utilization and degradation (catalysed by one or more polyphosphatases) existed. However, in our monograph we cited more than once a number of observations (see, for example, ref. 255) that under some conditions polyphosphatases function as transferases, i.e. they possess both hydrolytic and phosphotransferase activity.

Furthermore, several workers[24,215,216] based their denial of the function of polyphosphates as energy reserves on the fact that, when labelled polyphosphates are degraded, the ^{32}P which they contain is found in the first instance as orthophosphate, and only subsequently in other compounds.

It is possible, however, that in some cases, utilization of polyphosphates does not involve hydrolysis to orthophosphate, but rather transfer of phosphate without the loss of the energy of the phosphoric anhydride bond to very labile phosphorus compounds which readily undergo hydrolysis to orthophosphate during the course of fractionation. This seems to us a highly probable course of events. It seems very unlikely that the energy stored in the polyphosphates should be dissipated without being utilized for energy-requiring processes.

It is much more logical to suppose that polyphosphates do not function solely as depots for phosphorus, but also as a reserve of phosphate which has been activated by energy-releasing processes, and which is in consequence highly reactive and readily mobilized.

HIGH-MOLECULAR-WEIGHT POLYPHOSPHATES AS RESERVES OF ACTIVATED PHOSPHATE WHICH CAN BE UTILIZED IN VARIOUS BIOCHEMICAL AND PHYSIOLOGICAL PROCESSES

Sufficient data have been presented in this book to demonstrate that a universal function of all the polyphosphate fractions in the vital processes of contemporary organisms is to provide substantial reserves of activated phosphate within the cell. This function of polyphosphates is particularly important in the cells of microorganisms, as a result of their unusual dependence on external conditions. It must be advantageous for microorganisms to possess intracellular reserves of activated phosphate which, when conditions suitable for growth and development arise, can immediately be utilized to initiate a variety of biochemical and physiological processes.

It is, however, most significant that the various polyphosphate fractions evidently constitute specific reserves of activated phosphate, which are utilized by the cells of microorganisms only under certain culture conditions to carry out specific biochemical and physiological processes. Support for this view is provided, for example, by the results obtained by us in an investigation into the metabolism of the various polyphosphate fractions during phosphorus starvation in the yeast-like fungus *Endomyces magnusii.*[383]

In one experiment, cells of this fungus were grown initially for 12 h on a medium containing all of the components needed for growth, following which they were cultured for a further 2 h on a medium containing the usual amounts of sucrose and nitrogen sources, but which was devoid of phosphorus-containing compounds. In other words, these cells were maintained in conditions under which they could grow only by virtue of their internal sources of phosphorus. It was found that the cell biomass in this experiment increased by 20%. Other cells were cultured at the same time under conditions which precluded growth, i.e. on a medium containing, in addition to 0.01 M maleate buffer of pH 5.6, only 1 M rhamnose and 1.5% of albumin. The results obtained are shown in Fig. 63.

In both cases, during the 2-h period of phosphorus starvation, a substantial decrease in the total content of polyphosphate occurred.

The different fractions were however consumed to widely varying extents. In both parts of the experiment the acid-soluble polyphosphates were very largely or almost completely consumed, whereas the salt-soluble polyphosphates were consumed only by those cells which continued to grow. In the non-growing cells, the level of the latter fraction after 2 h phosphorus starvation was almost unchanged. The behaviour of the acid-insoluble, most highly polymerized polyphosphates was the reverse of this: they were scarcely consumed during phosphorus starvation of the growing culture, but their level fell substantially in the absence of growth. These findings may be rationalized as follows. The acid-soluble polyphosphates constitute the most readily mobilized general reserve of the activated phosphate required by the cell when phosphorus is deficient in the medium. The salt-soluble and acid-insoluble fractions

180

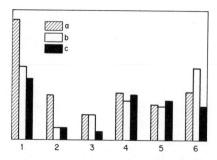

Figure 63. Effect of phosphorus and nitrogen starvation on the content of high-molecular-weight polyphosphate fractions, nucleoside polyphosphates, and orthophosphate, in cells of *E. magnusii.*[383] Incubation for 2 h at 28 °C. Results are given in arbitrary units. 1, Total polyphosphate; 2, acid-soluble polyphosphates; 3, salt-soluble polyphosphates; 4, acid-insoluble polyphosphates; 5, nucleoside polyphosphates; 6, orthophosphate. (a) Initial cells; (b) on a medium without P and N; (c) on a medium without P but with N

apparently constitute much more specialized reserves of phosphorus and energy which are drawn on to sustain certain specific biochemical and physiological processes.

It appears likely that the salt-soluble polyphosphates, which are utilized only when growth of the culture is prolonged, may provide a reserve of active phosphate which is specific for the synthesis of nucleic acids. This view is in full accordance with our own observation that this polyphosphate fraction is present in the cell nucleus, in which nuclei acid biosynthesis is known to take place.

The acid-insoluble polyphosphates on the other hand, which are localized somewhere in the vicinity of the cell boundary, are maintained in the growing cell in amounts close to their initial values. The fact that, under conditions of phosphorus starvation, growing cells maintain high levels of acid-insoluble polyphosphates is perhaps related to their involvement in the uptake of sugars, which takes place only in this type of experiment. This suggestion is based, firstly, on literature reports[104,616,690–694] demonstrating that polyphosphates are able to participate in the uptake of sugars by yeast cells, secondly, on its good agreement with our own experimental observation that the more highly polymerized polyphosphates are located peripherally in the cells of yeasts and other fungi, and lastly, on the convincing fact that, as our work has shown, there is a close correlation between the changes in the initial rate of active glucose transport in *N. crassa* cells and changes in polyphosphatase activity. This latter enzyme, which terminally hydrolyses polyphosphate to orthophosphate, is localized in *N. crassa* at the cell periphery, as are the polyphosphates.

The specific function of this most highly polymerized polyphosphate fraction therefore appears to be to participate in the mechanism of the active transport of glucose and other sugars by the cells of yeasts and other organisms. It is interesting that when *E. magnusii* is grown in a phosphorus- and sugar-free medium, the most highly polymerized polyphosphate fraction, localized at the cell periphery, does not accumulate to any significant extent. It may be that under these growth conditions, the polyphosphates could be utilized by the cell, apparently not for the transport of sugars, but for other purposes. It is possible that the highly polymeric, acid-insoluble polyphosphates are depolymerized, move to towards the centre of the cell, and are only then utilized in biochemical processes.

It can therefore be concluded that, depending on the culture conditions and the mode of metabolism adopted by the organism, the various polyphosphate fractions will be utilized to widely varying extents. On some occasions it is more important for the cell to retain most of the polyphosphates which it synthesizes at the cell periphery in the form of highly polymerized molecules, while on other occasions it is more desirable for these acid-insoluble polyphosphates to be broken down to lower molecular weight acid-soluble or salt-soluble fractions which are utilized in the cell in processes for which the acid-insoluble polyphosphates are unsuitable.

Thus, in addition to its general functions, each polyphosphate fraction appears to perform in the cell its particular specific function. This may be the reason for the presence in microorganisms (especially fungi) of a whole range of depolymerizing polyphosphatases which cleave the higher polyphosphates not to orthophosphate, but to shorter fragments.

Consequently, the physiological significance of the degradation of inorganic polyphosphates may reside, as has already been remarked, in the movement of these compounds from one part of the cell to another, with a parallel change in their functions.

An evident indication of this high specificity in the utilization of the various polyphosphate fractions is provided by our observation that the different fractions are attacked to differing extents by the polyphosphatases present in the cell. It is relevant to point out that this differing susceptibility of the polyphosphate fractions may be related to their state within the cell, as follows.

Firstly, some of them occur in the cell in vacuoles and the nucleus, and perhaps in other cell organelles, which are isolated from the cell contents by the appropriate membranes, and secondly, certain polyphosphate fractions occur in the cell in the form of complexes with other compounds such as nucleic acids and proteins, which must also hinder their hydrolysis by enzymes.

It is also possible that a specific type of vacuole, or 'vesicle', containing high-molecular-weight polyphosphates may function not only as a reserve of activated phosphate but also as a carrier of polyphosphate from the cell boundary, where it is synthesized, to the site where it is to be utilized. Such inclusions (carriers of low-molecular-weight metabolites) have recently been observed in fungi, algae, and other vegetable organisms.[191,238,486,525,816] These inclusions, known as lomasomes, have been shown to transport monosaccharides from the interior of the cell to its periphery, thereby delivering material for the construction of the polysaccharides of which the cell wall is composed.

The presence of such carrier inclusions is in itself of great interest. On the one hand it enables reactive metabolites, which could affect the osmotic pressure and pH of the cell, to be isolated from the general cytoplasmic constituents, and on the other hand it prevents their being attacked by hydrolytic and other enzymes. The complete or partial segregation of salt-soluble and acid-soluble polyphosphates, and their transport and localization in the cell in the form of granules, is obviously especially desirable in view of the fact that they are active ion exchangers,[740,790] which, by combining with cations, could have an extremely deleterious effect on a variety of cellular reactions.[297,298,693] Finally, the converse is also possible, for example in some cases the cell could with great advantage to itself utilize the ion-exchange properties of polyphosphates to regulate the course of certain enzymatic reactions.[107,230,635,778] In order to evaluate the extent to which microorganisms take advantage of the ion-exchange properties of polyphosphates, however, further experiments will be required.

The mechanisms involved in the utilization of polyphosphates in specific reactions are not yet clear. In some of the more primitive organisms, in particular the *Actinomycetes*, they may be utilized together with polyphosphate hexokinase for the direct phosphorylation of glucose and other hexoses. In other organisms they may be utilized by means of other phosphotransferase reactions, with the formation of ATP and perhaps other high-energy compounds for use in specific biochemical processes. Lastly, in some cases they may be hydrolysed by polyphosphatase with the liberation of free orthophosphate.

A very interesting example of the utilization of activated phosphoryl residues from polyphosphate, accompanied by the formation of free orthophosphate, is provided by the transport of sugars in *N. crassa*. It has already been pointed out that the uptake of each molecule of glucose by the mycelium of this fungus is accompanied by the cleavage of one molecule of orthophosphate from polyphosphate mediated by polyphosphatase (see Fig. 58). In all probability, the energy of hydrolysis of the phosphoric anhydride bond by means of polyphosphatase in this system is not dissipated as heat, but is utilized to transport the sugar across the plasmatic membrane and into the cell. Unfortunately, the actual mechanism by which this energy is utilized for the transport of sugars is not yet known, but the very fact that it is possible for polyphosphate to be utilized for various energy-requiring processes mediated by polyphosphatase must be taken into account in any consideration of possible metabolic reactions involving these high-energy phosphorus compounds.

The involvement of polyphosphates in the active transport of sugars in *N. crassa* is an excellent example of the utilization of these compounds in specific physiological processes. Another example is provided by their undoubted participation in sporogenesis in fungi.

As Kritsky and co-workers[360,361,409] have shown, during the process of sporogenesis in the fruiting bodies of the mushroom, very large amounts of relatively low-molecular-weight polyphosphates accumulate at the actual site of spore formation (the basidia). These polyphosphates are not transported into the spores, but they do participate in some way in the actual formation of the spores (or they are formed during this process).

It is also possible that the accumulation of large amounts of low-molecular-weight polyphosphates in the basidial cells of fungal fruiting bodies substantially increases their osmotic pressure. This in turn could be utilized as a force for the ejection and dispersal of the spores from the basidia. However, high-molecular-weight polyphosphates are utilized much more extensively in biochemical processes.

But what precisely are the biochemical processes in which the activated phosphate, which is accumulated in the cells of many organisms as polyphosphate, could be preferentially utilized? Our own experiments, and those of other workers, have given definite indications that the process primarily involved here is the biosynthesis of nucleic acids. The relationship between polyphosphate and nucleic acid metabolism may be extremely complex.

Firstly, polyphosphates may provide a source of the phosphorus present in nucleic acids. This was first shown by us in the mould fungus *Aspergillus niger*, using ^{32}P.[386] This observation was subsequently confirmed by other workers.[145, 529,681,834] It was shown that polyphosphates are utilized in yeasts primarily for the biosynthesis of RNA. It was shown using mycelia of the mould fungus *Penicillium chrysogenum*[354,391] and the mycelia and fruiting bodies of the Basidiomycete *Lentinus tigrinus*[354] that there was a clear correlation between the content of acid-insoluble polyphosphates (including the salt-soluble fraction) and that of RNA. This is well shown by results obtained[391] in an investigation into the accumulation of polyphosphate, RNA, and protein during the development of the fungus *Pencicillium chrysogenum* (Fig. 64).

In the preceding chapters, other examples have been cited which demonstrate clearly the relationship that exists between the behaviour of specific polyphosphate fractions and that of RNA during the development of various microorganisms. In all of the cases which have been studied, the same relationship has been found, i.e. a direct correlation between the accumulation of acid-insoluble (mainly salt-soluble) polyphosphates (fraction PP_2) and that of RNA, together with an inverse relationship between the content of RNA and that of the acid-soluble polyphosphates (fraction PP_1). From these findings, the impression is gained that the biosynthesis of RNA (and perhaps DNA) mainly involves the consumption of activated phosphate derived from the acid-soluble polyphosphates, which are localized in the cytoplasm partially or wholly in the form of volutin granules. Salt-soluble polyphosphates, however, which are largely located in the nucleus, are perhaps themselves formed during the biosynthesis of nucleic acids, perhaps as a result of the production of pyrophosphate during this process (see Fig. 28). The salt-soluble polyphosphates also appear to be indirectly involved in nucleic acid biosynthesis in that they remove from the sphere of reaction pyrophosphate, which is a product of the biosynthesis of these biopolymers, thereby shifting the reaction in the required direction.

The most highly polymerized polyphosphate fractions (PP_5, PP_4, and PP_3), which are localized at the cell periphery, may also be involved indirectly in nucleic acid biosynthesis.

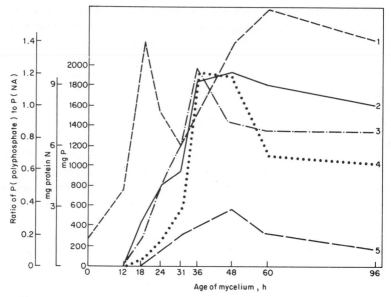

Figure 64. Changes in the content of pyrophosphate, nucleic acids, and protein, during the development of *P. chrysogenum* Q-176, on a medium without maize extract (per 100 ml of medium).[391] 1, Ratio of acid-insoluble polyphosphate phosphorus to nucleic acid phosphorus; 2, acid-insoluble polyphosphate phosphorus; 3, nucleic acid phosphorus; 4, protein nitrogen; 5, salt-soluble polyphosphate phosphorus

From the work of van Steveninck and co-workers[104,690–694] and ourselves[376,377,762] there is reason to beleive that, in yeasts and other organisms, these polyphosphate fractions participate in the transport of glucose and other sugars in the cell. Since glucose is essential for the synthesis of RNA, being the source of ribose-5-phosphate, the participation of the more highly polymerized polyphosphates in its uptake by the cell may also stimulate indirectly the biosynthesis of RNA. Although in this case polyphosphates themselves do not in general participate in RNA synthesis, they stimulate non-specifically the course of many biosynthetic processes, including the formation of RNA. It may be that these more highly polymerized polyphosphate fractions are utilized for the biosynthesis of nucleic acids by means of polyphosphate kinase and other kinases which generate ATP and other nucleoside triphosphates.

In discussing the interaction of polyphosphate metabolism with that of nucleic acids, it is pertinent also to mention that polyphosphates may participate in nucleic acid metabolism not merely as donors of active phosphate, but also as unique regulators of RNA and protein biosynthesis, functioning as repressors or in some other way in the process of transcription of genetic information. Our own experiments[60,387] and those of many other workers have shown that RNA readily forms complexes with polyphosphates.[141,142,144,153]

It may be that such RNA–polyphosphate complexes, bonded via metal ions or in some other way, are present in the cell, especially in the nucleus. It is possible that the formation of such complexes could facilitate the removal of newly synthesized RNA from its DNA matrix. RNA synthesized in the nucleus could in principle be transported into the cytoplasm in the form of such (or even more complicated) complexes. The RNA in these complexes would be protected from attack by ribonuclease and other enzymes. It may be suggested further that polyphosphates function in the cell as derepressors. The possibility of the formation of complexes between polyphosphates and proteins has long been debated.[300,475,665] Polyphosphates form complexes particularly readily with the basic proteins, the histones, and it is now well known that histones are involved in the regulation of the gene activity of chromosomes, bonding to DNA and thus protecting it from deletion of genetic information. In this connection, Ashmarin and co-workers[14,322] have also shown that polyphosphate which readily forms complexes with histones, has a marked effect on their ability to form complexes with DNA.

It is reasonable to suppose that polyphosphates, being polyanions, are present in the cell in association with histones, which may be regulators of gene activity, and thereby participate in the synthesis of RNA and proteins. Nevertheless, these hypotheses, although most intriguing and plausible, have insufficient experimental support and cannot yet be regarded as proved. With respect to the relationship of the metabolism of polyphosphates to that of nucleic acids in microorganisms, the work of Kohiyama and Saito[319] must not be overlooked. This demonstrated the powerful activating effect of polyphosphates on the genetic transformation of biochemical mutants of *Bacillus subtilis*, which is as yet poorly understood.

In concluding this section, it is emphasized that although very little is known at present concerning the mechanisms by which polyphosphates participate in nucleic acid metabolism, nevertheless the existence of an extremely close connection between the metabolism of these compounds has been firmly established.

It may be that polyphosphates are present in significant amounts only in the lower organisms in consequence of their very high rates of nucleic acid and protein synthesis, which lead in turn to very rapid cell division. Polyphosphates, which concentrate large amounts of phosphorus and energy, perhaps provide for the high rates of cell division in these organisms.

HIGH-MOLECULAR-WEIGHT POLYPHOSPHATES AS REGULATORS OF METABOLIC PROCESSES

Recent years have seen much discussion concerning the possible role of high-molecular-weight polyphosphates in regulating the cellular levels of ATP, ADP, and other high-energy phosphorus compounds, together with ortho-phosphate, thereby controlling a wide variety of metabolic processes. [73–75,157,158, 215,219,220,247,248,289,376,377,382,383,407,433]

The maintenance of constant levels of orthophosphate, ADP, and ATP in the cell is currently regarded as being of major importance in the regulation of the cellular metabolism.[16,17] This is due both to the fact that that these compounds are important metabolites which are of major significance both for the bioenergetics of the cell and for nucleic acid biosynthesis, and to their functions as allosteric activators and inhibitors in a large number of enzymatic reactions.[17] It follows that highly sensitive mechanisms for controlling the levels of these highly important phosphorus compounds in the cell are of vital importance.

The participation of polyphosphates in the regulation of the levels of cellular ATP and ADP was first suggested by Hoffman-Ostenhof and co-workers,[247,248] and received support from Bukhovich and Belozersky.[74,75] More recent work, however, has shown (see Chapter 7) that not all organisms contain the enzyme polyphosphate kinase, which links polyphosphate metabolism with the ADP–ATP system. Further, in several microorganisms which have been studied from this point of view, polyphosphate kinase has a much greater affinity for ADP than for ATP, or *vice versa*.

On the basis of currently available information, it can be concluded that in yeast, for example, the synthesis of ATP from ADP by means of polyphosphate kinase predominates almost completely over the reverse reaction.[166,167] In *Corynebacterium xerosis* and *Aerobacter aerogenes*, on the other hand, the equilibrium of this enzyme is shifted almost completely towards the biosynthesis of high-molecular-weight polyphosphates.[221,530] Even in these cases, however, owing to the action of polyphosphate kinase in the cell, regulation of the intracellular pool of ATP and ADP takes place nonetheless.

The detection in several organisms of 1,3-DPGA–polyphosphate phosphotransferase[394-396,426,437,541] raises the question of the possible participation of high-molecular-weight polyphosphates in the regulation of the rate of glycolysis in these organisms. It is interesting that the activity of this enzyme increases substantially when the synthesis of ATP and ADP in mycelia of *N. crassa* is strongly inhibited by the inclusion of 8-azaadenine in the medium.[426] It appears that the functioning of 1,3-DPGA–polyphosphate phosphotransferase at this time leads both to an acceleration of glycolysis and to the accumulation of the more readily mobilized acid-soluble polyphosphates.[510] It has been shown in our laboratory that both the above-mentioned phosphotransferase and the acid-soluble polyphosphates are localized in *N. crassa* cells in the same place, i.e. in the volutinlike granules.

Inorganic polyphosphates appear to fulfil an equally important role in controlling the intracellular concentration of pyrophosphate, which is formed in many biochemical reactions. On the basis of data and arguments which are as yet indirect, we have postulated the possible existence in various cell organelles, for instance the nucleus and membrane, of a reaction which transfers activated phosphate from pyrophosphate to inorganic polyphosphates.[377,407,433] In the papers cited, the hypothesis was put forward that such a reaction is mediated *in vivo* by means of polyphosphatases bonded to the membranes. The validity of these suggestions awaits experimental verification.

If, however, our hypothesis is confirmed, it will be apparent that the biosynthesis of high-molecular-weight polyphosphates provides the mechanism by which many biochemical processes are controlled, including the biosynthesis of polysaccharides, nucleic acids, and proteins. In fact, the removal of pyrophosphate, which is formed in numerous biosynthetic processes, from the sphere of reaction by the enzymatic transfer of one of its phosphoryl-groups to high-molecular-weight polyphosphate could be a very effective mechanism for controlling the rates of these processes. Still more likely is the possibility that high-molecular-weight polyphosphates, in the organisms in which they are present, function as the principal regulators of the level of inorganic orthophosphate.[220,289,382,383,541,542]

In all of the cases, in fact, in which rapid uptake of orthophosphate into the cell is taking place, and in which the pathway by which it is utilized is inhibited in some way, accumulation of high-molecular-weight polyphosphate takes place, the intra-cellular concentration of orthophosphate remaining at a low level.[213,289] The same effect is observed when free phosphate accumulates in the cell as a result of degradative processes occurring therein, especially the degradation of nucleic acids.[213]

The 'hypercompensation' effect (excess accumulation of polyphosphates when growth takes place on a phosphorus-rich medium following phosphorus starvation) may also be regarded as a mechanism for the regulation of the level of intracellular orthophosphate. It is well known that phosphorus starvation results in derepression of phosphatases localized at the cell surface, and of the permeases involved in the uptake of phosphate.[220,223,540] When the phosphorus-starved cells are placed in a medium containing this element, the phosphatases begin to cleave the organic phosphorus compounds present in the medium to orthophosphate. The orthophosphate thus liberated is accumulated by the cell in large amounts with the assistance of the permeases, and in order to maintain a sensibly constant level of orthophosphate, it is converted into polymeric condensed phosphates.[157,158,215]

In the series of experiments that we carried out on a range of organisms, which have already been described in this book, convincing evidence was obtained for the accumulation of polyphosphate in cells providing a highly efficient mechanism for the regulation of the level of intracellular orthophosphate.

In one of these experiments,[383] an investigation was carried out into the behaviour of various polyphosphate fractions and of orthophosphate under conditions of phosphorus enrichment in the yeast *E. magnusii*, following phosphorus starvation (i.e. under 'hypercompensation' conditions). Figure 65 presents results showing the changes in the content of the various polyphosphate fractions and of orthophosphate in the yeast under these growth conditions. It can be seen that, when the phosphorus-starved cells are transferred to a medium containing this element, very rapid accumulation of all of the polyphosphate fractions takes place with, as would be expected, the more highly polymerized acid-insoluble fractions being synthesized at the greatest rate. The acid-soluble and salt-soluble fractions were synthesized less rapidly,

188

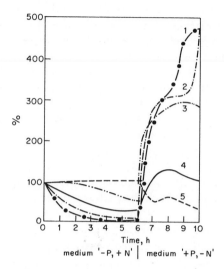

Figure 65. Effect of the composition of the growth medium on the content of inorganic polyphosphate fractions and other phosphorus compounds in cells of *E. magnusii* ('hypercompensation effect'), expressed as a percentage of the initial amounts.[383] 12-hr culture. 1, Acid-soluble fraction; 2, salt-soluble fraction; 3, acid-insoluble fraction; 4, orthophosphate; 5, labile phosphorus in acid-soluble nucleotides (nucleoside polyphosphates)

although still at a fairly high rate. It was the changes in orthophosphate concentration in this experiment which were, however, most worthy of note. Initially, the concentration of orthophosphate in the phosphorus-enriched cells also increased rapidly, but when the orthophosphate regained its initial levels (those present in normal cells), its rate of accumulation fell rapidly. Two hours after the beginning of enrichment, when the amount of orthophosphate was in excess of its normal level, its accumulation began to be strongly retarded, resulting finally in restoration of the initial level of orthophosphate in the cell.

The behaviour of the various polyphosphate fractions during this time demonstrates very clearly that the maintenance of the normal level of orthophosphate in yeast cells under hypercompensation conditions is brought about by the accumulation of certain of these polyphosphate fractions. In the first few minutes of enrichment, the phosphate entering the cell accumulates in the highly polymerized, acid-insoluble polyphosphates which are localized, as we have indicated, close to the surface of the cell. Phosphorus from these fractions, following degradation, gradually passes into the cell in the form of the less highly polymerized acid-soluble and salt-soluble polyphosphates.

One hour after the commencement of phosphorus enrichment, the acid-insoluble polyphosphate content evidently reaches its highest possible level, and degradation is then accelerated, together with the accumulation of the acid-soluble, and especially the salt-soluble fractions. It is interesting that the final restoration of orthophosphate to its initial level, which constitutes a unique 'tuning' process, is effected mainly by the accumulation of salt-soluble polyphosphates which, we suggest, are localized in the cell largely in the vacuoles. The accumulation of this fraction during phosphorus enrichment is described by a sigmoid curve, with a small plateau during the period when orthophosphate and acid-insoluble polyphosphates cease to accumulate in the cell. The curve for the accumulation of acid-soluble polyphosphates, on the other hand, is essentially rectilinear, and does not show such a plateau.

These results show that all of the polyphosphate fractions are involved to varying extents in maintaining constant (very low) levels of orthophosphate in the cell, but that the extent to which the different fractions are involved in regulating orthophosphate levels may vary during the culturing process.

This experiment shows clearly that the polyphosphate fractions are closely related both to each other and to orthophosphate, and control very closely the level in the cell of the latter. This function of inorganic polyphosphates is of the highest importance, at least in contemporary organisms.

The importance of maintaining a constant, low level of orthophosphate in the cell is relevant to several considerations. Firstly, the concentration of orthophosphate in the cell is in turn a powerful controlling factor on the course of biochemical processes. Secondly, the accumulation of any significant amounts of orthophosphate in the cell would result in a considerable change in its osmotic pressure and pH. It is also possible that the free phosphate ion is as toxic to the cell as ammonia, and it is therefore likely that, to some extent, the function of inorganic polyphosphates in the phosphorus metabolism of microorganisms is analogous to that of the amides asparagine and glutamine in the nitrogen metabolism of higher plants.[605]

In fact, it is likely that the inorganic polyphosphates present in microorganisms represent a detoxification product of the phosphate which enters the cell, in the same way as asparagine and glutamine in higher plants are a means of detoxification of ammonia. We make this suggestion on the basis of a number of investigations in which we studied the mechanism of the uptake of orthophosphate by microorganisms, and the state in which it is present in the cell.[369–371,397,412,415–418,421,431,436,568–572]

It is important that in the above investigations Kulaev and co-workers[418,419,568–572,574,575] found a new form of orthophosphate in several organisms, namely polymeric orthophosphates of various metals. A proposed structural formula for one of these compounds [a polymeric iron (III) orthophosphate] is shown in Fig. 66.

It was established that the molecular weight of these complexes could exceed 400 000. In other words, these compounds, like the condensed polyphosphates, constitute a macromolecular reserve of substantial amounts of orthophosphate in the cell. However, these polymeric metal orthophosphates and the condensed polyphosphates, in addition to carrying out their general functions of storing and detoxifying substantial amounts of phosphate ion, and participat-

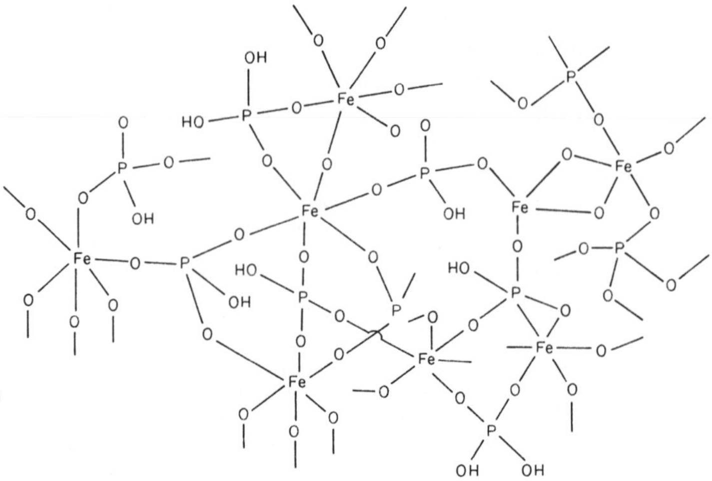

190

Figure 66. Proposed structure for polymeric iron orthophosphate[572]

ing in the regulation of the level of free phosphate in the cell, may also possess their own specific functions in the cellular metabolism.

It has already been pointed out that high-molecular-weight polyphosphate functions in the cell as a reserve of activated phosphate, i.e. the cell accumulates in the polyphosphate molecules, in addition to phosphate, a quantity of potential energy. The polymeric metal (iron, magnesium, calcium, manganase, etc.) phosphates, on the other hand, constitute reserves of phosphate ions and of the ions of various metals which are of considerable biological importance.

In other words, polyphosphates are able to accumulate in the cells of different organisms only if these cells possess an 'energy surplus'. Their accumulation must in some way be associated with energy-liberating processes. In the accumulation of phosphate in the form of polymeric metal orthophosphates, this 'energy surplus' is not required, but what is essential is the presence in the culture medium of some cation which is capable of accumulating in the cell, and of passing through the cell membrane.

Depending on the state of a given organism, its nature, and the conditions under which it is grown, phosphate may accumulate preferentially either as condensed polyphosphates or as polymeric metal orthophosphates.

Although in the yeast-like fungus *Endomyces magnusii* the excess of phosphate under the conditions which we used (see Fig. 65) appeared mainly in the form of condensed polyphosphates, in *P. chrysogenum*, in which energetic processes appear to proceed much less intensively than in *E. magnusii*, the surplus of phosphate taken up accumulated for the most part as a polymeric phosphate of iron and magnesium (see Fig. 67). It should be mentioned that in the experiment, the results of which are shown in Fig. 67 (carried out by Okorokov *et al.*[573]), the growth medium for *Penicillium chrysogenum* contained substantial amounts of iron and magnesium ions. Examination of Fig. 67 shows

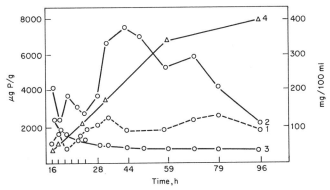

Figure 67. Content of (1) free orthophosphate, (2) $PO_{Fe} + PO_{Mg}$ phosphate, and (3) acid-insoluble PP during (4) the growth *P. chrysogenum.*[573] The amount of phosphorus is expressed in $\mu g/g$ dry weight of mycelium, and the growth of the fungus in mg per 100 ml

that here the leading role in the regulation of the level of free phosphate ions is played by the polymeric metal orthophosphate, rather than by the condensed polyphosphate which is also present in the fungal mycelium.

The simultaneous presence of polymeric metal phosphates and condensed polyphosphates demonstrates clearly how, by duplicating the control system for a given function, the cell increases the reliability of control of such an important aspect of its metabolism as the concentration of free phosphate. One must obviously always bear in mind the possibility that the intracellular phosphate level is controlled in several ways, such as polyphosphate metabolism, the formation of polymeric metal phosphates, and by other mechanisms. In a recent paper, Lichko and Okorokov[452] have shown in particular, that the compartmentalization of phosphates and cations in the yeast cell is a most important factor in regulating their intracellular levels. Thus, they showed that in the cytoplasm itself, the concentrations of phosphate and magnesium ions were maintained at constant, very low levels during the growth of cultures of *Saccharomyces carlsbergensi's*. These levels were held constant, in particular by the uptake of phosphate and magnesium into the cytoplasm from the vacuoles, in which the concentration of these ions is, firstly about ten times greater than in the cytosole, and secondly, varies greatly during the course of growth of this yeast. It is interesting that the phosphate and magnesium are present in yeast vacuoles largely in the form of polymeric magnesium phosphate.[574]

The discovery of polymeric metal phosphates and the study of their physiological function in the cell once again emphasizes the fact that in the condensed polyphosphates we are not dealing simply with a reserve of phosphate, as was believed earlier by several workers, but rather with a reserve of activated phosphate. Consequently, the metabolism of high-molecular-weight condensed polyphosphates in general cannot, it seems be considered without taking into account the energy-releasing and energy-requiring processes with which it is associated. It must always be borne in mind that the metabolism of polyphosphates, like that of adenosine polyphosphates and, apparently, pyrophosphate, constitutes an important component part of bioenergetics.

In concluding this discussion of the physiological role of inorganic polyphosphates in the metabolism of contemporary organisms, it must again be emphasized that they are to be found to a significant extent only in microorganisms. This in itself appears to us to be highly significant in establishing the role of these interesting phosphorus compounds in the metabolism of those organisms in which they occur. It is also well known that microorganisms are in general unusually dependent on their environment, and that they are capable of extremely high rates of growth and development under suitable conditions. It is with these features that the accumulation of such large amounts of polyphosphate in the cells of microorganisms must primarily be associated, and it is in these features too that the key to the puzzle of the physiological role of these compounds in microorganisms must be sought. High-molecular-weight polyphosphates provide reserves of both phosphorus and energy, which are essential for the development of all organisms.

The ability to store large amounts of activated phosphate in a polymeric form renders the cell less dependent on changes in external conditions. It is for this reason that polyphosphate is stored in large amounts in microorganisms, but is present in very small amounts in the cells of higher animals, in which the vital processes are much less dependent on external factors.

Further, the metabolism of microorganisms is much less balanced than that of the cells of higher organisms. The cells of microorganisms stand in greater need of 'buffer' compounds which are able to maintain within the cell the balance between the more important cell metabolites. There is reason to suppose that inorganic polyphosphates fulfil this function in respect of the phosphorus metabolism of microorganisms.

Finally, the ability of the cells of microorganisms to accumulate large amounts of activated phosphate in the form of polyphosphate renders them capable of simultaneously synthesizing large amounts of nucleic acids and proteins, which in turn provides the basis for the very rapid growth and development of these organisms when favourable conditions are restored.

In the cells of higher organisms, such simultaneous synthesis of large amounts of nucleic acids and proteins is not normally necessary, and therefore they do not need to accumulate significant amounts of polyphosphate. However, the cells of higher animals and plants do retain the capacity to synthesize high-molecular-weight polyphosphates, which indicates the important part played by these phosphorus compounds in some of the metabolic processes which take place in these organisms.

Chapter 11

Inorganic Polyphosphates in Chemical and Biological Evolution

The achievements of contemporary science have enabled biochemical investigations to be extended beyond the biochemistry of contemporary organisms, and it is now possible to pose, and to answer, many questions concerning evolutionary and comparative biochemistry. The present stage of the development of the science of biochemistry is characterized by the steadily increasing interest of research workers in a wide variety of problems in evolutionary biochemistry. These investigations, especially those carried out over the last 15–20 years, have provided a wealth of experimental material from which far-reaching conclusions may be drawn concerning the origin and development of life on earth.

These investigations have been mainly concerned with those aspects of chemical evolution which must have preceded the appearance of life on earth.[52,178,180,206,374,375,378,427,428,449,470,578–582,587,603,607,608,650,651,653,688] It is worthy of note that the experimental data so far obtained represent a great triumph for the theory of the origin of life on earth put forward by Academician A. I. Oparin half a century ago.

Despite the outstanding achievements of evolutionary biochemistry, however, many problems still await solution. Amongst these unsolved and relatively little-investigated problems of evolutionary biochemistry are the role of phosphorus compounds in the chemical evolution which preceded the appearance of life on earth, and the evolution of phosphorus metabolism from primitive organisms to contemporary living creatures.

Phosphorus, being one of the most important constituent elements of living cells, must without doubt have played an important part in the very earliest stages of the emergence and evolution of life.[578,579] Miller and Parris[517] consider that, even when the earth possessed a reducing atmosphere, phosphorus must have been present as phosphate rather than in a reduced form such as phosphite. Phosphites are known to be highly unstable compounds.[517,651] It appears that even at the very earliest stages of life on earth, phosphorus was taken up by primitive living organisms from the 'primaeval soup' in the form of phosphate or its derivatives.

Condensed inorganic phosphates could have arisen on the primitive earth by a wide variety of abiogenic processes. They could have been formed by the condensation of inorganic phosphates at high temperatures (for instance, during the eruption of volcanoes),[645–648] in the reaction between calcium phosphate and cyanide,[517] and by the action of heat on mixtures of ammonium phosphate and urea.[582] They could therefore easily have been present at the time life first appeared on earth.

On the other hand, pyrophosphate (the lowest member of this homologous series of compounds) could also have been formed on the primitive earth. It could have arisen from orthophosphate, either by inorganic redox reactions, or following preliminary activation of the phosphate by cyanogen, cyanate, or dicyandiamide. This has been demonstrated in model experiments by Miller and Parris,[517] Degani and Halmann,[103] and Steinman et al.[688] These activating agents could apparently have existed on the primaeval earth.[180] Further, from the work of Orgel and co-workers,[44,470] the Miller–Parris reaction (i.e. the conversion of hydroxyapatite into calcium pyrophosphate in the presence of cyanates) can take place under aqueous conditions.

The pyrophosphate formed in these, or other ways could be, in the words of Lipmann,[463] 'the simplest compound present on the primaeval earth to be involved in the accumulation and transfer of energy-rich bonds'. That inorganic pyrophosphate is an energy-rich compound has recently been shown conclusively by Flodgaard and Fleron,[172] who showed that hydrolysis of the phosphoric anhydride bond in pyrophosphate liberated, according to conditions, 4.5–5 kkal of energy per mole.

In model experiments, Lowenstein[472,473,731] and later Le Port et al.[449] were able to find conditions for the non-enzymatic synthesis of pyrophosphate, by the phosphorylation of orthophosphate in the presence of certain cations, and by means of ATP and inorganic tripolyphosphate, respectively. Kulaev and co-workers[427,428,493] were able to show experimentally that this reaction proceeded much more rapidly with inorganic polyphosphate, as follows:

$$PP_n + [^{32}P]\text{orthophosphate} \xrightarrow{\text{M}^{2+}} [^{32}P]\text{pyrophosphate} + PP_{n-1}$$

We have observed that radioactive pyrophosphate is formed non-enzymatically in substantial amounts when ^{32}P-labelled orthophosphate is phosphorylated by incubation with polyphosphate ($\bar{n} = 40$) in aqueous solution at 37 °C for 15 h at pH 9, in the presence of certain divalent cations.

As shown by the results presented below, 16–30% of the initial high-molecular-weight polyphosphate was utilized in the phosphorylation, and of the cations tested (Mg^{2+}, Mn^{2+}, Ca^{2+}, Cd^{2+}, and Ba^{2+}), the greatest amount of incorporation was achieved with Ba^{2+} (33%) and the least with Mg^{2+} (16%):

Cation	Mg^{2+}	Ca^{2+}	Mn^{2+}	Cd^{2+}	Ba^{2+}	—
$\dfrac{\text{Pyrophosphate}}{\text{Polyphosphate}} \times 100 (\%)$	16	22	23	26	33	0

It is interesting that, when organic and inorganic tripolyphosphates were employed under the same conditions, Lowenstein and Le Port *et al.* found that only 0.2–0.6% of the phosphate donor was utilized.

Our experimental findings thus lead to the conclusion that, as the earth cooled and a hydrosphere formed on its surface, a variety of transphosphorylation reactions became possible in the primaeval ocean, in particular the phosphorylation of orthophosphate by polyphosphate to give pyrophosphate.

Many model experiments (in particular those of Schramm and co-workers,[645–648] Ponnamperuma and co-workers,[180,602,603,607] Fox and co-workers,[174–176, 211] Schwartz and co-workers,[652,653] and others[581,587,644,848] have shown that high-molecular-weight polyphosphates, in contrast to pyrophosphate, could have functioned on the primaeval earth as condensing agents in reactions leading to the formation of nucleosides, nucleotides (including adenosine triphosphate), simple polynucleotides, polypeptides, and even primitive protein-like materials. It was observed that condensation reactions in which high-molecular-weight polyphosphate functioned as an activating agent could be carried out either at high temperatures in non-aqueous media, or at room temperature in aqueous solution. This gives reason to suppose that these reactions could have been involved in the synthesis of macromolecules which were subsequently incorporated into living cells, both before and after the appearance of a hydrosphere on earth,[374,375,427,428] whereas transphosphorylation by high-molecular-weight polyphosphates to form pyrophosphate, as we have shown, only takes place in aqueous solution at 37 °C.[427,428] It follows that it must be assumed that the formation of substantial amounts of pyrophosphate by transphosphorylation could take place only after the formation of a primaeval terrestrial hydrosphere containing previously-synthesized macromolecules. It is possible that the primary macromolecules might have arisen spontaneously at this time, in the form of primitive structures such as coacervates[578–580] or microspheres.[176,211] This factor may have been of decisive importance in the appearance and further evolution of the biochemical functions of pyrophosphate.

Model experiments, however, have not yet provided any reliable information concerning the functions of high-molecular-weight polyphosphates and pyrophosphate in the earliest living creatures, although some conclusions as to the role of these primitive high-energy compounds in the metabolism of protobionts may be drawn from comparative biochemistry.

By studying the metabolism of the more ancient, comparatively primitive forms of contemporary organisms, there may be discerned, as Lipmann[463] has said, 'antediluvian' metabolic features and 'fossil' biochemical reactions, which have been preserved since ancient times.

Investigations in this area of biochemistry, which could be termed "biochemical palaeontology' could lead, and have indeed led, to the detection of archaic metabolic features which in all probability derive from primitive life forms. Thus, Baltschevskii and co-workers[31,33,772] and Keister and Yike[304,306] have shown that in the phylogenetically ancient and primitive photosynthesizing bacterium *Rhodospirillum rubrum*, photosynthetic phosphorylation results in the production of high-energy phosphate much more in the form of pyrophosphate than in the form of ATP. The synthesis of pyrophosphate can proceed in the chromatophores of this bacterium even when the formation of ATP is totally suppressed.

Kulaev and co-workers[411,425] have confirmed the findings of Baltschevskii and co-workers and Keister and Yike, and in particular they were able to show that, in *Rh. rubrum*, pyrophosphate accumulates only in the light. They were unable to detect it in a dark culture, either in the whole cells, or in the chromatophores.[25-31] The light-dependent synthesis of pyrophosphate was recently also observed in our laboratory in the chloroplasts of higher plants.[411,620]

It was shown by Baltschevskii and co-workers[25-31] that the energy stored in the pyrophosphate molecule could be utilized both for reversed electron transport and for the active transport of ions through the chromatophore membranes in this bacterium.

Keister and Minton[302,303] observed that in the chromatophores of this organism, ATP is synthesized by the utilization of the energy liberated by the hydrolysis of pyrophosphate mediated by the appropriate pyrophosphatase. It is interesting that the pyrophosphatase in *Rh. rubrum* is exclusively a regulatory enzyme.[313]

Results obtained by Liberman and Skulachev[669] give reason to suppose that the energy of the pyrophosphate in chromatophores of *Rh. rubrum* is utilized via the intermediate development of a membrane potential ($\Delta \Psi$).

The reactions observed in the chromatophores of the photosynthesizing bacterium are summarized in Fig. 68. It is of interest that the ability to utilize the energy of pyrophosphate, which is liberated when the latter is cleaved by pyrophosphatase, for the synthesis of ATP, is not unique to the chromatophores of *Rh. rubrum*. Mansurova and co-workers[487,489,490] have shown that the same process occurs in animal and yeast mitochondria. Also, Mansurova and co-workers[489-491] have shown that, in addition to the utilization of pyrophosphates, synthesis of the latter can also take place in animal and yeast

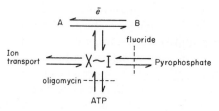

Figure 68. Conversion of pyrophosphate in *Rh. rubrum*[27]

mitochondria. Figure 69 shows that pyrophosphate is synthesized in rat liver mitochondria, together with ATP. Calculations show, however, that in the rat liver mitochondria isolated by us, pyrophosphate was synthesized at about one tenth the rate of the synthesis of ATP and ADP (AMP was present in the incubation medium). In experiments aimed at determining the effect of inhibitors of the respiratory chain [rotenone (2 μg/mg protein), antimycin (1 μg/mg protein), and cyanide (10^{-3} M)], together with the uncoupler 2,4-DNP (4×10^{-4} M), it was shown that the biosynthesis of pyrophosphate in these mitochondria occurred by transphosphorylation. These experiments showed that both the uncoupler and the inhibitors of electron transport in the respiratory chain at all three points of coupling completely suppressed the biosynthesis of pyrophosphate in rat liver mitochondria. This was shown both in intact mitochondria and in fragments of the internal mitochondrial membrane.[658] It is possible that pyrophosphate is formed in the mitochondria as a secondary product, by cleavage of part of the ATP to AMP and pyrophosphate. However, experiments on the effect of oligomycin (2 μg/mg protein), which inhibits the formation of ATP in mitochondria, showed that under these conditions the production of pyrophosphate, far from falling, increased substantially. These findings suggest that pyrophosphate is synthesized in animal mitochondria during the functioning of the respiratory chain independently of ATP, and to a certain extent in opposition to it. Similar results were obtained with yeast mitochondria.[489,490]

The lack of dependence of pyrophosphate synthesis in animal and yeast mitochondria on ATP was even more firmly established in experiments with mitochondria which had been depleted in ADP and ATP by pre-incubation with glucose (4×10^{-2} M), hexokinase (0.1 mg/ml), and oligomycin (1 μg/mg protein).

Figure 70 shows that synthesis of pyrophosphate in yeast mitochondria proceeds in the total absence of ADP and ATP. It is increased by adding orthophosphate.

It is also important to note that, as in *Rh. rubrum*, pyrophosphatase appears

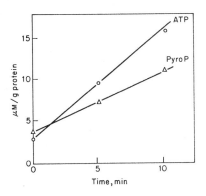

Figure 69. Synthesis of inorganic pyrophosphate by rat liver mitochondria[491]

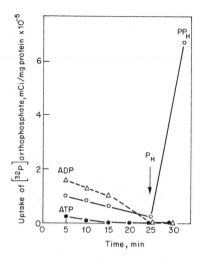

Figure 70. Synthesis of inorganic pyrophosphate by yeast mitochondria depleted in ADP and ATP[489]

to be involved in the biosynthesis and utilization of pyrophosphate in animal and yeast mitochondria, since in the presence of sodium fluoride (10^{-2} M), pyrophosphate biosynthesis is inhibited while that of ATP is increased. Mansurova and co-workers[490] have recently isolated from beef heart and characterized highly purified preparations of two pyrophosphatases (PPase I and PPase II). Both of these are Mg-dependent enzymes which are highly specific towards pyrophosphate, and possess the same molecular weight of approximately 75 000 and optimum pH of 8.0. The value of K_m for PPase I is $9.0.10^{-5}$ M and for PPase II $2.2.10^{-5}$ M. It was shown that PPase II is more firmly bound to the internal mitochondrial membrane than PPase I, and is a lipoprotein.

In the same investigation,[490] a reconstruction of the synthesis of pyrophosphate was carried out, using fragments of rat liver internal mitochondrial membrane free from pyrophosphatase activity and a highly purified preparation of PPase II. The results demonstrated conclusively that membrane PPase II mediates the synthesis of pyrophosphate, and is involved in the coupling of electron transport in the respiratory chain with the synthesis of pyrophosphate in mitochondria. It appears that the scheme proposed earlier for the chromatophores of *Rh. rubrum* (Fig. 68) may also used to interpret the relationship of the biosynthesis of pyrophosphate to that of ATP in animal mitochondria. In chromatophores of *Rh. rubrum*, however in consequence of the operation of the electron transport chain, much more pyrophosphate than ATP is synthesized, whereas in animal mitochondria substantially more ATP is formed. Also, in yeast mitochondria, under some conditions of incubation, the quantity of pyrophosphate synthesized may approximate to their ATP content.[489,490]

Comparative biochemistry thus suggests that, since ancient times, and perhaps from the appearance of the earliest organisms, pyrophosphate was involved in the energetic processes taking place at the membranes, and above all with photosynthetic phosphorylation and phosphorylation in the respiratory chain. In the opinion of Oparin,[578–580] however, the earliest process to provide energy for the vital processes of the first living organisms to appear on earth, even before the appearance of oxygen, was the anaerobic fermentation of hexoses to lactic acid and ethanol. This suggestion was based firstly on the very reasonable assumption that when life originated on earth the atmosphere did not contain oxygen, but possessed reducing properties, and that a variety of organic substances were present in abundance on the earth's surface. Secondly, all of the remaining and most essential energy-providing mechanisms encountered in living organisms today (the cleavage of hexoses during respiration, and the pentose phosphate and photosynthetic cycles) involve anaerobic fermentation reactions. From these considerations, it may be deduced with a reasonable degree of certainty that in primitive living organisms the principal, and perhaps the only, energy-providing process was the anaerobic fermentation of hexoses which were, it seems, already present in the 'primaeval soup'.

On the basis of results obtained by Uryson, Szymona, and Bobyk, Kulaev has suggested[370,374,375] that the energy-providing processes involved in glycolysis were mediated in the earliest organisms by high-molecular-weight polyphosphates, rather than by ATP and pyrophosphate.

As has already been pointed out, we have established that in certain contemporary organisms, for instance bacteria and fungi,[378,394–396,437,541] an enzyme is present which catalyses the transfer of phosphate, activated by glycolytic phosphorylation, from 1,3-diphosphoglyceric acid not to ADP to form ATP, as one would expect from the Meyerhof–Embden–Parnas scheme, but directly to polyphosphate, thereby lengthening its chain by one phosphate residue.

It seemed to us to be of prime importance from the point of view of comparative evolution to carry out parallel determinations in different microorganisms of the two enzymes involved in the formation of the same product, glucose-6-phosphate, from ATP on the one hand, and polyphosphate on the other. To this end, as has already been mentioned, we determined the activity of ordinary ATP hexokinase and polyphosphate hexokinase in more than 60 species of organisms of widely differing types. Other workers have carried out similar investigations using eleven further species of microorganisms. It was found that ATP hexokinase is present in all of the species examined, whereas polyphosphate hexokinase was detected only in the phylogenetically ancient organisms which are closely related to each other, being assigned in Krasil'nikov's classification[346,347] to the distinct class of Actinomycetes. The results obtained are summarized in Fig. 34 and Table 26. It can be seen from the latter that in the more ancient representatives of this group of microorganisms (in the opinion of Kluyver and van Niel) such as the Micrococci, Tetracocci, Mycococci, and the propionic bacteria, polyphosphate hexokinase

200

activity exceeded that of ATP hexokinase several-fold, whereas in the phylogenetically younger representatives of this class, the true Actinomycetes, ATP hexokinase activity was substantially greater than that of polyphosphate hexokinase.

These results show that the utilization of high-molecular-weight polyphosphate as a donor of active phosphate in the phosphorylation of glucose is apparently more ancient from the evolutionary point of view than the utilization of ATP for this purpose.

These experimental findings provide confirmation of the view of Belozersky[52] who, as long ago as 1957, suggested that in the earliest organisms high-molecular-weight polyphosphates fulfilled the functions carried out in contemporary organisms by ATP. This suggestion has also found support from other workers,[462,579] and our findings have filled in some details. They indicate that a high-molecular-weight polyphosphates may be primarily involved in protobionts in the coupling of glycolysis with the phosphorylation of sugars, for instance in a sequence of reactions such as that shown in Fig. 71.

High-molecular-weight polyphosphates are able to phosphorylate glucose to glucose-6-phosphate which, following glycolysis, is converted into 1,3-diphosphoglyceric acid. The latter may give rise to the synthesis of polyphosphate.

Thus, in the earliest stages of the evolution of the energy systems of living organisms, the function of linking exoenergetic and endoenergetic processes, which is normally accomplished in contemporary organisms by ATP, could apparently be carried out to some extent by the more primitively structured high-energy compounds, inorganic polyphosphates.

It may be that this function of polyphosphates when life first appeared on earth was their main, although by no means their only function.

In general, it is pointed out that polyfunctionality in a compound was, we believe, an important criterion in the selection of compounds as components of living cells. It appears likely, in fact, that only those compounds were selected for this purpose which the cell was able to utilize for a variety of processes, rather than one only.

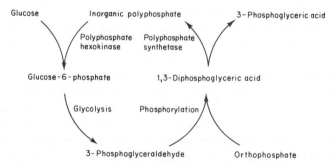

Figure 71. Proposed scheme for the participation of high-molecular-weight polyphosphate in the coupling of energy-liberating and energy-requiring processes[373]

It is possible, for instance, that in protobionts polyphosphates were not merely coupling compounds, but also provided a comparatively long-term depot for phosphorus and energy which enabled the organisms to become independent of their environment to some extent. They were able to fulfil the functions of detoxifying free orthophosphoric acid and its salts, which could accumulate in large amounts in the cell, thereby markedly influencing the reaction of the medium and the osmotic pressure within the cell. Being excellent ion exchangers, polyphosphates could also have fulfilled an important function as regulators of cation metabolism in the earliest organisms, since by binding one cation and liberating another they could regulate the functioning of many enzymes. Further, in the opinion of Gabel,[178,179,181] high-molecular-weight polyphosphates may have been involved in the formation of the first cell membranes.

At some stage of evolution, however, high-molecular-weight polyphosphates ceased to satisfy fully the requirements of the organism. More specific compounds were needed which had structures capable of carrying out an even greater variety of functions, and having the ability to interact more precisely and specifically with other cell components.

In general, it is very likely that selection for the ability to interact with other cell components has also been an important factor in the evolution of the metabolism of living organisms. Thus polyphosphates, which possess a monotonic macromolecular, essentially linear, structure without any special features, could, at a certain stage of complexity of structure and metabolism in the development of the cell, have become somewhat unsatisfactory compounds. The limited capacity of polyphosphates to make contact precisely and very specifically with other cellular metabolites resulted in an inconsistency with their function of coupling exo- and endoenergetic processes. For this purpose, one of the most vital in the metabolism of the cell, the compound selected was ATP, which has a much more specific and therefore more readily recognized structure. Further, ATP was able to fulfil in addition many other functions which could not be accomplished by high-molecular-weight polyphosphates. The fact that it was adenosine triphosphate which was selected in the evolutionary process instead of polyphosphates rather than some other compound was also not, it appears, a chance occurrence.

Numerous investigations, carried out for the most part by Oro[581] and Ponnamperuma and co-workers,[602,603,607,653] have demonstrated the comparative ease of formation of adenine, in comparison with other nitrogenous bases, under conditions similar to those which probably obtained in the prebiological stage of the earth's development. Oro has also shown that adenine is in addition much more stable to the various types of radiation which undoubtedly played an important part in the processes taking place on the primaeval earth. That adenosine, and adenosine polyphosphates, could have been formed under conditions similar to those which existed on earth at the time when life first appeared and during the early stages of its development has been demonstrated in a series of brilliant investigations carried out by

Ponnamperuma and co-workers.[603,607,653] It is very important to note that in these investigations the abiogenic formation of ATP took place only in the presence of ethyl polyphosphates, or of labile inorganic polyphosphates. This makes even more plausible the hypothesis that inorganic polyphosphates, rather than ATP, were the first to appear and to function as acceptors and donors of phosphate in primitive organisms.

Concerning the sequence of appearance and utilization as cell metabolites of the other nucleoside polyphosphates which are important in the metabolism of contemporary organisms, the suggestion can be put forward that they arose and took up their places in metabolism much later than the adenine nucleotides. This may have occurred when, during the course of continuing improvement and increasing metabolic complexity, the need arose for specific coenzymes, regulated by specific metabolic components.

It may have been at approximately this stage of evolution, which led, as a result of a great increase in the complexity of the metabolism of primaeval organisms, to the appearance of all of the presently known nucleoside polyphosphates, that there arose a need for the formation of polymeric, highly specific nucleic acids. Their appearance must have resulted from the development in early organisms of the need to conserve and to reproduce exactly in their offspring the most favourable and advantageous organization of their metabolism, which had arisen during the course of evolution. It is not intended to discuss in any further detail the evolution of the nucleic acids, since this subject has been treated in depth in numerous papers by Belozersky and co-workers.[53,55,56,61]

Returning once more to inorganic polyphosphates, it should be emphasized that in organisms possessing a relatively low degree of organization they have by no means been totally replaced by ATP and the other nucleoside polyphosphates, and they play, as we have seen, a very important part in these organisms.

It is natural to pose the question: why have high-molecular-weight phosphates not been displaced completely by ATP during the course of evolutionary development, but are present, sometimes in large amounts, in contemporary (especially lower) living organisms?

In our opinion, this is primarily due to the fact that contemporary organisms with a low degree of organization are still heavily dependent on their environment, and for this reason they need to possess a mechanism for conserving large amounts of active phosphate. One such mechanism for the storage of activated phosphate in contemporary lower organisms, especially microorganisms, remains, as in the protobionts, the accumulation of high-molecular-weight polyphosphates, which renders them independent of changes in their external environment to a substantial extent. Further, an important factor leading to the retention in contemporary organisms of the ability to accumulate high-molecular-weight polyphosphate appears to be the lack of balance of their metabolism. The accumulation of polyphosphates in these organisms appears to be, as we have already said, an efficient means of detoxifying free orthophosphate, the accumulation of which to any significant extent in the cell could markedly affect the pH and change the osmotic pressure.

It is also most important that phosphate is held in reserve in contemporary organisms in an activated form, as polyphosphate, so that when favourable conditions arise for their development they can be utilized for rapid growth. In this case the phosphorus and energy contained in the polyphosphates must greatly facilitate the rapid and simultaneous synthesis of the large amounts of nucleic acids required at an early stage in the rapid growth and multiplication of the cells.

The cells of the higher, multicellular organisms are no longer dependent to such an extent on the medium, and as a result the ability to accumulate large amounts of inorganic polyphosphate could have been lost to a great extent during the evolutionary process. In certain highly specialized organs in animals, in particular in the muscles, the requirement for a long-term depot of activated phosphate again arose during the course of evolution, because of the rhythmic nature of muscular contraction. In this case, still more specialized forms of storage of phosphorus and energy arose in the shape of phosphagens (arginine phosphate and creatine phosphate). However, the active phosphate stored in the phosphagens is both accumulated and utilized in muscle tissue only via ATP. In this respect, the phosphagens differ essentially from the inorganic polyphosphates, which can also be utilized and synthesized, as we have seen, by other routes.

It is therefore reasonable to suppose that phosphagens, as distinct from inorganic polyphosphates, arose in the course of evolution subsequent to ATP.

As we have mentioned elsewhere in this book, polyphosphates are also found in small amounts in contemporary higher plants and animals. In particular, they have been found and well characterized in many animal tissues.[179,181,422] For example, high-molecular-weight polyphosphates occur in the brain of mammals, i.e. at a site where usually a typical phosphagen, creatine phosphate, is also found.

It appears that in the higher organisms high-molecular-weight polyphosphates function as a reserve of activated phosphate for certain highly specific processes. It is possible, for example, that they are involved in nucleic acid metabolism in higher organisms. This is suggested by the fact that high-molecular-weight polyphosphates have been reliably detected in the nuclei of cells of higher animals,[194] i.e. at those sites at which nucleic acid synthesis occurs most rapidly.

The function of detoxification of orthophosphate has in most cases, however, evidently ceased to be necessary as a result of the evolutionary development in the higher organisms of a very delicately balanced metabolism, and of mechanisms for its precise control. In many cases, however, the need to eliminate surplus orthophosphate in a form which is nontoxic to the cell has again arisen in certain tissues in higher organisms.

Such a case is apparently the accumulation in the seeds of some plants of significant amounts of phytin, the calcium and magnesium salts of inositol-hexaphosphoric acid.

A number of investigators, in particular Sobolev,[672] have in fact observed that phytin is formed in large amounts during the ripening of the seeds of higher plants, in parallel with the accumulation of reserve substances, such as starch and fats. The accumulation of phytin under these conditions, in the light of our hypothesis, can be regarded as a means of detoxifying the orthophosphate which is liberated during the synthesis of starch and fats, and which cannot be removed from the ripening seeds by any other means.

This method of detoxification of orthophosphate, i.e. its deposition as phytin, may be regarded in a certain sense, however, as much more advanced from the evolutionary point of view. This follows from work carried out by Kulaev and co-workers[436,775] on the metabolism of phosphorus during the germination of the seeds of cotton. It was shown in particular that one of the products of the cleavage of phytin during germination was 3-phosphoglyceric acid. From the evolutionary point of view this is obviously a greatly improved process, as a result of which large amounts of 3-phosphoglyceric acid accumulate during the germination of the seed. This compound is known to occupy a central position in the Calvin photosynthetic cycle and other energetic and plastic processes in higher plants.

However, high-molecular-weight polyphosphates are also present at the sites at which large amounts of phytin accumulate in the higher plants (e.g., the seeds),[12,13,773,775] which is highly significant. The ability to synthesize high-molecular-weight polyphosphates is apparently of great importance even in the higher animals and plants, since their cells are thereby rendered less dependent on external factors.

This could well be the principal reason for the continued ability of contemporary organisms, in varying degrees, to synthesize high-molecular-weight polyphosphates, which apparently played such an important part in primitive living creatures, although there may also be other reasons for this which are as yet undiscovered.

The experimental data so far obtained, especially those derived from our own work, make it possible even now to envisage a role for high-molecular-weight polyphosphates in chemical and biochemical evolution.[378]

The results which we have obtained, and our deductions therefrom, are summarized in the hypothetical scheme shown in Fig. 72. When the earth was still very hot, phosphate occurred largely in the form of polyphosphates. As the temperature fell and a hydrosphere formed, the polyphosphates were able to participate in abiogenic transphosphorylation reactions, with the formation of pyrophosphate. At some time during this period, according to the findings of Ponnamperuma and co-workers, ATP could have been formed in the primaeval ocean.

During the development of the earliest organisms, the functions which were initially largely carried out by pyrophosphate, for example participation in redox reactions taking place in membranes, and by high-molecular-weight

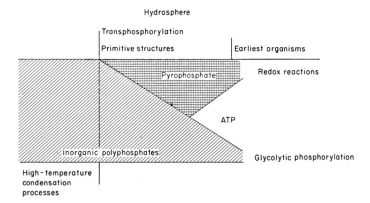

Figure 72. Hypothetical scheme for the formation and reactions of high-molecular-weight polyphosphates, pyrophosphate, and ATP, at different stages of the appearance of life on earth[428]

polyphosphate in reactions occurring in solution, were gradually transferred to ATP.

In the metabolism of contemporary organisms, however, neither pyrophosphate nor polyphosphates have been totally superseded. Alongside ATP, they continue to play an important part in the energetic and plastic metabolism of, it would appear, all contemporary organisms.

References

1. Afanas'eva, T. P., 'Some properties and possible modes of regulation of the activity of *Endomyce*? ...ᵤₛ ᵗⁱⁱ polyphosphate phosphohydrolase' (in Russian), in *Tezisy Dokl. na Vsesoyu...oi Konferentsii 'Regulyatsiya Biokhimicheskikh Protsessov u Mikroorganismov' (Abstracts of the All-Union Conference on 'The Regulation of Biochemical Processes in Microorganisms')*, Pushchino-na-Oke, 1972, p. 4.
2. Afanas'eva, T. P., and Kulaev, I. S., 'Polyphosphate phosphydrolase from *Endomyces magnusii*', *Biochim. Biophys. Acta*, **321**, 336 (1973).
3. Afanas'eva, T. P., and Kulaev, I. S., 'Polyphosphate phosphohydrolase from *Endomyces magnusii*', *Proceedings of the 3rd Int. Spec. Symposium on Yeasts*, *Otaniemi/Helsinki* 1973, Part I, p. 137.
4. Afanes'eva, T. P., Kulaev I. S., Mansurova, S. E. and Polyakov, V. F., 'Nucleotides and other phosphorus compounds in mitochondria of the yeast *Endomyces magnusii*' (in Russian), *Biokhimiya*, **33**, 1245 (1968).
5. Afanas'eva, T. P., and Rotschanez, V. V., 'Some properties of polyphosphatase from *Endomyces magnusii*, In *Reaction Mechanisms and Control Properties of Phosphotransferases, Joint Biochemical Symposium UdSSR–DDR, Reinhardsbrunn* Akademie-Verlag, Berlin, 1973, p. 57.
6. Afanas'eva, T. P., Uryson, S. O., Kulaev, I. S., 'The detection of multiple forms of polyphosphate phosphohydrolase in *Endomyces magnusii*' (in Russian), *Biokhimiya*, **41**, 1078 (1976).
7. Aitchison, P. A. and Butt V. S., 'The relation between the synthesis of inorganic polyphosphate and phosphate uptake by *Chlorella vulgaris*', *J. Exp. Bot.*, **24**, 497 (1973).
8. Akhromenko, A. I., and Shestakov, V. A., 'The influence of microorganisms in the rhizosphere on the uptake and excretion of phosphorus and sulphur by the roots and seedlings of woody plants' (in Russian), *Trudy II Mezhdunarodnoi Konferentsii po mirnomu ispol'zovaniyu atomnoi energii* (*Proceedings of the Second International Conference on the Peaceful Uses of Atomic Energy*), Reports by Soviet Scientists, Atomizdat, Moscow, **6**, 306 (1959).

9. Albaum, H. G. Schatz, A., Hutner, S. T., and Harschfeld, A., 'Phosphorylated compounds in *Euglena*', *Arch. Biochem.*, **29**, 210 (1950).
10. Allfrey, V. G., Mirsky, A. E., and Stern, H., 'The chemistry of the cell nucleus', *Advan. Enzymol.*, **16**, 411 (1955).
11. Arloing, F., and Richard, G., 'Les corpuscules metachromatiques des Corynebacteries. Cytologie experimentale et comparé, *Rev. Gen. Bot.*, **33**, 88 (1921).
12. Asamov, D. K., 'The metabolism of phytin and other phosphorus compounds in ripening seeds of cotton' (in Russian), *Author's Abstract, Candidate's Thesis*, Tashkent, 1972.
13. Asamov, D. K., and Valikhanov, M. N., 'A study of phosphorus compounds in ripening seeds of cotton' (in Russian), *Uzb. Biol. Zh.*, No. 2, 3 (1972).
14. Ashmarin, I. P., and Komkova, A. I., 'The reaction of histones with phosphoproteins and other phosphorus-containing compounds' (in Russian), *Biokhimiya*, **32**, 640 (1967).
15. Ascoli, A., 'Uber die Plasminsaure', *Z. Physiol. Chem.*, **28**, 426 (1899).
16. Atkinson, D. E., 'Regulation of enzyme activity', *Ann. Rev. Biochem.*, **35**, 85 (1966).
17. Atkinson, D. E., 'The adenylate control hypothesis', *Abstr. VII Int. Congr. Biochem. Tokyo*, Symp. V2, 249 comm. V–1, 6 (1967).
18. Axelrod, B., 'A new mode of enzymatic phosphate transfer', *J. Biol. Chem.*, **172**, 1 (1948).
19. Bachman, B. J., and Bonner, D. M., 'Protoplasts from *Neurospora crassa*', *J. Bacteriol.*, **78**, 550 (1959).
20. Bajaj, V., Damle, S. P., and Krishnan, P. S., 'Phosphate metabolism of mold spores. I. Phosphate uptake by the spores of *Aspergillus niger*', *Arch. Biochem. Biophys.*, **50**, 451 (1954).
21. Bajaj, V., and Krishnan, P. S., 'A note on phosphate partition in trichloroacetic acid. Extract of mold tissue', *Arch. Biochem. Biophys.*, **47**, 1 (1953).
22. Baker, A. L., and Schmidt, R. R., 'Intracellular distribution of phosphorus during synchronous growth of *Chlorella pyrenoidosa*', *Biochim. Biophys. Acta*, **74**, 75 (1963).
23. Baker, A. L., and Schmidt, R. R., 'Induced utilization of polyphosphate during nuclear division in synchronously growing *Chlorella*', *Biochim. Biophys. Acta*, **93**, 180 (1964).
24. Baker, A. L., and Schmidt, R. R., 'Further studies on the intracellular distribution of phosphorus during synchronous growth of *Chlorella pyrenoidosa*', *Biochim. Biophys. Acta*, **82**, 336 (1964).
25. Baltschevskii, H., 'Inorganic pyrophosphate and the evolution of biological energy transformation', *Acta Chem. Scand.*, **21**, 1973 (1967).
26. Baltschevskii, H. 'Biochemistry of electron transport and photophosphorylation', in *Progress in Photosynthesis Research*, Ed. Metzner, H., Springer Verlag, Tübingen, 1969, vol. 3, p. 1420.
27. Baltschevskii, H., 'Inorganic pyrophosphate and the origin and evolution of biological energy transformation', in *Chemical Evolution and the Origin of Life*, Eds. Buvet, R., and Ponnamperuma, C., North-Holland, Amsterdam, 1971, p. 466.
28. Baltschevskii, H., Baltschevskii, M., and Stedingk, L. V., 'Inorganic pyrophosphate, bacterial photophosphorylation and evolution of biological energy transformation,' in: *Progress in Photosynthesis Research*, Ed. Metzner, H., Springer Verlag, Tübingen, 1969, Vol. 3, p. 1313.
29. Baltschevskii, H., and Stedingk, L. V., 'Bacterial photophosphorylation in the absence of added nucleotide. A second intermediate stage of energy transfer in light-induced formation of ATP,' *Biochem. Bioph. Res. Commun.*, **22**, 722 (1966).
30. Baltschevskii, H., Stedingk, L. V., Heldt, H. W., and Klingenberg, M., 'Inorganic

208

pyrophosphate: formation in bacterial photophosphorylation,' *Science, N. Y.,* **153**, 1120 (1966).

31. Baltschevskii, M., 'Inorganic pyrophosphate and ATP energy donors in chromatophores from *Rhodospirillum, Nature, Lond.,* **216**, 241 (1967)

32. Baltschevskii, M., 'Inorganic pyrophosphate: its hydrolysis and function as energy donor in chromatophores from *Rhodospirillum rubrum', Abstr. VII Int. Congr. Biochem. Tokyo,* Symp. V, **H-67**, 897 (1967).

33. Baltschevskii, M., 'Inorganic pyrophosphate as an energy donor in photosynthetic and respiratory electron transport phosphorylation systems', *Biochem. Biophys. Res. Commun.,* **28** 270 (1967).

34. Baltschevskii, M., 'Reversed energy transfer in *Rhodospirillum rubrum* chromatophores', in *Progress in Photosynthesis Research,* Ed. Metzner, H., Springer-Verlag, Tübingen, 1969, Vol. 3, p. 1306.

35. Baltschevskii M., 'Energy conversion-linked changes of carotenoid absorbence in *Rhodospirillum rubrum* chromatophores', *Arch. Biochem. Biophys.,* **130**, 646 (1969).

36. Baltschevskii, M., 'Reversed energy conversion reactions of bacterial photophosphorylation', *Arch. Biochem. Biophys.,* **133**, 46 (1969).

37. Baltschevskii, M., Baltschevskii, H., and Stedingk, L. V., 'Light-induced energy conversion and the inorganic pyrophosphatase reactions in chromatophores from *Rhodospirillum rubrum',* in *Energy Conversion and Photosynthetic Apparatus,* Ed. Uption M. J., Brookhaven National Laboratory, Brookhaven, 1967, p. 246.

38. Ramann, R., and Heumüller, E., 'The activation of phosphatases by different metal ions', *Naturwissenschaften,* **28**, 353 (1940).

39. Barker, H. A., and Kornberg, A., 'The structure of the adenosine triphosphate of *Thiobacillus thiooxydans', J. Bacteriol.,* **68**, 655 (1954).

40. Bartholomew, J. W., and Levin, R., 'The structure of *Saccharomyces carlsbergensis* and *S. cerevisiae* as determined by ultra-thin sectioning methods and electron microscopy', *J. Gen. Microbiol.,* **12**, 473 (1955).

41. Bashirelashi, N., and Dallam, D., 'Nuclear metabolism, I. Nuclear phosphorylation', *Arch. Biochem. Biophys.,* **141**, 329 (1970).

42. Baslavskaya, S. S., and Bystrova, E. I., 'The action of light on phosphorus metabolism in protococcal algae' (in Russian), *Dokl. Akad. Nauk SSSR,* **155**, 1220 (1964).

43. Baslavskaya, S. S., Kulikova, R. F., Markarova, E. N., and Savchenko, R. V., 'A study of phosphorus metabolism in protococcal algae' (in Russian), *Nauchn. Dokl. Vyssh. Shkoly, Biol. Nauk* (*High School Science Reports, Biological Sciences*), No. 2, 144 (1966).

44. Beck, A., and Orgel, L. E., 'The formation of condensed phosphate in aqueous solution', *Proc. Nat. Acad. Sci. USA,* **54**, 664 (1965).

45. Bekker, Z. E., *Fiziologiya Gribov i ikh Prakticheskoe ispol'zovanie* (*The Physiology and Practical Uses of Fungi*), Izd. MGU, Moscow, 1963.

46. Belyavskaya, E. A., and Pinevich, V. V., 'The effect of trace element nutrition on the metabolism of polyphosphates and other phosphorus-containing compounds in *Chlorella pyrenoidosa*' (in Russian), *Trudy XIII Nauchnoi Konferentsii Dal'GU* (*Proceedings of the Thirteenth Scientific Far-Eastern State University Conference*), Part 3, 1960, p. 163.

47. Belousova, A. I., and Popova, L. A., 'The effect of inorganic phosphorus on tetracycline biosynthesis and the composition of phosphorus fractions in *Actinomyces aureofaciens* with respect to the conditions of culture and growth of the mycelia' (in Russian), *Antibiotiki,* **4**, 302 (1961).

48. Belozersky, A. N., 'The dependence of the composition of the protoplasm of cells of *Spirillum volutans* on culture age' (in Russian), *Mikrobiologiya,* **10**, 185 (1941).

49. Belozersky, A. N., 'The chemical nature of volutin' (in Russian), *Mikrobiologiya*, **14**, 29 (1945).
50. Belozersky, A. N., 'The metaphosphate–nuclein complexes of yeast, and the chemical nature of volutin' (in Russian), in *III Mezhdunarod. Biokhim. Kongr. (Third International Biochemical Congress)*, Summary of papers, Brussels 1955, p. 78.
51. Belozersky, A. N., 'The formation and functions of polyphosphates in the developmental processes of some lower organisms' (in Russian), *Soobshch. i. Dokl. IV Mezhdunarod. Biokhim. Kongr. (Communications and Reports to the Fourth International Biochemical Congress)*, Vienna, 1958, refs. 3–12.
52. Belozersky, A. N., Contribution to the discussion at the Symposium on 'The Origin of Life on Earth' (in Russian), Izd. Akad. Nauk SSSR, Moscow, 1959, p. 370.
53. Belozersky, A. N., 'Nucleoproteins and nucleic acids in plants, and their biological significance' (in Russian), *XIV. Bakhovskoe Chtenie (Fourteenth Bach's Reading)*, Izd. Akad. Nauk SSSR, Moscow, 1959.
54. Belozersky, A. N., 'Nucleic acids of microorganisms', *Onzième Conseil de Chimie*, Institut International de Chimie Solvay, Brussells, 1959, p. 199.
55. Belozersky, A. N., 'Nucleic acids and their links with the evolution, physiology, and classification of organisms' (in Russian). *Trudy II-go Vsesoyuznogo biokhimicheskogo s'ezda. Plenarnaya Lektsiya (Proceedings of the Second All-Union Biochemical Congress, Plenary Lecture)*, Izd. FAN Uzbekskoi SSR, Tashkent, 1969.
56. Belozersky, A. N., Antonov, A. S., and Mednikov, B. M., *Stroenie DNK i Polozhenie Organismov v Sisteme (The structure of DNA and the Classification of Organisms)*, Izd. Mosovskogo Universiteta, Moscow, 1972, p. 3.
57. Belozersky, A. N., Korchagin, V. B., and Smirnova, T. I., 'Changes in the chemical composition of diphtheria bacteria with the age of the culture' (in Russian), *Dokl. Akad. Nauk SSSR*, **71**, 89 (1950).
58. Belozersky, A. N., and Kulaev, I. S., 'The significance of polyphosphates in the development of *Aspergillus niger*' (in Russian), *Biokhimiya*, **22**, 29 (1957).
59. Belozersky, A. N., and Kulaev, I. S., 'The correlation of the composition of free nucleotides and nucleotide-containing coenzymes with the course of biochemical processes and environmental conditions in fungi' (in Russian), in *Problemy Evolyutsionnoi i Technicheskoi Biokhimii (Problems of Evolutionary and Technical Biochemistry)*, Izd. Nauka, Moscow, 1964.
60. Belozersky, A. N., and Kulaev, I. S., 'The polyphosphate–ribonucleic acid complexes from yeasts' (in Russian), in *Biokhimiya i Fiziologiya Bol'nogo i Zdorovogo Rasteniya (The Biochemistry and Physiology of Diseased and Healthy Plants)*, Izd. MGU, Moscow, 1970, p. 42.
61. Belozersky, A. N., and Spirin, A. S., 'Nucleic acid composition and systematics' (in Russian), *Izv. Akad. Nauk SSSR, Ser. Biol.*, No. 1, 64 (1960).
62. Berg, G. G., 'Inorganic polyphosphatase in the developing of frog', *J. Cell. Comp. Physiol.*, **45**, 435 (1955).
63. Berg, G. G., 'Ascending chromatography of polyphosphates', *Analyt. Chem.*, **30**, 213 (1957).
64. Berg, G., 'Histochemical demonstration of acid trimetaphosphatase', *J. Histochem. Cytochem.*, **8**, 92 (1960).
65. Bergeron, J. A., and Singer, M., 'Metachromasy: an experimental and theoretical reevaluation', *J. Biophys. Biochem. Cytol.*, **4**, 433 (1958).
66. Biberacher, G., 'Die Papierchromatographie der Amido- und Imidophosphate', *Z. anorg. allg. Chem.*, **285**, 86 (1956).
67. Boulle, A. L., 'Les phosphates condensés minéraux. Aspects structuraux', *Coll. nat. du CNRS 'Composés organiques du phosphore'*, Toulouse, CNRS, Paris, 1965, p. 279.

210

68. Bowdes, F. C., and Pirie, N. W., 'The virus content of plants suffering from tobacco mosaic', *Br. J. Exp. Pathol.*, **27**, 81 (1946).
69. Bresler, S. E., and Nidzyan, E., 'The enzymatic transfer of phosphate from adenosinetriphosphoric to ribonucleic acid' (in Russian), *Dokl. Akad. Nauk SSSR*, **75**, 79 (1950).
70. Bresler, S. E., and Rubina, Kh. M., 'The enzymatic transfer of phosphate residues from ribonucleic acid to fructose monophosphate' (in Russian), *Biokhimiya*, **20**, 6 (1955).
71. Bresler, S. E., Rubina, Kh. M., and Vinokurov, Yu. A., 'The enzymatic transfer of phosphate from ribonucleic acid to creatine' (in Russian), *Biokhimiya*, **22**, 794 (1967).
72. Bringmann, G., 'Vergleichende Licht- und Electron-mikroskopische Unter-suchungen an Oszillatorien', *Planta*, **38**, 541 (1950).
73. Bukhovich, E., 'Polyphosphate metabolism in yeasts' (in Russian), *Candidate's Thesis*, Moscow, 1958.
74. Bukhovich, E., and Belozersky, A. N., 'Some aspects of the formation of poly-phosphates in yeast cells' (in Russian), *Biokhimiya*, **23**, 254 (1958).
75. Bukhovich, E., and Belozersky, A. N., 'Some aspects of the mode of synthesis and utilization of polyphosphates in yeasts' (in Russian), *Dokl. Akad. Nauk SSSR*, **124**, 1147 (1959).
76. Burkard, G., Weil, J.–H., and Ebel, J. P., 'Mise en evidence d'un effect inhibiteur des polyphosphates sur l'activité acceptrice de l'acide ribonucleique soluble de levure', *Bull. Ser. Chim. Biol.*, **47**, 561 (1965).
77. Bush, N., and Ebel, J. P., 'Recherces sur les poly- et metaphosphates. VI. Resul-tats complementaires sur le fractionment des polyphosphates par les composés organiques basiques', *Bull. Soc. Chim. Fr.*, **23**, 758 (1956).
78. Bush, N., Ebel, J. P., and Blanck, M., 'Recherches sur les poly- et metaphosphates. VII. Application de la chromatographie par échange d'ions à l'isolement de certains polyphosphates: tetra-, penta-, hexa- et octaphosphates', *Bull. Soc. Chim. Fr.*, **24**, 486 (1957).
79. Boutin, J. -P., and Roux, L., 'Etude cinetique de l'hydrolyse du pyrophosphate et du tripolyphosphate par les de tomate cultivees *in vitro*', *C. R. Acad. Sci. Paris*, **277**, 49 (1973).
80. Butt, W. D., and Lees, H., 'The biochemistry of nitrifying organisms. VII. The phosphates compounds of nitrobacter and the uptake of orthophosphate by the organisms', *Can. J. Biochem. Physiol.*, **38**, 1295 (1960).
81. Carr, N. G., and Sandhu, G. R., 'Endogenous metabolism of polyphosphates in two photosynthetic microorganisms', *Biochem. J.*, **99**, 29 (1966).
82. Cartier, P., 'Le role de l'ATP et des pyrophosphates dans l'ossification', *Colloq. Int. CNRS*, **106**, 626 (1962).
83. Chaloupka, J., and Babicky, A., 'Komplex ribonukleova kyselina-polyfosfat à jehe deleni, *Čsl. Microbiol.*, **2**, 371 (1957).
84. Chaloupka, J., and Babicky, A., 'The ribonucleic acid–polyphosphate complex and its separation', *Folia Biol.* **4**, 233 (1958).
85. Chayen, R., Chayen, S., and Roberts, E. R., 'Observations on nucleic acid and polyphosphate in torulopsis utilis', *Biochim. Biophys. Acta*, **16**, 117 (1955).
86. Chernysheva, E. K., 'A study of the metabolism of high-molecular polyphos-phates, and their enzymatic depolymerization, in *Neurospora crassa*' (in Russian), *Candidate's Thesis*, Moscow, 1972.
87. Chernysheva, E. K., and Mel'gunov, V. I., 'Determination of the correlation between the rates of accumulation of various polyphosphate fractions and RNA as a possible approach to investigating the regulation of polyphosphate meta-bolism' (in Russian), in *Tezisy Dokl. na Vsesoyuznoi Konferentsii Regulyatsiya Biokhimicheskikh Protsessov u Mikroorganizmov' (Abstracts of the All-Union*

Conference on 'The Regulation of Biochemical Processes in Microorganisms'), Pushchino-na-Oke, 1972, pp. 178.

88. Chernysheva, E. K., Kritsky, M. S., and Kulaev, I. S., 'The determination of the degree of polymerization of different inorganic polyphosphate fractions from mycelia of *Neurospora crassa*' (in Russian), *Biokhimiya*, **36**, 138 (1971).

89. Cole, J. A., and Hughes, D. E., 'The metabolism of polyphosphates in *Chlorobium thiosulfatophillium*', *J. Gen. Microbiol.*, **38**, 65 (1965).

90. Cordonnier, E., 'Recherches sur les dipeptidases de *Saccharomyses cerevisiae ellipsoideus*. Coix des substrats d'étude activation par un phosphate condensé', *C. R. Acad. Sci. Paris*, **253**, 748 (1961).

91. Correll, D. L., 'Ribonucleic acid-polyphosphate from algae. III. Hydrolysis studies'. *Plant Cell Physiol.* (*Tokyo*), **6**, 661 (1965).

92. Correll, D. L., 'Imidonitrogen in *Chlorella* 'polyphosphate', *Science, N.Y.*, **151**, 819 (1966).

93. Correll, D. L., and Tolbert, N. E., 'Ribonucleic acid–polyphosphate from algae. I. Isolation and physiology', *Plant. Physiol.*, **37**, 627 (1962).

94. Correll, D. L., and Tolbert, N. E., 'Ribonucleic acid–polyphosphate from algae. II. Physical and chemical properties of the isolated complexes', *Plant Cell Physiol.*, **5**, 171 (1964).

95. Corska-Brylass, A., 'Inorganic polyphosphates in pollen tubes', *Acta Soc. Bot. Pol.*, **35**, 3 (1966).

96. Crowther, J. P., 'Filter-paper chromatographie analyses of phosphate mixtures', *Nature, Lond.*, **173**, 486 (1954).

97. Curnutt, S. G., and Schmidt, R. R. Possible mechanisms controlling the intra-cellular level of inorganic polyphosphate during synchronous growth of *Chlorella pyrenoidosa*', *Biochim. Biophys. Acta*, **86**, 201 (1964).

98. Curnutt, S. G., and Schmidt, R. R., 'Possible mechanisms controlling the intra-cellular level of inorganic polyphosphate during synchronous growth of *Chlorella pyrenoidosa*. Endogenous respiration', *Exp. Cell. Res.*, **36**, 102 (1964).

99. Damle, S. P., and Krishnan, P. S., 'Studies on the role of metaphosphate in moids. I. Quantitative studies on the metachromatic effect of metaphosphate', *Arch. Biochem. Biophys.*, **49**, 58 (1954).

100. Davis, J. C., and Mudd, St., 'The cytology of a strain of *Corynebacterium dipheriae*', *J. Bacteriol.*, **69**, 372 (1955).

101. Davis, J. C., and Mudd, St., 'Cytological effects of ultraviolet radiation and azaserine on *Corynebacterium diphtheriae*, *J. Gen. Microbiol.*, **14**, 527 (1956).

102. Dawes, E. A., and Senior P. T., 'The role and regulation of energy reserve polymers in microorganisms', *Adv. Microbiol. Physiol.*, **10**, 135 (1973).

103. Degani, Cn., and Halmann M., 'Mechanism of the cyanogen induced phos-phorylation of sugars in aqueous solution', in *Molecular Evolution, Vol. 1, Chemical Evolution and the Origin of Life*, Eds. Buvet, R., and Ponnamperuma, C., North-Holland, Amsterdam, 1971, p. 224.

104. Deierkauf, F. A., and Booij, H. L., 'Changes in the phosphatide pattern of yeast cells in relation to active carbohydrate transport', *Biochim. Biophys. Acta*, **150**, 214 (1968).

105. Dietrich, S. M. C., 'Presence of polyphosphate of low molecular weight in zygomycetes', *J. Bacteriol.*, **127**, 1408 (1976).

106. Dirheimer, G., 'Recherches sur les phosphates condensés chez les microorgan-ismes. Étude de leur nature et de quelques enzymes intervenants dans leur utilisation metabolique', *Thése Doct. Sci. Phys.*, Strasbourg, 1964.

107. Dirheimer, G., and Ebel, J. P., 'Étude de l'inhibition des réactions hexokinasique et phosphohexokinasique par les polyphosphates inorganiques', *Bull. Soc. Chim. Biol.*, **38**, 1337 (1956).

108. Dirheimer, G., and Ebel, J. P., 'Sur la présence de metaphosphates cycliques dans la levure', *Bull. Soc. Chim. Biol.*, **39**, suppl. 2, 89 (1957).

109. Dirheimer, G., and Ebel J. P., 'Mise en evidence d'une polyphosphate–glucose–phosphotransferase dans *Corynebacterium xerosis*', *C. R. Acad. Sci. Paris*, **254**, 2850 (1962).

110. Dirheimer, G., and Ebel, J. P., 'Sur le metabolisme des polyphosphates inorganiques chez *Corynebacterium xerosis*', *C. R. Soc. Biol.*, **158**, 1948 (1964).

111. Dirheimer, G., and Ebel J. P., 'Séparation des acides ribonucleiques et des polyphosphates inorganiques. IV. Séparation par filtration sur gel de dextrane (Sephadex G-200)', *Bull. Soc. Chim. Biol.*, **46**, 396 (1964).

112. Dirheimer, G., and Ebel, J. P., 'Caracterisation d'une polyphosphate–AMP–phosphotransferase dans *Corynebacterium xerosis*', *C.R. Acad. Sci. Paris*, **260**, 3787 (1965).

113. Dirheimer, G., Muller-Felter, S., Weil, J. H., Jacoub, M., and Ebel, J. P., 'Étude des complexes entre acides ribonucleiques et polyphosphates inorganiques dans la levure. II. Essais de séparation des acides ribonucleiques et des polyphosphates dans les complexes naturels', *Bull. Soc. Chim. Biol.*, **45**, 875 (1963).

114. Dmitrieva, S. V., and Bekker, Z. E. 'The nature of volutin granules in *Penicillium chrysogenum*' (in Russian), *Tsitologiya*, **4**, 691 (1962).

115. Domanski-Kaden, J., and Simonis, W., 'Veranderungen der Phosphatfraktionen besonders des Polyphosphates, bei synchronisierten Ankistrodesmus braunii-Kulturen', *Arch. Mikrobiol.*, **87**, 11 (1972).

116. Dounce, A., and Kay, E., 'Attempted enzymic phosphorylation of ribonucleic acid', *Proc. Soc. Exp. Biol. Med.*, **83**, 321 (1953).

117. Drews, G., 'Die granulären Einschlüsse der Mycobacterium', *Arch. Mikrobiol.*, **28**, 369 (1958).

118. Drews, G., 'Untersuchungen über Granulabildung und Phosphateinbau in wachsenden Kulturen von *Mycobacterium phlei*', *Arch. Mikrobiol.*, **31**, 16 (1958).

119. Drews, G., 'Der Einfluss von Ribonuclease auf den Phosphatstoffwechsel von *Mycobacterium phlei*', *Monatsber. Deut. Akad. Wiss., Berlin*, **1**, 607 (1959).

120. Drews, G., 'Die Bildung metachromatische Granula in washsenden und ruhenden Kulturen von *Mycobacterium phlei*'. *Naturwissenschaften*, **46**, 87 (1959).

121. Drews, G., 'Der Einfluss von 2,4-Dinitrophenol auf den Phosphorstoffwechsel und die Bildung der metachromatischen Granula bei *Mycobacterium phlei*', *Z. Naturforsch.*, **14b**, 265 (1959).

122. Drews G., 'Electronenmicroscopische Untersuchungen an *Mycobacterium phlei* (Struktur und Bildung det metachromatischen Granula)', *Arch. Microbiol.*, **35**, 53 (1960).

123. Drews, G., 'Untersuchungen zum Polyphosphatstoffwechsel und der Bildung metachromatischer granula bei *Mycobacterium phlei*', *Arch. Mikrobiol.*, **36**, 387 (1960).

124. Drews, G., 'The cytochemistry of polyphosphates', in *Acides Ribonucleiques et Polyphosphates. Structure, Synthèse et Fonctions, Colloq. Int. CNRS, Strasbourg, 1961*, CNRS, Paris, 1962, p. 533.

125. Drews, G., and Niklowitz, W., 'Beiträge zur Cytologie der Blaualgen. III. Mitteilung. Untersuchungen über die granulären Einschlüsse der Hormogenales', *Arch. Mikrobiol.*, **25**, 335 (1957).

126. Duguid, J. P., Smith, J. W., and Wilkinson, J. F., 'Volutin production in *Bacterium aerogenes* due to development of an acid reaction', *J. Path. Bacteriol.*, **67**, 289 (1954).

127. Ebel, J. P., 'Sur le dosage des metaphosphates dans les microorganismes par hydrolyse differentielle technique et application aux levures', *C.R. Acad. Sci. Paris*, **226**, 2184 (1948).

128. Ebel, J. P., 'Participation de l'acide metaphosphorique à la constitution des bacteries et des tissus animaux', *C. R. Acad. Sci. Paris*, **228**, 1312 (1949).

129. Ebel, J. P., 'Recherches chimiques et biologiques sur les poly- et metaphosphates', *Thése Doct. Sci. Phys.*, Strasbourg, 1951.

130. Ebel, J. P., 'Recherches sur les polyphosphates contenus dans diverses cellules vivantes. I. Mise au point d'une méthode d'extraction', *Bull. Soc. Chim. Biol.*, **34**, 321 (1952).

131. Ebel, J. P., 'Recherches sur les polyphosphates contenus dans diverses cellules vivantes. II. Étude chromatographique et potentiometrique des polyphosphates de levure', *Bull. Soc. Chim. Biol.*, **34**, 330 (1952).

132. Ebel, J. P., 'Recherches sur les polyphosphates contenus dans diverses cellules vivantes. III. Recherche et dosage des polyphosphates dans les cellules de diverses microorganismes et animaux superieures', *Bull. Soc. Chim. Biol.*, **34**, 491 (1952).

133. Ebel, J. P., 'Recherches sur les polyphosphates contenus dans diverses cellules vivantes. IV. Localisation cytologique et role physiologique des polyphosphates dans la cellule vivante', *Bull. Soc. Chim., Biol.*, **34**, 498 (1952).

134. Ebel, J. P., 'Recherches sur les poly- et metaphosphates. I. Mise au point d'une méthode de separation par chromatographie sur papier', *Bull. Soc. Chim. Fr.*, **20**, 991 (1953).

135. Ebel, J. P., 'Recherches sur les poly- et metaphosphates. II. Essai d'application quantitative de la technique de separation des poly- et metaphosphates par chromatographie sur papier', *Bull. Soc. Chim. Fr.*, **20**, 998 (1953).

136. Ebel, J. P., 'Separation par chromatographie sur papier des oxyacides de polyphosphates', *Microchim. Acta*, **6**, 679 (1954).

137. Ebel, J. P., 'Étude chromatographique de quelques poly- et metaphosphates synthetiques et biologiques', *Colloq. Münster. Westf.*, *Sept.* 1954.

138. Ebel, J. P., 'Action physiologique et toxicologique des phosphates condensés', *Ann. Nutr. Aliment.*, **12**, 59 (1958).

139. Ebel, J. P., and Bush, N., 'Separation de polyphosphates par chromatographie sur colonne', *C. R. Soc. Biol.*, **242**, 647 (1956).

140. Ebel, J. P., and Dirheimer, G., 'Relations metaboliques entre polyphosphates inorganiques et adenosine di- et triphosphates', *C.R. Soc. Biol.*, **151**, 979 (1957).

141. Ebel, J. P., Dirheimer, G., and Jacoub, M., 'Sur l'existance d'une combinaison acide ribonucleique–polyphosphates dans la levure', *Proc. 4th Int. Congr. Biochem.*, *Vienna*, **17**, 31 (1958).

142. Ebel, J. P., Dirheimer, G., and Jacoub M., 'Sur l'existence d'une combinaison acide ribonucleique–polyphosphate dans la levure', *Bull. Soc. Chim. Biol.*, **40**, 738 (1958).

143. Ebel, J. P., Dirheimer, G., Jacoub, M., and Muller-Felter, S., 'Separation des acides ribonucleiques et des polyphosphates inorganiques. I. Essais de separation par electrophores et par chromatographie', *Bull. Soc. Chim. Biol.*, **44**, 1167 (1962).

144. Ebel, J. P., Dirheimer, G., Stahl, A. J. C., and Jacoub, M., 'Étude des complexes entre acides ribonucleiques et polyphosphates inorganiques dans la levure. I. Présence de polyphosphates inorganiques dans les préparations d'acides ribonucleiques de haut poids moleculaire extractés de levure', *Bull. Soc. Chim. Biol.*, **45**, 865 (1963).

145. Ebel, J. P., Dirheimer, G., Stahl, A., and Felter, S., 'Quelques aspects nouveaux de la biochimie des polyphosphates inorganiques', *Colloq. Nat. du CNRS Composés Organiques du Phosphore, Toulouse*, CNRS, Paris, 1966, p. 289.

146. Ebel, J. P., and Colas, J., 'Combinaisons entre poly- et metaphosphates et diverse composés organiques basiques', *C.R. Soc. Biol.*, **239**, 173 (1954).

147. Ebel, J. P., and Colas, J., 'Étude de la précipitation de quelques composés

basiques biologiques par de polyphosphates de longue chaine', *Bull. Soc. Chim. Biol.*, **37**, 445 (1955).

148. Ebel, J. P., Colas, J., and Bush N., 'Recherches sur le poly- et metaphosphates. V. Reaction de characterisation des poly- et metaphosphates par précipitation par divers composés organiques basiques', *Bull. Soc. Chim. Fr.*, **5**, 1087 (1955).

149. Ebel, J. P., and Mehr, U., 'Sur l'activité poly- et metaphosphatasique du sérum et ses variations pathologiques', *Bull. Soc. Chim. Biol.*, **39**, 1935 (1957).

150. Ebel, J. P., Colas, J., and Muller, S., 'Recherches cytochimiques sur les polyphosphates inorganiques contenus dans les organismes vivants. II. Mise au point de méthodes de détection cytochimiques specifiques des polyphosphates', *Exp. Cell. Res.*, **15**, 28 (1958).

151. Ebel, J. P., Colas, J., and Muller, S., 'Recherches cytochimiques sur les polyphosphates inorganiques contenus dans les organismes vivants. III. Présence de polyphosphates chez divers organismes inférieurs', *Exp., Cell. Res.*, **15**, 36 (1958).

152. Ebel, J. P., and Muller, S., 'Recherches cytochimique sur les polyphosphates inorganiques contenus dans les organismes vivants. I. Étude de la reaction de métachromasie et des colorations au vert de méthyle et à la pyronine en fonction de la longueur de chaîne des polyphosphates', *Exp. Cell. Res.*, **15**, 21 (1958).

153. Ebel, J. P., Stahl, A., Dirheimer, G., and Felter-Muller, S., 'Relations de structure entre acide ribonucleique et polyphosphates', in *Acides Ribonucleiques et Polyphosphates. Structure, Synthése et Fonctions, Colloq. Int. CNRS, Strasbourg, 1961*, CNRS, Paris, 1962, p. 545.

154. Ebel, J. P., and Volmer, Y., 'Chromatographie sur papier des ortho- pyro-, meta- et polyphosphates', *C.R. Soc. Biol.*, **233**, 415 (1951).

155. Echoly, H., Garen, A., Garen, S., and Torriani, A., 'Genetic control of repression of alkaline phosphatase in *E. coli.*, *J. Mol. Biol.*, **3**, 425 (1961).

156. Egorov, S. N., and Kulaev, I. S., 'The isolation and properties of tripolyphosphatase from the fungus *Neurospora crassa*' (in Russian), *Biokhimiya*, **41**, 1958 (1976).

157. Ehrenberg, M., 'Die Anfangstadien des ^{32}P-Einbaues in P-Zuführ', *Naturwissenschaften.*, **47**, 607 (1960).

158. Ehrenberg, M., 'Der Phosphorstoffwechsel von *Saccharomyces cerevisiae* in Abhängigkeit von intra- und extracellulärer Phosphatkonsentration', *Arch. Microbiol.*, **40**, 126 (1961).

159. Eigener, U., and Bock, E. 'Auf- und Abbau der Polyphosphat-fraction in Zellen von *Nitrobacter winogradskyi* Buch'., *Arch. Mikrobiol.*, **81**, 367 (1972).

160. Esposito, E., and Wilson, P., 'Calcium and polymetaphosphate synthesis in *Azotobacter vinelandii*', *Biochim. Biophys. Acta*, **22**, 186 (1956).

161. Evtodienko, Yu. V., and Mokhova, E. N., 'The quantitative determination of the proportions of cytochromes in mitochondria' (in Russian), in *Mekhanismy Dykhaniya, Fotosinteza, i Fiksatsii Azota (Mechanisms of Respiration, Photosynthesis, and Nitrogen Fixation)*, Izd. Nauka, Moscow, 1967, p. 35.

162. Fedorov, V. D., 'Polyphosphates in photosynthesizing bacteria' (in Russian), *Dokl. Akad. Nauk SSSR*, **126**, 406 (1959).

163. Fedorov, V. D., 'Phosphorus metabolism in green sulphur bacteria in relation to the photoassimilation of carbon dioxide', (in Russian), *Candidate's Thesis*, Moscow, 1961.

164. Felter, S., 'Recherche sur la distributions, la structure et le metabolisme des polyphosphates biologiques', *Thése Doct. Sci. Phys.*, Strasbourg, 1964.

165. Felter, S., Dirheimer, G., and Ebel, J. P., 'Recherches sur les polyphosphatases de levure de boulangerie', *Bull. Soc. Chim. Biol.*, **52**, 433 (1970).

166. Felter, S., and Stahl, A. J. C., 'Recherches sur les enzymes du metabolisme des polyphosphates dans la levure: 1. Étude de la variation des activites enzymatiques

en fonction de la concentration intracellulaire en polyphosphates', *Bull. Soc. Chim. Biol.*, **52**, 75 (1970).

167. Felter, S., and Stahl, A. J. C., 'Enzymes du metabolisme des polyphosphates dans la levure. III. Purification et propertieles de la polyphosphate–ADP–phosphotransferase', *Biochimie*, **55**, 245 (1973).
168. Felter, S., and Stahl, A., 'Recherches sur la polyphosphate-synthetase de *Saccharomyces cerevisiae*', *C.R. Acad. Sci. Paris*, **280D**, 1903 (1975).
169 Ferdman, D. L., 'The chemistry and biochemistry of condensed phosphates', (in Russian), *Ukr. Biokhim. Zh.*, **32**, 452 (1960).
170. Fisher, C. J., and Stetten, M. R., 'Parallel changes *in vivo* in microsomal inorganic pyrophosphatase, pyrophosphate-glucose phosphotransferase and glucose 6-phosphatase activities', *Biochim. Biophys. Acta*, **121**, 102 (1966).
171. Fleitmann, T., and Hennenberg, W., 'Über phosphorsäure Salze', *Justus Liebigs Ann. Chem.*, **65**, 304 (1848).
172. Flodgaard, H., and Fleron, P., 'Thermodynamic parameters for the hydrolysis of inorganic pyrophosphate at pH 7.4 as a function of Mg^{2+}, K^+ and ionic strength determined from equilibrium studies of the reaction', *J. Biol. Chem.*, **249**, 3465 (1974).
173. Friedberg, J., and Avigad, G., 'Structures containing polyphosphate in *Micrococcus lysodeikticus*', *J. Bacteriol.*, **96**, 544 (1968).
174. Fox, S. W., and Harada, K., 'Thermal copolymerization of amino acids to a product resembling protein', *Science, N.Y.*, **128**, 1214 (1958).
175. Fox, S. W., and Harada, K., 'Thermal copolymerization of amino acids common to protein', *J. Am. Chem. Soc.*, **82**, 3745 (1960).
176. Fox, S. W., Yuki, A., Wachneldt, T. V., and Lacey, Y. C., 'The primordial sequence, ribosomes and the genetic code', in *Molecular Evolution, Vol. 1, Chemical Evolution and the Origin of Life*, Eds. Buvet, R., and Ponnamperuma, C., North-Holland, Amsterdam, 1971, p. 252.
177. Fuhs, G. W., 'Bau, Verhalten und Bedeutung der Kernäquivalent in Structuren bei *Oscillatoria amoena* (Küts.)', *Arch. Microbiol.*, **28**, 270 (1958).
178. Gabel, N. W., 'Excitability and the origin of life: a hypothesis', *Life Sci.*, **4**, 2085 (1965).
179. Gabel, N. W., 'Excitability, polyphosphates and precellular organization', in *Molecular Evolution, Vol. 1, Chemical Evolution and the Origin of Life*, Eds. Buvet, R., and Ponnamperuma, C., North-Holland, Amsterdam, 1971, p. 369.
180. Gabel, N. W., and Ponnamperuma, C., 'Primordial organic chemistry', in *Exobiology*, Ed. C. Ponnamperuma, North-Holland, Amsterdam, 1972, p. 95.
181. Gabel, N. W., and Thomas, V., 'Evidence for the occurrence and distribution of inorganic polyphosphates in vertebrate tissues', *J. Neurochem.*, **18**, 1229 (1971).
182. Gasior, E., and Szymona, M., 'Comparative study of low and high molecular weight polyphosphates. Relation to the inorganic polyphosphates', *Bull. Acad. Polon. Sci., Ser. Sci. Biol.*, **11**, 231 (1963).
183. Gezelius, K., Felter, S., and Stahl, A., 'Étude des polyphosphates pendant le development multicellulaire de *Dictyostellium discoideum*', *C.R. Acad. Sci. Paris*, **276D**, 117 (1973).
184. Gierer, A., and Schramm, G., 'Infectivity of ribonucleic acid from tobacco mossaic virus', *Nature, Lond.*, **177**, 702 (1956).
185. Glauert, A. M., and Brieger, E. M., 'The electron-dense bodies of *Mycobacterium phlei*', *J. Gen. Microbiol.*, **13**, 310 (1955).
186. Glonek, Th., Lunde, M., Mudgett, M., and Myery, T. C. 'Studies of biological polyphosphate through the use of phosphorus-31 nuclear magnetic resonance', *Arch. Biochem. Biophys.*, **142**, 508 (1971).
187. Goodman, M., Benson, A., and Calvin, M., 'Fractionation of phosphates from *Scenedesmus* by anion exchange', *J. Am. Chem. Soc.*, **77**, 4257 (1955).

216

188. Goodman, E. M., Sauer, H. W., and Rusch, H. P., 'Changes in polyphosphate in *Physarum polycephalum*', *J. Cell. Biol.*, **39**, 530 (1968).
189. Gotovtseva, V. A., 'Phosphorus compounds in the fungus *Penicillium chrysogenum*' (in Russian), *Antibiotiki*, **1**, 18 (1956).
190. Graham, T., 'Researches on the arseniates, phosphates and modifications of phosphoric acid', *Phil. Trans. R. Soc. London*, **123**, 253 (1833).
191. Grawley, J. C. W., 'A cytoplasmic organelle associated with the cell wall of *Chara* and *Nitella*', *Nature, Lond.*, **205**, 200 (1965).
192. Green, H. A., and Meyerhof, O., 'Synthetic action of phosphatase. III. Transphorphorylation with intestinal and semen phosphatase', *J. Biol. Chem.*, **197**, 347 (1952).
193. Griffin, J. B., and Penniall, B. R., 'Studies of phosphorus metabolisms by isolated nuclei. IV. Labeled components of the acid- insoluble fraction', *Arch. Biochem. Biophys.*, **114**, 67 (1966).
194. Griffin, J. B., Davidian, N. M., and Penniall, R., 'Studies of phosphorus metabolism by isolated nuclei. VII. Identification of polyphosphate as a product', *J. Biol. Chem.*, **240**, 4427 (1965).
195. Griffith, E. J., 'The chemical and physical properties of condensed phosphates, in *Abstracts of 2nd IUPAC Symposium on Inorganic Phosphorus Compounds, Prague, Sept. 1974*, p. 26.
196. Griffith, E. J., Beeton, A., Spencer, J. M., and Mitchell, D. T., *Environmental Phosphorus Handbook*, Wiley-Interscience, New York. 1973.
197. Grigor'eva, S. P., Vorob'eva, G. I., Vysloukh, V. A., Maksimova, G. N., and Kulaev, I. S., Polyphosphates, polysaccharides, and nucleic acids in cells of *Candida quillermondii* assimilating petroleum under different conditions of culture' (in Russian), *Prikl. Biokhim. Mikrobiol.*, **9**, 805 (1973).
198 Grossman, D., and Lang, K., 'Inorganic poly- and metaphosphatases as well as polyphosphates in the animal cell nucleus', *Biochem. Z.*, **336**, 351 (1962).
199. Grula, E. A., and Hartsell, S. E., 'Intracellular structures in *Caulobacter vibroides*', *J. Bacteriol.*, **68**, 498 (1954).
200. Grunze, H., and Thilo, E., *Die Papierchromatographie der kondensierten Phosphate*, 2 Aufl., Academie Verlag, Berlin, 1955.
201. Guberniev, M. A., Listvinova, S. N., Torbochkina, L. I., and Kats, L. N., 'Polyphosphates of the chlortetracycline producer *Actinomyces aureofaciens*' (in Russian), *Antibiotiki*, **4**, 40 (1959).
202. Guberniev, M. A., Torbochkina, L. I., and Kats, L. N., 'Polyphosphates of *Actinomyces aureofaciens*' (in Russian), *Antibiotiki*, **4**, 24 (1959).
203. Guberniev, M. A., and Torbochkina, L. I., 'Phosphorus compounds of some *Actinomycetes*, and their connection with antibiotic activity' (in Russian), *Antibiotiki*, **6**, 752 (1961).
204. Guberniev, M. A., Torbochkina, L. I., and Bondareva, N. S., 'The polyphosphate constitution of volutin granules in *Actinomyces erythreus*', *Antibiotiki*, **6**, 5 (1961).
205. Guillory, R. J., and Fisher, R. R., 'Measurement of simultaneous synthesis of inorganic pyrophosphate and adenosine triphosphate', *Analyt. Biochem.*, **39**, 170 (1971).
206. Gulik, A., 'Phosphorus as a factor in the origin of life', *Am. Sci.*, **43**, 479 (1955).
207. Guttes, E., Guttes, S., and Rusch, H. R., 'Morphological observations on growth and differentiation of *Physarum polycephalum* grown in pure culture', *Develop. Biol.*, **3**, 588 (1961).
208. Hagedorn, H., 'Elektronenmikroskopische Untersuchungen an *Streptomyces griseus* (Krainsky)', *Zentralbl. Bakt. Parasitenkd. Infektionskr. Hyg.*, *Abt. 2*, **113**, 234 (1959).
209. Halmann, M., *Analytical Chemistry of Phosphorus Compounds*, Wiley-Interscience, New York, 1972.

217

210. Hanes, S., and Isherwood, F., 'Separation of the phosphoric esters on the filter paper chromatogram', *Nature, Lond.*, **164**, 1107 (1949).
211. Harada, K., and Fox, S., 'The thermal polycondensation of free amino acids with polyphosphoric acid' (in Russian), in *Proiskhozhdenie Predbiologicheskikh Sistem* (*The Origins of Prebiological Systems*) (Translated into Russian), Mir, Moscow, 1966, p. 292.
212. Harold, F. M., 'Annumulation of inorganic polyphosphates in *Neurospora crassa*', *Fed. Proc.*, **18**, Part 1, 242 (1959).
213. Harold, F. M., 'Accumulation of inorganic polyphosphate in mutants of *Neurospora crassa*', *Biochim. Biophys. Acta*, **45**, 172 (1960).
214. Harold, F. M., 'Binding of inorganic polyphosphate to the cell wall of *Neurospora crassa*', *Biochim. Biophys. Acta*, **57**, 59 (1962).
215. Harold, F. M., 'Depletion and replenishment of the inorganic polyphosphate pool in *Neurospora crassa*', *J. Bacteriol.*, **83**, 1047 (1962).
216. Harold, F. M., 'Accumulation of inorganic polyphosphate in *Aerobacter aerogenes*. I. Relationship to growth and nucleic acid synthesis', *J. Bacteriol.*, **86**, 216 (1963).
217. Harold, F. M., 'Inorganic polyphosphate of high molecular weight from *Aerobacter aerogenes*', *J. Bacteriol.*, **86**, 885 (1963).
218. Harold, F. M., 'Enzymic and genetic control of polyphosphate accumulation in *Aerobacter aerogenes*', *J. Gen. Microbiol.*, **35**, 81 (1964).
219. Harold, F. M., 'Regulatory mechanism in the metabolism of inorganic polyphosphate in *Aerobacter aerogenes*', Colloq. *Int. CNRS, Marseilles*, **124**, 307 (1965).
220. Harold, F. M., 'Inorganic polyphosphates in biology: structure, metabolism and functions', *Bacteriol. Rev.*, **30**, 772 (1966).
221. Harold, R. L., and Harold, F. M., 'Mutants of *Aerobacter aerogenes* blocked in the accumulation of inorganic polyphosphate', *J. Gen. Microbiol.*, **31**, 241 (1963).
222. Harold, F. M., and Harold, R. L., 'Degradation of inorganic polyphosphates in mutants of *Aerobacter aerogenes*', *J. Bacteriol.*, **89**, 1262 (1965).
223. Harold, F. M., Harold, R. L., and Abrams, A., 'A mutant of *Streptococcus faecalis* defective in phosphate uptake', *J. Biol. Chem.*, **240**, 3145 (1965).
224. Harold, F. M., Harold, R. L., and Abrams A., 'Isolation and characterization of phosphate transport mutants of *Streptococcus faecalis*', *Proc. VI Int. Congr. Biochem., New York, 1964*. Symp. 111–125 (1964).
225. Harold, F. M., and Miller A., 'Intracellular localization of inorganic polyphosphate in *Neurospora crassa*', *Fed. Proc.*, **I** (abstr.) 135 (1961).
226. Harold, F. M., and Miller, A., 'Intracellular localization of inorganic polyphosphate in *Neurospora crassa*', *Biochim. Biophys. Acta*, **50**, 261 (1961).
227. Harold, F. M., and Sylvan, S., 'Accumulation of inorganic polyphosphate in *Aerobacter aerogenes*. II. Environmental control and the role of sulphur compounds', *J. Bacteriol.*, **86**, 222 (1963).
228. Hase, E., Miyachi, S., and Mihara, S., 'A preliminary note on the phosphorus compounds in chloroplasts and volutin granules isolated from *Chlorella* cells', in *Microalgae and Photosynthetic Bacteria*, 1963, p. 619, Ed H. Tamiya, Tokyo University of Tokyo Press.
229. Hattler, H., 'Zur Papierchromatographie der Phosphorverbindungen. I. Anorganische Phosphorverbindungen', *J. Chromatog.*, **1**, 389 (1958).
230. Hayashibe, M., 'Alcoholic fermentation by intact cells of baker's yeast. I. Inhibition of fermentation by phosphate', *J. Biochem.* (*Tokyo*), **44**, 543 (1957).
231. Heinonen, J., 'Intracellular concentration of inorganic pyrophosphate in the cells of *Escherichia coli*. A method for its determination', *Analyt. Biochem.*, **59**, 366 (1974).

218

232. Heldt, H. W., and Klingenberg, M., 'Endogenous nucleotides of mitochondria participating in phosphate transfer reactions as studies with ^{32}P-labelled ortho-phosphate and ultra-microscale ion-exchange chromatography', *Biochem. Z.*, **343**, 433 (1965).

233. Heller, J., 'O zwiarkach fosforowych wysekiej energii', *Postepy Biochim.*, **1**, 5 (1953).

234. Heller, J., 'Podstawowe reakcje w oddychanin rollin i swierat', *Postepy Biochem.*, **2**, 44 (1954).

235. Heller, J., Karpiak, S., and Zubikowa, J., 'Pyrofosforany w cicle tluszczowym wilczomleczka', *Spraw. Wroclaw. Tow. Nauk.*, **5**, 5 (1950).

236. Heller, J., Karpiak, St., and Subikowa, J., 'Inorganic pyrophosphate in insect tissue', *Nature, Lond.*, **166**, 187 (1950).

237. Heller, J., Karpiak, S., and Zubikowa, J., 'Pyrofosforany nieorganiczne u motyla wilczomleczka', *Spraw. Wroclaw. Tow. Nauk.*, **6**, 80 (1951).

238. Hendy, R. J., 'Resemblance of lomasomes of *Pythium debaryanum* to structures recently described in *Chara* and *Nitella*' *Nature, Lond.*, **209**, 1258 (1966).

239. Henneberg, W., 'Über das "Volutin" der die "metachromatischen Körperchen" in der Hefezelle', *Wochenschr. Brau.*, **4**, 36 (1915).

240. Hermann, E. C., and Schmidt, R. R., 'Synthesis of phosphorus-containing macro-molecules during synchronous growth of *Chlorella pyrenodosa*', *Biochim. Biophys. Acta.*, **95**, 63 (1965).

241. Herwerden, van M., 'Über des Volutin und seine chemische Zusammensetzung', *Chem. Zentralbl.*, **1**, 216 (1918).

242. Hoffman-Ostenhof, O., 'Metaphosphate and other phosphorus fractions in yeasts' (in Russian), *Ukr. Biokhim. Zh.*, **27**, 488 (1955).

243. Hoffman-Ostenhof, O., 'Some biological functions of the polyphosphates', *Colloq. Int. CNRS*, **106**, 640 (1962).

244. Hoffman-Ostenhof, O., Keck, K., and Kenedy, J., 'Zur Kenntniss des Phosphat-stoffwechsels der Hefe. II', *Mh. Chem.*, **86**, 616 (1955).

245. Hoffman-Ostenhof, O., Kenedy, J., Keck, K., Gabriel, O., and Schönfellingen, H. S., 'Ein neues Phosphat-übertragen- des Ferments aus Hefe', *Biochim. Biophys. Acta*, **14**, 285 (1954).

246. Hoffman-Ostenhof, O., Klima, A., Kenedy, J., and Keck, K., 'Zur Kenntniss des Phosphatstoffwechsels der Hefe. I. **86**, 604 (1955).

247. Hoffman-Ostenhof, O., and Slechta, L., 'Transferring enzymes in the metabolism of inorganic polyphosphates and pyrophosphate', *Proc. Int. Symp. Enzyme Chem.*, *Tokyo and Kyoto, 1957*, Pergamon Press, Oxford, 1958, Vol. 2, p. 180.

248. Hoffman-Ostenhof, O., and Weigert, W., 'Über die Mögliche Funktion des polymeren Metaphosphats als Speicher energie-reichen Phosphate in der Hefe', *Naturwissenschaften.*, **39**, 303 (1952).

249. Hohne, W. E., and Heitmann, P., 'Tripolyphosphate as a substrate of the inor-ganic pyrophosphatase from baker's yeast; the role of divalent metal ions', *Acta Biol. Med. Ger.*, **33**, 1 (1974).

250. Holzer, H., 'Photosynthese und Atmungakettenphosphorylierung', *Naturwissen-schaften*, **66**, 424 (1951).

251. Holzer, H., and Lynen, F., 'Uber den aeroben Phosphatbedarf der Hefe. III. Labil an die Struktur gebundenes Phosphat in lebender Hefe', *Justus Liebigs Ann. Chem.*, **569**, Heft 2, 128 (1950).

252. Khomlyak, M. N., and Grodzinskii, D. M., 'The heterogeneity of reserves of phosphorus metabolites in tomato leaves' (in Russian), in *Fiziologiya i Biokhimiya Kulturnykh Rastenii* (*The Physiology and Biochemistry of Cultivated Plants*), Naukova Dumka, Kiev, 1970, Vol. 2, No. 4.

253. Khomlyak, M. N., and Grodzinskii, D. M., 'The possible regulatory function of the heterogeneous nature of cellular reserves of phosphorus compounds' (in

Russian), in *Tezisy Dokl. na Vsesoyuznoi Konferentsii 'Regulyatsija Biokhimicheskikh Protsessov u Mikroorganizmov'* (*Abstracts of the All-Union Conference on 'The Regulation of Biochemical Processes in Microorganisms'*), Pushchino-na-Oke, 1972, p. 171.

254. Houlahan, M. B., and Mitchell, H. K., 'The accumulation of acid labile, inorganic phosphate in mutants of *Neurospora*', *Arch. Biochem.*, **19**, 257 (1948).

255. Huennekens, F. M., and Whiteley, H. R., 'Phosphoric acid anhydrides and other energy-rich compounds', in *Comparative Biochemistry*, Eds. Florkin, M., and Mason, H. S., Academic Press, New York, 1960, p. 107.

256. Hughes, D. E., Conti, S. E., and Fuller, R. C., 'Inorganic polyphosphate metabolism in *Chlorolium thiosulfatophilium*', *J. Bacteriol.*, **85**, 577 (1963).

257. Hughes, D. E., and Muhammed, A., 'The metabolism of polyphosphate in bacteria', in *Acides Ribonucleiques et Polyphosphates. Structure, Synthese et Fonctions. Colloq. Int. CNRS, Strasbourg, 1961*, CNRS, Paris, 1962, p. 591.

258. Iida, T., and Yamable, T., 'Chromatographic behaviour and structural units of condensed phosphates. II. Influences of developing solvents', *J. Chromatog.*, **54**, 413 (1971).

259. Indge, K. J., 'Metabolic lysis of yeast protoplasts', *J. Gen. Microbiol.*, **51**, 433 (1968).

260. Indge, K. J., 'The isolation and properties of the yeast cell vacuole', *J. Gen. Microbiol.*, **51**, 441 (1968).

261. Indge, K. J., 'Polyphosphates of the yeast cell vacuole', *J. Gen. Microbiol.*, **51**, 447 (1968).

262. Ingelman, B., 'Isolation of a phosphorus-rich substance of high molecular weight from *Aspergillus niger*', *Acta Chem. Scand.*, **1**, 776 (1947).

263. Ingelman, B., 'Isolation of polymetaphosphate of high molecular weight from *Aspergillus niger*', *Sven. Kem. Tidskr.*, **60**, 222 (1948).

264. Ingelman, B., and Malmgren, H., 'Enzymatic breakdown of polymetaphosphate. I', *Acta Chem. Scand.*, **1**, 422 (1947).

265. Ingelmann, B., and Malmgren, H., 'Enzymatic breakdown of poly-metaphosphate. II', *Acta Chem. Scand.*, **2**, 365 (1948).

266. Ingelman, B., and Malmgren, H., 'Enzymatic breakdown of polymetaphosphate. III', *Acta Chem. Scand.*, **3**, 157 (1949).

267. Ingelman, B., and Malmgren, H., 'Investigation of metaphosphats of high molecular weight isolated from *Aspergillus niger*', *Acta Chem. Scand.*, **4**, 478 (1950).

268. Inhülsen, D., und Niemeyer, R. 'Kondensierte Phosphate in *Lemna minor* L. und ihre Beziehugen zu den Nucleinsäuren', *Planta*, **124**, 159 (1975).

269. Ipata, F. L., and Felicioli, R. A., 'Formazione enzimatica di pyrofosfato nel lievito per fosforolisi dei polifosfati inorganici', *Bull. Soc. Ital. Biol. Sperim.*, **39**, 85 (1963).

270. Ivanov, V. B., 'A cytochemical investigation of the proteins of maize root tips' (in Russian), *Candidate's Thesis*, Moscow, 1964.

271. Iwamura, T., and Kuwashima, S., 'Formation of adenosine-5'-triphosphate from polyphosphate by a cell-free extract from *Chlorella*', *J. Gen. Appl. Microbiol.*, **10**, 83 (1964).

272. James, A. W., and Casida, L. E., 'Accumulation of phosphorus compounds by *Mucor racemosus*', *J. Bacteriol.*, **37**, 150 (1964).

273. Janakidevi, K., Dewey, V. C., and Kidder, C. W., 'The nature of the inorganic polyphosphates in a flagellated protozoa', *J. Biol. Chem.*, **240**, 1754 (1965).

274. Jeener, R., 'L'hetérogénité des granules cytoplasmiques dennés complementaires fournies par leur fractionnement en solution saline concentrée', *Biochim. Biophys. Acta*, **2**, 633 (1948).

275. Jeener, R., and Brachet, J., 'Recherches sur la synthese de l'acide pentosenucleique par les levures, relations avec la fermentation et la respiration', *Bull. Cl. Sci. Acad. Roy. Belg.*, **5**, 476 (1943).

276. Jeener, R. R., and Brachet, J., 'Recherches sur l'acide ribonucleique des levures', *Enzymologia*, **11**, 222 (1944).

277. Jeffrey, D. W., 'The formation of polyphosphate in *Banksia ornate*, an Australian heath plant', *Aust. J. Biol. Sci.*, **17**, 845 (1964).

278. Jensen, T. E., and Sicko, L. M., 'Phosphate metabolism in blue–green algae. 1. Fine structure of the "polyphosphate overplus" phenomenon in *Plectonema boryanum*', *Can. J. Microbiol.*, **20**, 1235 (1974).

279. Jordanov, J., Grigorov, J., and Bojadjieva-Michaijlova, A., 'Über die cytochemie der Polyphosphate bei *Aspergillus oryzae*', *Acta Histochem.*, **13**, 165 (1962).

280. Jordanov, J., Grigorov, J., and Bojadjieva-Michaijlova, A., 'An investigation into the cytochemical characteristics and metabolism of polyphosphate in *Aspergillus oryzae*' (in Bulgarian), *Izv. Inst. Morf. Bulg. Akad. Nauk*, **5**, 69 (1962).

281. Josan, V., Tavari, K. K., and Krishnan, P. S., 'Occurrence of acid-soluble and acid-insoluble polyphosphates in the rabbit liver', *Indian J. Biochem.*, **2**, 134 (1965).

282. Jost, K. H., 'Zur Struktur des Kurrollschen Silbersalzes $(AgPO_3)_x$. Vorläufige Mitteilung', *Z. Anorg. Allg. Chem.*, **296**, 154 (1958).

283. Jungnickel, F., 'Untersuchungen über die Aufnahme und Verwertung von Phosphatverbindungen durch Phosphatmangel-zellen von *Candida utilis*. 1. Die Aufnahme von Pyrophosphat', *Z. Allg. Mikrobiol.*, **11**, 367 (1971).

284. Jungnickel, F. 'Orthophosphorsäuremonoester-Phosphohydrolaseaktivität der repressiblen Pyrophosphat-Phosphohydrolase 1 von *Candida utilis*', *Z. Allg. Mikrobiol.*, **12**, 451 (1972).

285. Jungnickel, F., 'Significance of repressible polyphosphate phosphohydrolases in yeast and higher plants', in *Reaction Mechanisms and Control Properties of Phosphotransferases, Joint Biochemical Symposium UdSSR–DDR, Reinhardsbrunn*, Akademie-Verlag, Berlin, 1973, p. 87.

286. Juni, E., Kamen, M. D., Reiner, J. M., and Spiegelman, S., 'Turnover and distribution of phosphate compounds in yeast metabolism', *Arch. Biochem.*, **18**, 387 (1948).

287. Juni, E., Kamen, M. D., Spiegelman, S., and Wiame, J. M., 'Physiological heterogenity of metaphosphate in yeast', *Nature, Lond.*, **160**, 717 (1947).

288. Kaltwasser, H., and Schlegel, H. G., 'Nachweis und quantitative Bestimmung der Polyphosphate in wasserstoff-oxydierenden Bakterien', *Arch. Mikrobiol.*, **34**, 76 (1959).

289. Kaltwasser, H., 'Die Rolle der Polyphosphate im Phosphatstoffwechsel eines Knallgazbakterium (Hydrogenemonas Stamm 20)', *Arch. Mikrobiol.*, **41**, 282 (1962).

290. Kaltwasser, H., Vogt, G., and Schlegel, N. C., 'Polyphosphatsynthese während der Nitrat-Atmung von *Micrococcus denitrificans* Stamm 11', *Arch. Microbiol.*, **44**, 259 (1962).

291. Kamen, M. D., and Spiegelman, S., 'Studies on the phosphate metabolism of some unicellular organisms', *Cold Spring Harbor Symp. Quant. Biol.*, **13**, 151 (1948).

292. Kanai, R., and Simonis, W., 'Einbau von ^{32}P in Verschiedene Phosphatfraktionen besonders polyphosphate bei einzelligen Gründalgen (*Ankistrodesmus braunii*) im Licht und im Dunkeln', *Arch. Mikrobiol.*, **62**, 56 (1968).

293. Kanai, R., Miyachi, S., and Miyachi, S., 'Light-induced formation and mobilization of polyphosphate "C" in *Chlorella* cells', in *Microalgae and Photosynthetic Bacteria*, 1963, p. 613. Ed. H. Tamiya, University of Tokyo Press.

294. Kandler, O., 'Über die Beziehungen zwischen Phosphataushalt und Photosynthese' *Naturforsch.*, **128**, 271 (1957).
295. Karbe, K., and Jander, G., 'The metaphosphates', *Kolloid-Beih.*, **54**, 1 (1942).
296. Karl-Kroupa, E., 'Use of paper chromatography for differential analysis of phosphate mixture', *Analyt. Chem.*, **28**, 1091 (1956).
297. Katchman, B. J., and Fetty, W. O., 'Phosphorus metabolism in growing cultures of *Saccharomyces cerevisiae*', *J. Bacteriol.*, **69**, 607 (1955).
298. Katchman, B. J., Fetty, W. O., and Busch, K. A., 'Effect of cell division inhibition on the phosphorus metabolism of growing cultures of *Saccharomyces cerevisiae*', *J. Bacteriol.*, **77**, 331 (1959).
299. Katchman, B. J., and Smith, H. E., 'Diffusion of synthetic and natural polyphosphates', *Arch. Biochem. Biophys.*, **75**, 396 (1958).
300. Katchman, B. J., and van Wazer, J. R., 'The "soluble" and "insoluble" polyphosphates in yeast', *Biochim. Biophys. Acta*, **14**, 445 (1954).
301. Keck, K., and Stich, H., 'The widespread occurrence of polyphosphate in lower plants', *Ann. Bot.*, **21**, 611 (1957).
302. Keister, D. L., and Minton, N. J., 'Energy-linked reactions in photo-synthetic bacteria', *Arch. Biochem, Biophys.*, **147**, 330 (1972).
303. Keister, D. L., and Minton, N. J., 'ATP synthesis derived by inorganic pyrophosphate in *Rhodospirillum rubrum* chromatophores', *Biochem. Biophys. Res. Commun.*, **42**, 932 (1971).
304. Keister, D. L., and Yike, N. J., 'Studies on an energy-linked pyridine nucleotide transhydrogenase in photosynthetic bacteria', *Biochem. Biophys. Res. Commun.* **24**, 510 (1966).
305. Keister, D. L., and Yike, N. J., 'Energy-linked reactions in photosynthetic bacteria', *Arch. Biochem. Biophys.*, **121**, 425 (1967).
306. Keister, D. L., and Yike, N. J., 'Energy-linked reactions in photosynthetic bacteria', *Biochemistry*, **6**, 3847 (1967).
307. Kingsbury, N., and Masters, C. J., 'On the determination of component molecular weights in complex protein mixtures by means of disk electrophoresis', *Analyt. Biochem.*, **36**, 144 (1970).
308. Kirby, K. S., 'A new method for the isolation of ribonucleic acids from mammalian tissues', *Biochem. J.*, **64**, 405 (1956).
309. Kita, D. A., and Peterson, W. H., 'Forms of phosphorus in the penicillin-producing mold *Penicillium chrysogenum* Q-176', *J. Biol. Chem.*, **203**, 861 (1953).
310. Kitasato, T., 'Über Metaphosphatase', *Biochem. Z.*, **197**, 257 (1928).
311. Kitasato, T., 'Weiters Untersuchungen über die Metaphosphatase', *Biochem. Z.*, **201**, 206 (1928).
312. Klein, R. M., 'Nitrogen and phosphorus fractions, respiration and structure of normal and grown gall tissues of tomato', *Plant Physiol.*, **27**, 355 (1952).
313. Klemme, J.-H., and Gest, H., 'Regulatory properties of an inorganic pyrophosphatase from the photosynthetic bacterium *Rhodospirillum rubrum*', *Proc. Nat. Acad. Sci. USA*, **68**, 721 (1971).
314. Klungsöyr, L., King, J., and Cheldelin, V., 'Oxidative phosphorylations in *Acetobacter suboxydans*', *J. Biol. Chem.*, **227**, 135 (1957).
315. Kluyver, A. J., and van Niel, C. B., 'Prospects for a natural system of classification of bacteria', *Zentralbl. Bakteriol. Parasitenkd. Infektionskr. Hyg.*, Abt. 2., **94**, 369 (1936).
316. Knappe, E., Drews, G., and Böckel, V., 'Untersuchungen zum Phosphorstoffwechsel der Mycobakterien', *Zentralbl. Bakteriol. Parasitenkd. Infektionstr. Hyg.*, Abt. 2, **112**, 1 (1959).
317. Knaysi, G., 'Chemical composition of the granules of *Mycobacterium thamnopheos* with special reference to their biological identity and the chemical nature of volution', *J. Bacteriol.*, **77**, 532 (1959).

222

318. Koike, M., and Takeya, K., 'Fine structures of intracytoplasmic organelles of *Mycobacteria*', *J. Biophys. Biochem. Cytol.*, **9**, 597 (1961).
319. Kohiyama, M., and Saito, A., 'Stimulation of genetic transformation in *Bacillus subtilis* by biologically inactive DNA and polyphosphate', *Biochim. Biophys. Acta*, **41**, 280 (1960).
320. Kokurina, N. A., Kulaev, I. S., and Belozersky, A. N., 'Phosphorus compounds in some strains of *Actinomycetes*' (in Russian), *Mikrobiologiya*, **30**, 15 (1961).
321. Kölbel, H., 'Untersuchungen am *Mycobacterium tuberculosis*. V. Über die Vermehrung der Phosphatgranula', *Zentralbl. Bakteriol. Parasitenkd. Infektionstr. Hyg., Abt. 1*, **171**, 486 (1958).
322. Komkova, A. N., Pirogova, E. A., and Ashmarin, I. P., 'The formation and properties of complexes of some phosphorus-containing compounds with DNA and histones' (in Russian), *Biokhimiya*, **133**, 465 (1968).
323. König, H., and Winkler, A., 'Über Einschlusse in Bakterien und ihre Veränderung im Elektronenmikroskop', *Naturwissenschaften*, **35**, 136 (1948).
324. Konovalov, S. A., 'The conversion of phosphorus compounds in yeasts at various stages of alcoholic fermentation' (in Russian), *Mikrobiologiya*, **29**, 661 (1960).
325. Konovalov, S. A., and Grebeshkova, R. N., 'An investigation into certain phosphorus compounds in yeasts' (in Russian), *Mikrobiologiya*, **28**, 838 (1959).
326. Konovalova, L. V., and Vorob'eva, L. I., 'The effect of polymyxin M on the accumulation of lipids and polyphosphates by cells of *Propionibacterium shermamanii*' (in Russian), *Dokl. Vyssh. Shkoly, Ser. Biol.*, No. 7, 101 (1972).
327. Konoshenko, G. I., Umnov, A. M., Egorov, S. N., and Bobyk, M. A., 'The compartmentalization of polyphosphate-metabolizing enzymes in cells of *Neurospora crassa*' (in Russian), in *Tezisy Dokl. na Vsesoyuznoi Konferentsii po 'regulyatsii Biokhimicheskikh Protsessov u Mikroorganizmov* (*Abstracts of the All-Union Conference on 'The Regulation of Biochemical Processes in Microorganisms'*), Pushchino-na-Oke, 1972, p. 72.
328. Konoshenko, G. I., Umnov, A. M., and Kulaev, I. S., 'Characterization of *Neurospora crassa* polyphosphate phosphohydrolases', in *Reaction Mechanisms and Control Properties of Phosphotransferases, Joint Biochemical Symposium UdSSR–DRR, Reinhardsbrunn*, Akademie-Verlag, Berlin, 1973, p. 65.
329. Kornberg, A., 'The metabolic role of phosphorolysis and pyrophosphorolysis', in *Gorizonty Biokhimii* (*Horizons in Biochemistry*) (translation into Russian), Mir, Moscow, 1964, p. 191.
330. Kornberg, S. R., 'Tripolyphosphate and trimetaphosphate in yeast extracts', *J. Biol. Chem.*, **218**, 23 (1956).
331. Kornberg, S. R., 'Adenosine triphosphate synthesis from polyphosphate by an enzyme from *Escherichia coli*', *Biochim. Biophys. Acta*, **26**, 294 (1957).
332. Kornberg, S., and Kornberg, A., 'Inorganic tripolyphosphate and trimetaphosphate in yeast extracts', *Fed. Proc.*, **13**, 244 (1954).
333. Kornberg, A., Kornberg, S., and Simms, E., 'Metaphosphate synthesis by an enzyme from *Escherichia coli*', *Biochim. Biophys. Acta*, **20**, 215 (1956).
334. Kornberg, S., Lehman, L., Bessman, M., Simms, E., and Kornberg, A., 'Enzymatic cleavage of deoxyguanosine triphosphate to deoxyguanosine and trypolyphosphate', *J. Biol. Chem.*, **233**, 159 (1958).
335. Korchagin, V. B., 'Yeast metaphosphates and the chemical nature of volutin', (in Russian), *Candidate's Thesis*, Moscow, 1954.
336. Kossel, A., 'Über die Nucleinsäure', *Arch. Anat. Physiol.*, **23**, 157 (1893).
337. Kostlan, N. V., 'The properties of phosphorus compounds of the blue–green algae with respect to the nutrient status' (in Russian), in *Tezisy Dokl. na Vsesoyuznoi Konferentsii 'Regulyatsiya Biokhimicheskikh Protsessov u Mikroorganizmov'* (*Proceedings of the All-Union Conference on 'The Regulation of Biochemical Processes in Microorganisms'*), Pushchino-na-Oke, 1972, p. 74.

338. Kotyk, A., 'The influence of K$^+$ on phosphorylation processes in *Saccharomyces cerevisiae*' (in Russian), *Biokhimiya*, **23**, 737 (1958).

339. Kotyk, A., 'A contribution to the problem of phosphorylation in yeast', *Folia Microbiol.*, **4**, 23 (1959).

340. Kotyk, A., 'Molecular aspects of nonelectrolyte transport in yeasts', *Proceedings of the 3rd Int. Spec. Symp. on Yeasts, Otaniemi/Helsinki*, 1973, Part II, p. 103.

341. Kozel'tsev, V. L., 'The role of inorganic polyphosphates in the formation of adenyl nucleotides in a culture of *Brevibacterium ammoniagenus* ATSS-6872' (in Russian), *Author's Abstract, Candidate's Thesis*, Moscow, 1969.

342. Kozel'tsev, V. L., Debov, S. S., and Votrin, N. I., 'The role of inorganic polyphosphates in the formation of adenyl nucleotides in a culture of *Brevibacterium ammoniagenus*' (in Russian), *Vopr. Med. Khim.*, **5**, 602 (1969).

343. Krasheninnikov, I. A., Konoshenko, G. I., and Kulaev, I. S., 'Phosphorus compounds in the cell walls of mycelia of *Neurospora crassa*' (in Russian), *Dokl. Akad. Nauk SSSR*, **177**, 711 (1967).

344. Krasheninnikov, I. A., Konoshenko, G. I., Mansurova, S. E., and Kulaev, I. S., 'Localization of polyphosphates and polyphosphatases in *Neurospora crassa*', in *Reaction Mechanisms and Control Properties of Phosphotransferases, Joint Biochemical Symposium UdSSR–DDR, Reinhardsbrunn*, Akademie-Verlag, Berlin, 1973, p. 36.

345. Krasheninnikov, I. A., Kulaev, I. S., Konoshenko, G. I., and Biryuzova, V. I., 'Phosphorus compounds of *Neurospora crassa* mycelia' (in Russian), *Biokhimiya*, **33**, 800 (1968).

346. Krasil'nikov, N. A., *A Guide to Bacteria and Actinomycetes* (in Russian), Izd. Akad. Nauk SSSR, Moscow, 1949.

347. Krasil'nikov, N. A., *Rukovodstvo po Mikrobiologii, Klinike i Epidemologii Infektsiomykh Zabolevanii (A Handbook of the Microbiology, Treatment, and Epidemiology of Infectious Diseases)*, Izd. Meditsina, Moscow, 1962.

348. Krishnan, P., 'Apyrase, pyrophosphatase and metaphosphatase of *Penicillium chrysogenum*', *Arch. Biochem. Biophys.*, **37**, 224 (1952).

349. Krishnan, P., and Bajaj, V., 'The polyphosphatase of *Aspergillus niger*. I. The nonidentity of metaphosphatase with apyrase and pyrophosphatase', *Arch. Biochem. Biophys.*, **42**, 174 (1953).

350. Krishnan, P., and Bajaj, V., 'The polyphosphatases of *Aspergillus niger*. II. Metabolic role of the polyphosphatases', *Arch. Biochem. Biophys.*, **47**, 39 (1953).

351. Krishnan, P. S., Damle, S. P., and Bajaj, V., 'Studies on the role of "metaphosphate" in molds. II. The formation of "soluble" and "insoluble" metaphosphate" *Aspergillus niger*', *Arch. Biochem. Biophys.*, **67**, 35 (1957).

352. Kritsky, M. S., and Belozerskaya, T. A., 'Nucleic acids, nucleotides and polyphosphates in normally and abnormally developed sporophores of mushroom *Lentinus tigrinus*', *Abstr. VII Int. Congr. Biochem. Tokyo*, Symp. J5, 1020. comm. j–295 (1967).

353. Kritsky, M. S., and Belozerskaya, T. A., 'Acid-soluble nucleotides and nucleic acids in the fruiting bodies of the fungus *Lentinus tigrinus* with various morphogenetic features' (in Russian), *Biokhimiya*, **33**, 874 (1968).

354. Kritsky, M. S., Belozerskaya, T. A., and Kulaev, I. S., 'The interrelationship of RNA and polyphosphate metabolism in some fungi' (in Russian), *Dokl. Akad. Nauk SSSR*, **180**, 746 (1968).

355. Kritsky, M. S., and Chernysheva, E. K., 'The study of polyphosphates and polyphosphate depolymerases of *Neurospora crassa*', *Abstr. 7th Meeting FEBS, Varna*, 1971.

356. Kritsky, M. S., and Chernysheva, E. K., 'The study of polyphosphate depolymerase activity of *Neurospora crassa*', in *Reaction Mechanisms and Control*

224

Properties of Phosphotransferases, Joint Biochemical Symposium UdSSR–DDR, Reinhardsbrunn, Akademie-Verlag, Berlin, 1973, p. 49.

357. Kritsky, M. S., Chernysheva, E. K., and Kulaev, I. S., 'The correlation between certain inorganic polyphosphate fractions and RNA in Neurospora crassa' (in Russian), Dokl. Akad. Nauk SSSR, 192, 1166 (1970).

358. Kritsky, M. S., Chernysheva, E. K., and Kulaev, I. S., Polyphosphate depolymerase activity in cells of the fungus Neurospora crassa' (in Russian), Biokhimiya, 37, 983 (1972).

359. Kritsky, M. S., and Kulaev, I. S., 'Acid-soluble nucleotides of the fruiting bodies of the mushroom Agaricus bisporus' (in Russian), Biokhimiya, 28, 694 (1963).

360. Kritsky, M. S., Kulaev, I. S., Klebanova, L. M., and Belozersky, A. N., 'Two modes of phosphorus transport in fruiting bodies of the mushroom' (in Russian), Dokl. Akad. Nauk SSSR, 160, 949 (1965).

361. Kritsky, M. S., Kulaev, I. S., Maiorova, I. P., Fais, D. A., and Belozersky, A. N., 'The movement of phosphorus in fruiting bodies of the mushroom Agaricus bisporus' (in Russian), Biokhimiya, 30, 778 (1965).

362. Krüger-Thiemer, E., and Lembke A., 'Zur Definition der Mykobakteriengranula', Naturwissenschaften, 41, 146 (1954).

363. Kuhl, A., 'Die Biologie der Kondensierten anorganischen Phosphate', Ergeb. Biol., 23, 144 (1960).

364. Kuhl, A., 'Inorganic phosphorus uptake and metabolism', in Physiology and Biochemistry of Algae, Ed. Lewin, R. A., Academic Press, New York and London, 1962, p. 211.

365. Kulaev, I. S., 'The role of polyphosphates in the developmental processes of some mould fungi (in Russian), Candidate's Thesis, Moscow, 1956.

366. Kulaev, I. S., Polyphosphate–ribonucleic complexes from Aspergillus niger' (in Russian), Referaty Soobshch. na VIII Mendeleevskom s'ezde (Abstracts of Communications to the Eighth Mendeleev Congress), Izd. Akad. Nauk SSSR, Moscow, 1958, p. 82.

367. Kulaev, I. S., 'The physiological significance of polyphosphates' (in Russian), Referaty Nauchnoi Sessii po Nukleinovym Kislotam Rastenii (Abstracts of the Scientific Session on Plant Nucleic Acids), Ufa, Izd. Akad. Nauk SSSR, 1958, p. 6.

368. Kulaev, I. S., 'Polyphosphate metabolism in some fungi' (in Russian), Referat Sektsionnykh Soobshchenii V Mezhdunarodnogo Biokhimicheskogo Kongressa v Moskve (Abstracts of Section Papers at the Fifth International Biochemical Congress, Moscow), Sections 14–28 2, 37, (1961).

369. Kulaev, I. S., 'The uptake of orthophosphate from the medium by Penicillium chrysogenum mycelia' (in Russian), Dokl. Akad. Nauk SSSR, 155, 204 (1964).

370. Kulaev, I. S., 'Some features of the evolution of phosphorus and nucleic acid metabolism' (in Russian), Tezisy Dokl. na 1. Vsesoyuzn. Biokhim. s'ezde (Abstracts of the First All-Union Biochemical Congress), No. 1, 12 (1964).

371. Kulaev, I. S., 'Some aspects of the mode of uptake of orthophosphate by cells of lower organisms' (in Russian), Tezisy Dokl. na 1. Vsesoyuzn. Biokhim. s'ezde (Abstracts of the First All-Union Biochemical Congress), No. 1, 77 (1964).

372. Kulaev, I. S., 'On the mechanism of the orthophosphate absorption by the cells of some organisms', Abstr. 6th Int. Congr. Biochem., New York, X-37, 1964, p. 782.

373. Kulaev, I. S., 'Inorganic polyphosphates and the evolution of phosphorus metabolism' (in Russian), in Abiogenez i Nachalnye Stadii Evolyutsii (Abiogenesis and the Early Stages of Evolution), Izd. Nauka, Moscow, 1968, p. 97.

374. Kulaev, I. S., 'Inorganic polyphosphates in evolution of phosphorus metabolism', in Molecular Evolution, Eds. Buvet, R., and Ponnamperuma, C., North-Holland, Amsterdam, 1971, Vol. 1, p. 458.

375. Kulaev, I. S., 'The role of inorganic polyphosphates in abiogenesis and the

evolution of energetic systems' (in Russian), in *Evoyutsionnaya Biokhimiya* (*Evolutionary Biochemistry*) [in series *Novoe v Zhizni, Nauke, i Tekhnike* (*Recent Advances in Life, Science, and Technology*), Ser. *Biol.*, No. 4], Izd. Znanie, Moscow, 1973, p. 46.

376. Kulaev, I. S., 'The enzymes of polyphosphate metabolism in protoplasts and some subcellular structures of *Neurospora crassa*', in *Yeast Mould and Plant Protoplasts*, Ed. Villanueva, J., Academic. Press, London, 1973, p. 259.

377. Kulaev, I. S., 'Phosphotransferase reactions in the inorganic polyphosphate metabolism', in *Reaction Mechanisms and Control Properties of Phosphotransferases, Joint Biochemical Symposium UdSSR–DDR, Reinhardsbrun*, Akademie-Verlag. Berlin, 1973, p. 36.

378. Kulaev, I. S., 'The role of inorganic polyphosphates in chemical and biological evolution', in *The Origin of Life, Evolutionary Biochemistry*, Ed. Dose K. Fox S. W., Deborin G. A., Pavlovskaya, T. E., Plenum Press, New York and London, 1974, p. 271; Kulaev, I. S., 'Inorganic polyphosphates in chemical and biological evolution', (in Russian), in *Problemy Vozniknovenie i Sushchnosti Zhizni* (*Problems of the Origin and Nature of Life*), Ed. Oparin, A. I., Izd. Nauka, Moscow, 1973, p. 176.

379. Kulaev, I. S., *Neorgancheskie Polifosfaty i ib Fiziologicheskaya Rol'* (*The Physiological Role of Inorganic Polyphosphates*) XXX Bakhovskol chtenie (XXX Bach's Reading) Izd. Nauka, Moscow, 1975.

380. Kulaev, I. S., 'Biochemistry of inorganic polyphosphates', in *Review of Physiology, Biochemistry, and Pharmacology.*, Springer-Verlag, Berlin, Heidelberg and New York, **73**, 131, (1975), Ed. M. Holzer a F. Lyner.

381. Kulaev, I. S., Afanas'eva, T. P., Krasheninnikov, I. A., and Mansurova S. É., 'Phosphorus compounds in protoplasts of *Endomyces magnusii* and *Neurospora crassa* and their subcellular structures', *Acta Fac. Med. Univ. Brun.*, **37**, 81 (1970).

382. Kulaev, I. S., and Afanas'eva, T. P., 'Localization and physiological role of inorganic polyphosphates in the cells of yeasts', *Antonie van Lceuwenhoek, J. Microbiol. Serol.*, **35**, Suppl. 'Yeasts Symposium', 13 (1969).

383. Kulaev, I. S., and Afanas'eva, T. P., 'The physiological role of inorganic polyphosphates in the yeast *Endomyces magnusii* (in Russian), *Dokl. Akad. Nauk SSSR*, **192**, 668 (1970).

384. Kulaev, I. S., Afanas'eva, T. P., and Belikova, M. P., 'The localization of inorganic polyphosphates and nucleoside polyphosphates in cells of the yeast *Endomyces magnusii*' (in Russian), *Biokhimiya*, **32**, 539 (1967).

385. Kulaev, I. S., Afanas'eva, T. P., and Belozersky, A. N., 'Polyphosphate phosphohydrolases in *Endomyces magnusii*' (in Russian), *Fiziol. Rast.*, **19**, 984 (1972).

386. Kulaev, I. S., and Belozersky, A. N., 'An investigation using ^{32}P of the physiological role of polyphosphates in the development of *Aspergillus niger*' (in Russian) *Biokhimiya*, **22**, 587 (1957).

387. Kulaev, I. S., and Belozersky, A. N., 'An electrophoretic investigation into the polyphosphate–ribonucleic acid complexes from *Aspergillus niger*' (in Russian), *Dokl. Akad. Nauk SSSR*, **120**, 1080 (1958).

388. Kulaev, I. S., and Belozersky, A. N., 'Condensed inorganic phosphates in the metabolism of living organisms, Part 1', (in Russian), *Izv. Akad. Nauk SSSR, Ser. Biol.*, No. 3, 354 (1962).

389. Kulaev, I. S., and Belozersky, A. N., 'Condensed inorganic phosphates in the metabolism of living organisms, Part 2', (in Russian), *Izv. Akad. Nauk SSSR, Ser. Biol.*, No. 4, 502 (1962).

390. Kulaev, I. S., Belozersky, A. N., Kritsky, M. S., and Kokurina, N. A., 'Polyphosphates of the fruiting bodies of the mushroom and false morel' (in Russian), *Dokl. Akad. Nauk SSSR*, **130**, 667 (1960).

391. Kulaev, I. S., Belozersky, A. N., and Mansurova, S. É., 'Polyphosphate metabo-

226

lism in deepcultured *Penicillium chrysogenum'* (in Russian), *Biokhimiya*, **24**, 253 (1959).

392. Kulaev, I. S., Belozersky, A. N., and Ostrovskii, D. N., 'Acid-soluble phosphorus compounds in *Penicillium chrysogenum* Q-176 under different growth conditions' (in Russian), *Biokhimiya*, **26**, 188 (1961).

393. Kulaev, I. S., and Bobyk, M. A., 'The ferment of biosynthesis of high-polymer polyphosphates–1,3-diphosphoglycerate: polyphosphatephosphotransferase', *Proc. X Int. Congr. Microbiol., Mexico,* 1970, p. 157.

394. Kulaev, I. S., and Bobyk, M. A., 'The detection of a new enzyme, 1,3-phosphoglycerate–polyphosphate phosphotransferase, in *Neurospora crassa'* (in Russian), *Biokhimiya*, **36**, 426 (1971).

395. Kulaev, I. S., Bobyk, M. A., Nikolaev, N. N., Sergeev, N. S., and Uryson, S. O., 'The polyphosphate-synthesizing enzymes of some fungi and bacteria' (in Russian), *Biokhimiya*, **36**, 943 (1971).

396. Kulaev, I. S., Bobyk, A. M., Tobek, I., and Hoštalek, Z., 'The possible role of high-molecular polyphosphates in the biosynthesis of chlortetracycline by *Streptomyces aureofaciens'* (in Russian), *Biokhimiya*, **41**, 343 (1976).

397. Kulaev, I. S., Fais, D. A., Metlitskaya, A. Z., and Krasheninnikov, I. A., 'The rapid uptake of ^{32}P into mycelia of *Penicillium chrysogenum'* (in Russian), *Dokl. Akad. Nauk SSSR*, **159**, 198 (1964).

398. Kulaev, I. S., Grigor'eva, S. P., and Vorob'eva, G. I., 'Polyphosphate phosphohydrolase activity in cells of the petroleum hydrocarbon-assimilating yeast *Candida quillermondii* under different growth conditions' (in Russian), *Prikl. Biokhim. Mikrobiol.*, **10** p. 396 (1974).

399. Kulaev, I. S., and Konoshenko, G. I., 'The detection and properties of polyphosphatases of *Neurospora crassa* which hydrolyse inorganic polyphosphates to orthophosphate' (in Russian), *Biokhimiya*, **36**, 1175 (1971).

400. Kulaev, I. S., and Konoshenko, G. I., 'The localization of 1,3-diphosphoglycerate–polyphosphate phosphotransferase in cells of *Neurospora crassa'* (in Russian), *Dokl. Akad. Nauk SSSR*, **200**, 477 (1971).

401. Kulaev, I. S., Konoshenko, G. I., and Umnov, A. M., 'The localization of polyphosphatases which hydrolyse polyphosphate to orthophosphate, in subcellular structures of *Neurospora crassa'* (in Russian), *Biokhimiya*, **37**, 227 (1972).

402. Kulaev, I. S., Konoshenko, G. I., Chernysheva, E. K., and Kritsky, M. S., 'The localization and possible physiological role of polyphosphate depolymerase in cells of *Neurospora crassa'* (in Russian), *Dokl. Akad. Nauk SSSR*, **206**, 233 (1972).

403. Kulaev, I. S., Krasheninnikov, I. A., Afanas'eva, T. P., Mansurova, S. É, and Uryson, S. O., The absence of inorganic polyphosphates from mitochondria' (in Russian), *Dokl. Akad. Nauk SSSR*, **177**, 229 (1967).

404. Kulaev, I. S., Krasheninnikov, I. A., and Afanas'eva, T. P., 'The localization and nature of the polyphosphates and polyphosphatases present in fungal cells' (in Russian), *Dokl. Akad. Nauk SSSR*, **190**, 1238 (1970).

405. Kulaev, I. S., Krasheninnikov, I. A., and Kokurina, N. K., 'The localization of inorganic polyphosphates and nucleotides in mycelia of *Neurospora crassa'* (in Russian), *Biokhimiya*, **31**, 850 (1966).

406. Kulaev, I. S., Krasheninnikov, I. A., and Polyakov, V. Yu., 'Nuclear phosphorus compounds of *Neurospora crassa'* (in Russian), in *Trudy II Konferentsii po Kletochnym Yadram* (*Proceedings of the Second Conference on Cell Nuclei*), Ed. Zbarsky I. V., Izd. Nauka *Riga*, 1970.

407. Kulaev, I. S., Krasheninnikov, I. A., and Tyrsin, Yu. A., 'Nucleic acids, polyphosphates, and some polyphosphate-metabolizing enzymes in synchronous cultures of *Schizosaccharomyces pombe'* (in Russian), *Mikrobiologiya*, **38**, 613 (1973).

408. Kulaev, I. S., Krasheninnikov, I. A., and Urbanek, H., 'The localization and

biosynthesis of inorganic polyphosphates and nucleoside polyphosphates in mycelia of *Neurospora crassa'*(in Russian), *Tezisy Mezhdunarod. Mikrobiol. Kongressa* (*Abstracts of the International Microbiological Congress*), Moscow, 1966, p. 96.

409. Kulaev, I. S., Kritsky, M. S., and Belozersky, A. N., 'The metabolism of polyphosphates and other phosphorus compounds during the development of fruiting bodies of the mushroom' (in Russian), *Biokhimiya*, **25**, 735 (1960).

410. Kulaev, I. S., Mansurova, S. É. Afanas'eva, T. P., Krasheninnikov, I. A., Kholodenko, V. P., Konoshenko, G. I., and Uryson, S. O., 'Nucleoside polyphosphates and inorganic polyphosphates in mitochondria of different origins' (in Russian), *III Vsesoyuzni Simpozium po Mitokhondryam* (*Third All-Union Symposium on Mitochondria*), *Collected Reports*, Izd. Nauka, Moscow, 1968, p. 15.

411. Kulaev, I. S., Mansurova, S. É., Shadi, A., Rubtsov, P. M., Konoshenko, G. I., Tsydendambaev, V. D., Efremovich, N. V., and Gaziyants, S. M., 'The biosynthesis of pyrophosphate and high-molecular polyphosphates in the chromatophores and chloroplasts of lower and higher plants' (in Russian), *XII Mezhdunar. Botanicheskii Kongress, Leningrad* (*Twelfth International Botanical Congress, Leningrad*), 1975, Vol. 2, p. 427.

412. Kulaev, I. S., and Mel'gunov, V. I., 'Orthophosphate uptake by the adenine-deficient mutant of *Neurospora crassa*, 286–10 HSA' (in Russian), *Biokhimiya*, **32**, 913 (1967).

413. Kulaev, I. S., Mel'gunov, V. I., and Yakovleva, L. A., 'The effect of 8-azaadenine on phosphorus metabolism in adenine-deficient mutants of *Neurospora crassa*' (in Russian), *Dokl. Akad. Nauk SSSR*, **194**, 1202 (1970).

414. Kulaev, I. S., and Okorokov, L. A., 'On the primary acceptors of orthophosphate in *Penicillium chrysogenum* cells', *Abstr. 3rd Meeting, FEBS Warsaw*, M–60, 1966, p. 148.

415. Kulaev, I. S., and Okorokov, L. A., 'The initial steps in the entry of orthophosphate from the medium into the mycelia of *Penicillium chrysogenum*' (in Russian), *Biokhimiya*, **32**, 695 (1967).

416. Kulaev, I. S., and Okorokov, L. A., '"Pseudoorthophosphate" in *Penicillium chrysogenum* cells', *Abstr. VII Int. Congr. Biochem., Tokyo*, **5**, J–294, 1967, p. 1020.

417. Kulaev, I. S., and Okorokov, L. A., 'The variable nature of the uptake of phosphate from the medium by *Penicillium chrysogenum*' (in Russian), *Dokl. Akad. Nauk SSSR*, **179**, 982 (1968).

418. Kulaev, I. S., Okorokov, L. A., Fen-Yui-Tsyan, and Kuritsyna, G. A., 'Organophosphorus compounds in *Penicillium chrysogenum*, the phosphorus of which is determined as orthophosphate' (in Russian), *Dokl. Akad. Nauk SSSR*, **171**, 473 (1966).

419. Kulaev, I. S., Okorokov, L. A., Mansurova, S. É., and Uryson, S. O., 'The constitution of the phosphorus compounds present in cells' (in Russian), in *Mekhanismy Dykhaniya, Fotosinteza i Fiksatsii Azota* (*Mechanisms of Respiration, Photosynthesis, and Nitrogen Fixation*), Izd. Nauka, Moscow, 1967, p. 105.

420. Kulaev, I. S., Ostrovskii, D. N., and Belozersky, A. N., 'The problem of the initial products in the uptake of orthophosphate from the medium by mycelia of *Penicillium chrysogenum*' (in Russian), *Dokl. Akad. Nauk SSSR*, **135**, 467 (1960).

421. Kulaev, I. S., Polonskii, Yu. S., Khlebalina, O. I., and Chigirev, V. S., 'The mechanism of the uptake of orthophosphate from the medium by mycelia of *Penicillium chrysogenum*' (in Russian), *Biokhimiya*, **29**, 759 (1964).

422. Kulaev, I. S., and Rozhanets, V. V., 'Inorganic polyphosphates in the brain of the rat' (in Russian), *Dokl. Vyssh. Shkoly 'Biologicheskie Nauki'*, No. 11, 119 (1973).

423. Kulaev, I. S., Rozhanets, V. V., Kobylanskii, A., and Filippovich, Yu. V.,

'Inorganic polyphosphates in insects' (in Russian), *Evol. Biokhim. Fiziol.*, **10**, 147 (1974).

424. Kulaev, I. S., Rubtsov, P. M., Brommer, B., and Yazykov, A. I., 'High-molecular polyphosphates in the development of individual cells of *Acetabularia crenulata*' (in Russian), Fiziol. Rast. , **22**, 537 (1975).

425. Kulaev, I. S., Shadi, A., and Mansurova, S. É., 'Polyphosphates in the phototrophic bacterium *Rhodospirillum rubrum* under various culture conditions' (in Russian), *Biokhimiya*, **39**, 656 (1974).

426. Kulaev, I. S., Szymona, O. V., and Bobyk, M. A., 'The biosynthesis of inorganic polyphosphates in *Neurospora crassa*' (in Russian), *Biokhimiya*, **33**, 419 (1968).

427. Kulaev, I. S., and Skryabin, K. G., 'Reactions of abiogenic transphosphorylation involving high-polymer polyphosphates and their role in the phosphorus metabolism evolution', *Abstracts of Symposium 'Origin of Life and Evolutionary Biochemistry'*, *Varna*, 1971, p. 22.

428. Kulaev, I. S., and Skryabin, K. G., 'Abiogenic transphosphorylation reactions of high-molecular polyphosphates, and their role in the evolution of phosphorus metabolism' (in Russian), *Evol. Biokhim. Fiz.*, No. 2, 533 (1974).

429. Kulaev, I. S., and Urbanek, H., 'The dependence of ^{32}P uptake by an adenine-deficient mutant of *Neurospora crassa* on the adenine content of the medium and on the presence of 8-azaadenine' (in Russian), *Dokl. Akad. Nauk SSSR*, **169**, 958 (1966).

430. Kulaev, I. S., and Uryson, S. O., 'Free nucleotides and other phosphorus compounds in ergot, *Claviceps paspali*' (in Russian), *Biokhimiya*, **30**, 282 (1965).

431. Kulaev, I. S., and Vagabov, V. M., 'The effect of growth conditions on the metabolism of inorganic polyphosphates and other phosphorus compounds in *Scenedesmus obliquus*' (in Russian), *Biokhimiya*, **32**, 253 (1967).

432. Kulaev, I. S., Vagabov, V. M., and Tsiomenko, A. B., 'On the correlation of the accumulation of polysaccharides and condensed inorganic polyphosphates of the yeast', *Abstracts of International Symposium 'Yeasts as Models in Sciences and Techniques'*, Smolenice, Czechoslovakia, 1971, p. 229.

433. Kulaev, I. S., Vagabov, V. M., and Tsiomenko, A. B., 'The correlation between the accumulation of polysaccharides by the cell walls and some polyphosphate fractions in yeasts' (in Russian), *Dokl. Akad. Nauk SSSR*, **204**, 734 (1972).

434. Kulaev, I. S., Vagabov, V. M. and Tsiomenko, A. B., 'Features of the biosynthesis of cell-wall components in yeasts' (in Russian), in *Biokhimiya i Fiziologiya Mikroorganizmov (The Biochemistry and Physiology of Microorganisms)*, 1975, p. 63. Ed. Skryabin G. K., Pushchino na oke. Izd. Nauka.

435. Kulaev, I. S., Vagabov, V. M. Tsiomenko, A. B., and Shabalin, Yu. A., 'The action of inhibitors of protein and carbohydrate metabolism on the biosynthesis of polysaccharides and polyphosphates by yeasts protoplasts' (in Russian), *Tezisy Dokl. na XII Mezhdunarodnom Botanicheskom Kongresse Leningrad (Abstracts of the Twelfth International Botanical Congress, Leningrad)*, 1975, Vol. 2, p. 386.

436. Kulaev, I. S., Valikhanov, M. N., and Belozersky, A. N., 'Possible routes of conversion of phytin in higher plants' (in Russian), *Dokl. Akad. Nauk SSSR*, **159**, 668 (1964).

437. Kulaev, I. S., Vorob'eva, L. I., Konovalova, L. V., Bobyk, M. A., Konoshenko, G. I., and Uryson, S. O., 'Polyphosphate-metabolizing enzymes in the development of *Propionibacterium shermanii*, under normal conditions and in the presence of polymyxin M' (in Russian), *Biokhimiya*, **38**, 712 (1973).

438. Kulyash, Yu. V. 'The relationship of respiration to certain features of phosphorus metabolism in *Staphylococcus aureus* 209-p at various temperatures' (in Russian), in *Regulyatsiya Biokhimicheskikh Protsessov u Mikroorganizmov (The Regulation of Biochemical Processes in Microorganisms)*, 1972, p. 91. Ed. Kulaev I. S., Izd, Nauka, Pushchino na oke.

439. Kura, G., and Ohashi, S., 'Chromatographic separation of cyclic phosphates by means of an anion exchange dextran gel', *J. Chromatog.*, **56**, 111 (1971).

440. Lampen, J. O., Kuo, S.-C., and Liras, P., 'Synthesis of external glycoprotein enzymes by yeast', *Proceedings of the 3rd Int. Spec. Symp. on Yeasts. Otanienii/ Helsinki*, 1973, Part II, p. 129.

441. Langen, P., 'Über Polyphosphate in Ostesee-Algen', *Acta Biol. Med. Ger.* **1**, 368 (1958).

442. Langen, P., and Liss, E., 'Über Bindung und Umsatz der Polyphosphate der Hefe', *Biochem. Z.*, **330**, 455 (1958).

443. Langen, P., 'Vorkommen und Bedeutung von Polyphosphates in Organismen', *Biol. Rundsch.*, 145 (1965).

444. Langen, P., and Liss, E., 'Über die Polyphosphates der Hefe', *Naturwissenschaften*, **45**, 191 (1958).

445. Langen, P., and Liss, E., 'Über den Polyphosphat-Stoffwechsel der Hefe', *Proc. 4th Int. Congr. Biochem., Vienna*, **17**, 35 (1958).

446. Langen, P., and Liss E., 'Differenzierung des Orthophosphates der Hefezelle', *Biochem. Z.*, **322**. 403 (1960).

447. Langen, P., Liss, E., and Lohmann, K., 'Art, Bildung und Umsatz der Polyphosphate der Hefe', in *Acides Ribonucleiques et Polyphosphates. Structure, Synthese et Fonctions, Colloq. Int. CNRS, Strasbourg, 1961*, CNRS, Paris, 1962, p. 603.

448. Lee, L. P. K., and Kosicki, G. W., 'Citrate synthetase interaction with polyphosphate derivatives', *Biochim. Biophys. Acta*, **139**, 195 (1967).

449. Le Port, L., Etaix, E. Godin, F., Leduc, P., and Buvet, R., 'Archetypes of present day process of transphosphorylation, transacylation and peptide synthesis', in *Molecular Evolution, Vol. 1, Chemical Evolution and the Origin of Life*, Eds. Buvet, R., and Ponnamperuma, C., North-Holland, Amsterdam, 1971, p. 197.

450. Levchuk, T. P., Gershkovich, V. G., and Lozinov, A. B., 'Some features of phosphorus metabolism in *Candida* yeasts when grown on *n*-alkanes' (in Russian), *Mikrobiol. Sint.*, No. 7, 22 (1969).

451. Levinson, S. L., Jacons, L. H., Krulwich, T. A., and Li, H. C., 'Purification and characterization of a polyphosphate kinase from *Arthrobacter atrocyaneus*', *J. Gen. Microbiol.*, **88**, 65 (1975).

452. Lichko, L. P., and Okorokov, L. A., 'The compartmentalization of magnesium ions and orthophosphate in cells of *Saccharomyces carlsbergensis*' (in Russian), *Dokl. Akad. Nauk SSSR*, **227**, 756 (1976).

453. Liebermann, L., 'Über das Nuclein der Hefe und Künetliche Darstellung eines Nuclein's Eiweiss und Metaphosphorsäure', *Ber. Deut. Chem. Ges.*, **21**, 598 (1888).

454. Liebermann, L., 'Nachweis der Metaphosphorsäure im Nuclein der Hefe', *Pflüger's Arch.*, **47**, 155 (1890).

455. Liebermann, I., and Eto, W. H., 'Inorganic triphosphate synthesis by muscle adenylate kinase'. *J. Biol. Chem.* **219**, 307 (1956).

456. Li Hend Chun, and Brown, C., 'Orthophosphate and histone dependent polyphosphatkinase from *E. coli*', *Biochem. Biophys. Res. Commun.*, **53**, 875 (1973).

457. Lindeberg, G., and Malmgren, H., 'Enzymatic breakdown of polymetaphosphate. VI. Influence of nutritional factors on the polymetaphosphatase production of *Aspergillus niger*', *Acta Chem. Scand.*, **6**, 27 (1952).

458. Lindegreen, C. C., 'The origin of volutin on the chromosomes. Its transfer to the nucleolus and suggestions concerning the significance of this phenomenon', *Proc. Nat. Acad. Sci. USA*, **34**, 187 (1948).

459. Lindegreen, C. C., 'The relation of metaphosphate formation to cell division in yeast', *Exp. Cell. Res.*, **2**, 275 (1951).

460. Lindebaum, S., Peters, T., and Rieman, W., 'Analyses of mixtures of the con-

230

densed phosphates by ion-exchange chromatography', *Analyt. Chim. Acta*, **11**, 530 (1954).

461. Lindgren, D. C., Pfister, R. M., and Merrick, J. M., 'Structure of poly-β-hydroxy-buturic acid granules', *J. Gen. Microbiol.*, **34**, 441 (1964).

462. Lipmann, F., 'The contemporary stage of evolution of biosynthesis and the developments which preceded it' (translated into Russian), in *Proiskhozhdenie predbiologichkikh Sistem (The Origins of Prebiological Systems)*, Mir, Moscow, 1966, p. 261.

463. Lipmann, F., 'Gramicidin S and Tyrocidine biosynthesis: primitive process of sequential addition of amino acids on polyenzymes', in *Molecular Evolution, Vol. 1, Chemical Evolution and the Origin of Life* Eds. Buvet, R., and Ponnam-peruma, C., North-Holland, Amsterdam, 1971, p. 381.

464. Liss, E., and Langen, P., 'Zur Charakterisierung der Polyphosphate der Hefe', *Naturwissenschaften*, **46**, 151 (1959).

465. Liss, E., and Langen, P., 'Uber die hochmoleculares Polyphosphan der Hefe', *Biochem. Z.*, **333**, 193 (1960).

466. Liss, E., and Langen, P., 'Versuche zur Polyphosphate-Überkompensation in Hefezellen nach Phosphatverarmung', *Arch. Mikrobiol.*, **41**, 383 (1962).

467. Lohmann, K., 'Über das Vorkommen der Kondensierten Phosphate in Lebewe-sen', in *Kondensierte Phosphate in Lebensmitteln, Symposium, 1957, Mainz*, Springer-Verlag, Berlin, Göttingen and Heidelberg, 1958, p. 29.

468. Lohmann, K., and Langen, P., 'Über die kondensierten Phosphate in Organismen, in *10 Jahre der Deutschen Akademie der Wissenschaften zu Berlin*, Akademie verlag, Berlin 1956, p. 297.

469. Lohmann, K., and Langen, P., 'Untersuchungen an den kondensierten Phosphaten der Hefe', *Biochem. Z.*, **328**, 1 (1956).

470. Lohmann, R., and Orgel, L. E., 'Prebiotic synthesis: phosphorylation in aqueous solution', *Science, N.Y.*, **161**, 64 (1968).

471. Lorenc, J., Modrzejewska, A., Gniazdowski, M., and Filipowics, B., 'The effect of phosphorus starvation on the content of the phosphate fractions of yeast (*Saccharomyces cerevisiae)'*, *Abstr. 5th Meeting FEBS, Prague* 1968, p. 98.

472. Lowenstein, J. M., 'Transphorylations catalysed by bivalent metal ions', *J. Biochem.*, **70**, 222 (1958).

473. Lowenstein, J. M., 'Synergism of bivalent metals ions in transphorylation', *Nature, Lond.*, **187**, 570 (1960).

474. Lynn, W. S., and Brown, R. H., 'Synthesis of polyphosphate by rat liver mito-chondria', *Biochem. Biophys. Res. Commun.*, **7**, 367 (1963).

475. Lyons, J. W., and Siebenthal, Ch.D., 'On the binding of condensed phosphates by proteins', *Biochim. Biophys. Acta*, **120**, 176 (1966).

476. Lysek, G., and Simonis, W., 'Substrataufnahme und Phosphatstoffwechsel bei *Ankistrodesmus braunii*. I. Beteilligung der Polyphosphate an der Aufnahme von Glucose und 2-Desoxyglucose im Dunkeln und im Licht', *Planta*, **79**, 133 (1968).

477. MacFarlane, M. G., 'Phosphorylation in living yeast', *Biochem. J.*, **30**, 1369 (1936).

478. MacFarlane, M., 'The phosphorylation of carbohydrate in yeast.', *Biochem. J.*, **33**, 565 (1939).

479. Macrae, R. M., and Wilkinson, J. F., 'Poly-β-hydroxybutyrate metabolism in washed suspensions of *Bacillus cereus* and *Bacillus megaterium*', *J. Gen. Microbiol.*, **19**, 210 (1958).

480. Malmgren, H., 'Contribution to the physical chemistry of colloid metaphos-phates. I'. *Acta Chem. Scand.*, **2**, 147 (1948).

481. Malmgren, H., 'Enzymatic breakdown of polymetaphosphate. IV. The activa-tion and the inhibition of the enzyme', *Acta Chem. Scand.*, **3**, 1331 (1949).

482. Malmgren, H., 'Enzymatic breakdown of polymetaphosphate. V. Purification and specificity of the enzyme', *Acta Chem. Scand.*, **6**, 16 (1952).

483. Mann, T., 'Studies on the metabolism of mould fungi. I. Phosphorus metabolism in moulds', *Biochem. J.*, **38**. 339 (1944).
484. Mann, T., 'Studies on the metabolism of mould fungi. II', Isolation of pyrophosphate and metaphosphate from Aspergillus niger. *Biochem. J.*, **38**, 345 (1944).
485. Manocha, M. S., and Colvin, J. P., 'Structure and composition of the cell wall of *Neurospora crassa*', *J. Bacteriol.*, **94**, 202 (1967).
486. Manocha, M. S., and Shaw, M., 'Occurrence of lomasomes in mesophyll cell of Khapli wheat', *Nature, Lond.*, **203**, 1402 (1964).
487. Mansurova, S. É., Belyakova, T. N., and Kulaev, I. S., 'The role of inorganic polyphosphate in the energy metabolism of mitochondria' (in Russian), *Biokhimiya*, **38**, 223 (1973).
488. Mansurova, S. É., Ermakova, S. A., and Kulaev, I. S., 'The extramitochondrial energy-dependent synthesis of inorganic pyrophosphate in yeasts' (in Russian), *Biokhimiya*, **41**, 1716 (1976).
489. Mansurova, S. É., Ermakova, S. A., Zvyagil'skaya, R. A., and Kulaev, I. S., 'The synthesis of inorganic polyphosphate by mitochondria of the yeast-like fungus *Endomyces magnusii* linked to the operation of the respiratory system' (in Russian), *Mikrobiologiya*, **44**, 874 (1975).
490. Mansurova, S. É., Shakhov, Yu. A., Ermakova, S. A., Zvyagil'skaya, R. A., and Kulaev, I. S., 'The conjugation of inorganic pyrophosphate with respiration and the synthesis of inorganic pyrophosphate in mitochondria' (in Russian), in *Mitokhondrii, Akkumulatsiya Energii, i Regulyatsiya Fermentativnykh Protsessov (Mitochondria, Energy Accumulation, and the Control of Enzymatic Processes)*, Izd. Nauka, Moscow, 1977, p. 148.
491. Mansurova, S. É., Shakhov, Yu. A., and Kulaev, I. S., 'The coupling of inorganic pyrophosphate synthesis with respiration in animal tissue mitochondria' (in Russian), *Dokl. Akad. Nauk SSSR*, **213**, 1207 (1973).
492. Mansurova, S. É., Shama, A. M., Sokolovskii, V. Yu., and Kulaev, I. S., 'High-molecular polyphosphates of rat liver nuclei. Their function during liver regeneration' (in Russian), *Dokl. Akad. Nauk SSSR*, **225**, 717 (1975).
493. Mansurova, S. É., Skryabin, K. G., and Kulaev, I. S. 'On a possibility of non-enzymatic transphosphorylation by high-polymeric polyphosphate', in *Reaction Mechanisms and Control Properties of Phosphotransferases, Joint Biochemical Symposium UdSSR–DDR Reinhardsbrunn*, Akademie-Verlag, Berlin, 1973, p. 75.
494. Markarova, E. N., and Baslavskaya, S. S., 'The relationship between the biosynthesis of polyphosphates and oxidative phosphorylation in cells of *Scenedesmus obliquus* (Turpin)' (in Russian), *Dokl. Akad. Nauk SSSR*, **193**, 709 (1970).
495. Martinez, R. J., 'On the nature of the granules of the genus *Spirillum*', *Arch. Microbiol.*, **44**, 334 (1963).
496. Matsuhashi, M., 'Über die chromatographische Trennung von Kondensierten Phosphaten an Anionenaustauscherharen', *J. Biochem. (Tokyo)*, **44**, 65 (1957).
497. Matsuhashi, M., 'Die Trennung von Polyphosphaten durch Anionenaustausch-Chromatographie', *Z. Physiol. Chem.*, **333**, 28 (1963).
498. Mattenheimer, H., 'Die enzymatische Aufspaltung anorganischer Poly- und Metaphosphate durch Organextrakte und Trockenhefe', *Biochem. Z.*, **322** 36 (1951).
499. Mattenheimer, H., 'Die Substratspezifität "anorganischer" Poly- und Metaphosphatasen. I. Optimale Wirkungsbedeutungen für den enzymatischen Abbau von Poly- und Metaphosphaten', *Z. Physiol. Chem.*, **303**, 107 (1956).
500. Mattenheimer, H., 'Die Substratspezifität "anorganischer" Poly- und Metaphosphatasen. II. Trennung der Enzyme', *Z. Physiol. Chem.* **303**, 115 (1956).
501. Mattenheimer, H., 'Die Substratspezifität "anorganischer" Poly- und Metaphosphatasen. III. Papier-chromatographische Untersuchungen beim enzyma-

tischen Abbau von anorganischen Poly- und Metaphosphates', *Z. Physiol. Chem.*, **303**, 125 (1955).

502. Mattenheimer, H., 'Polyphosphate une Polyphosphatasen in Amöben', *Acta Biol. Med. Ger.*, **1**, 405 (1958).

503. Mattenheimer, H., 'Auf kondensierte Phosphate wirkende Enzyme', in *Kondensierte Phosphate in Lebensmitteln, Symposium, 1957, Mainz*, Springer-Verlag, Berlin, Göttingen and Heidelberg, 1958, p. 45.

504. McCullough, J. F., Van Wazer, J. R., and Griffith, E. J., 'Structure and properties of the condensed phosphates. XI. Hydrolytic degradation of Graham's salt', *J. Am. Chem. Soc.*, **78**, 367 (1956).

505. McElroy, W. D., 'Phosphate bound energy and bioluminescence', in *Phosphorus Metabolism*, Eds. McElroy, W. D., and Glass, B., Johns Hopkins Press, Baltimore, 1951, Vol. 1, p. 585.

506. Meissner, J., 'Die anorganischen Poly- und Metaphosphate und ihr biochemisches Verhalten', *Iber. Borstel, 1956/1957*, Springer-Verlag, Berlin, Göttingen, and Heildelberg, 1957, Vol. 4, p. 586.

507. Meissner, J., and Diller, W., 'Über die Aufnahme von Ortho- und Metaphosphate durch das *Mycobacterium tuberculosis*', *Z. Hyg. Infektionskr.*, **137**, 518 (1953).

508. Meissner, J., and Kropp, F., 'Über die Phosphatresorption bei Mycobacterium tuberculosis', *Z. Hyg. Infektionskr.*, **137**, 429 (1953).

509. Meissner, J. and Lemke, J., 'Über die Aufnahme 32-markierter kondensierter Phosphate bei BCG–und H 37–Stämmen von *Mycobacterium* tubercules', *Iber. Borstel 1954/55*, Springer-Verlag, Berlin, Göttingen, and Heidelberg, 1956, p. 3.

510. Mel'gunov, V. I., and Kulaev, I. S., 'The effect of 8-azaadenine on the accumulation of high-molecular inorganic polyphosphates in *Neurospora crassa*' (in Russian), *Dokl. Vyssh. Shkoly 'Biologicheskie Nauki'*, No. 3, 87 (1971).

511. Mel'gunov, V. I., and Kulaev, I. S., '8-Azaadenine and metabolism of phosphorus compounds in adenine deficient mutants and the wild type of *Neurospors crassa*', *Abstr. 7th Meeting FEBS, Varna*, 1971, Abstr. 1022, p. 343.

512. Merrick, J. M., Schwartz, T., and Wagner, B., 'Studies on the Metabolism of poly-β-hydroxybutyrate: depolymerization by an enzyme system from *Rhodosperillum rubrum*', *Abstr. VII Int. Congr. Biochem., Tokyo. Symp.* V. J-343, p. 1011. (1967).

513. Meyerhof, O., and Green, H., 'Synthetic action of phosphatase. II. Transphosphorylation by alkaline phosphatase in the absence of nucleotides', *J. Biol. Chem.*, **183**, 177 (1950).

514. Meyerhof, O., Shates, H., and Kaplan, A., 'Heat of hydrolysis of trimetaphosphate', *Biochim. Biophys. Acta*, **12**, 121 (1953).

515. Meisel', M. N., *Funktsional'naya Morfologiya Drozhzhevykh Organizmov (The Functional Morphology of Yeasts)*, Izd. Akad. Nauk SSSR, Moscow, 1950.

516. Mikhailovskaya, T., The biochemistry of phosphorus compounds in antibiotic-producing strains of *Aspergillus niger* and *Actinomyces violaceus*' (in Russian), *Candidate's Thesis*, Khar'kov, 1951.

517. Miller, S. L., and Parris, M., 'Synthesis of pyrophosphate under primitive earth conditions', *Nature, Lond.*, **204**, 1248 (1964).

518. Minck, R., and Minck, A., 'Constitution of diphtheria bacillus, nuclear apparatus and metachromatic granulations', *C.R. Acad. Sci. Paris*, **228**, 1313 (1949).

519. Mitchell, P., and Moyle, J. M., 'Transport of phosphate across the surface of *Micrococcus pyrogenes*. Nature of cell inorganic phosphate', *J. Gen. Microbiol.*, **9**, 273 (1953).

520. Miyachi, S., 'Inorganic polyphosphate in spinach leaves', *J. Biochem. (Tokyo)*, **50**, 367 (1961).

521. Miyachi, S., 'Turnover of phosphate compounds in *Chlorella* cells under phosphate deficiency in darkness', *Plant Cell Physiol.*, **3**, 1 (1961).

522. Miyachi, S., Kanai, R., Mihara, S., Miyachi, S., and Aoki, S., 'Metabolic roles of inorganic polyphosphates in *Chlorella* cells', *Biochim. Biophys. Acta*, **93**, 625 (1964).

523. Miyachi, S., and Miyachi, S., 'Modes of formation of phosphate compounds and their turnover in *Chlorella* cells during the process of life cycle as studied by the technique of synchonous culture', *Plant Cell Physiol.*, **2**, 415 (1961).

524. Miyachi, S., and Tamiya, H., 'Distribution and turnover of phosphate compounds in growing *Chlorella* cells', *Plant Cell Physiol.*, **2**, 405 (1961).

525. Moore, R. T., and McAlear, J. H., 'Fine structure of mycota lomasomes, previously uncharacterized hyphal structures', *Mycologia*, **53**, 194 (1961).

526. Moorison, J. F., and Enner, A. H., '*N*-Phosphorylated guanidines', in *The Enzymes*, Ed. Boyer, P. D., Lardy, H., and Myrbäck K. Academic Press, New York and London, 1960, Vol. 2, p. 89.

527. Mudd, S., 'Cytology of bacteria. Part I. The bacterial cell', *Ann. Rev. Microbiol.*, **8**, 1 (1954).

528. Mudd, S. H., 'Activation of methionine for transmethylation. VI. Enzyme-bound tripolyphosphate as an intermediate in the reaction catalyzed by the methionine-activation enzyme of baker's yeast', *J. Biol. Chem.*, **238**, 2156 (1963).

529. Mudd, S., Yoshida, S. A., and Koike, M., 'Polyphosphate as accumulator of phosphorus and energy', *J. Bacteriol.*, **75**, 224 (1958).

530. Muhammed, A., 'Studies on biosynthesis of polymetaphosphate by an enzyme from *Corynebacterium xerosis*', *Biochim. Biophys. Acta*, **54**, 121 (1961).

531. Muhammed, A., Rodgers, A., and Hughes, D. E., 'Purification and properties of a polymetaphosphatase from *Corynebacterium xerosis*', *J. Gen. Microbiol.*, **20**, 482 (1959).

532. Mühlradt, P. F., 'The formation of polyphosphate–RNA complexes in cell-free extracts of *Salmonella minnesota*' (in Russian), in *7i Mezhdunarodnyi Simpozium po Khimii Prirodnykh Soedinenii (Seventh International Symposium on the Chemistry of Naturally Occurring Compounds)*, Abstracts, Zinatne, Riga, 1970, p. 276.

533. Mühlradt, P. F., 'Synthesis of high molecular weight polyphosphate with partially purified enzyme from *Salmonella*', *J. Gen. Microbiol.*, **68**, 115 (1971).

534. Muller, S., and Ebel, J. P., 'Étude des phosphates condensés contenus dans divers champignons', *Bull. Soc. Chim. Biol.*, **40**, 1153 (1958).

535. Muller, S., and Ebel, J. P., 'Étude des phosphates condensés contenus dans divers champignons', *Proc. 4th Int. Congr. Biochem., Vienna*, **17**, 36 (1958).

536. Muller-Felter, S., and Ebel, T. P., 'Separation des acides ribonucleiques et des polyphosphates inorganiques. II. Mise au point d'une technique de separation par adsorbtion differentielle sur charbon', *Bull. Soc. Chim. Biol.*, **44**, 1175 (1962).

537. Nassery, H., 'Polyphosphate formation in the roots *Deschampsia flexiosa* and *Urtica dioica*', *New Phytol.*, **68**, 21 (1969).

538. Naumova, N. B., Streshinskaya, G. M., and Gololobov, A. D., 'Phosphorus-containing non-nucleic polymers in some species of yeasts grown on various media' (in Russian), *Dokl. Akad. Nauk SSSR*, **182**, 465 (1968).

539. Nesmeyanova, M. A., 'The function of membranes in regulating the activity and synthesis of bacterial enzymes' (in Russian), in *Biokhimiya in Fiziologiya Mikroorganizmov (The Biochemistry and Physiology of Microorganisms)*, Ed. Skryabin G. K., Izd. Nauka, Pushchino na oke 1975, p. 68.

540. Nesmeyanova, M. A., Dmitriev, A. D., Bobik, M. A., and Kulaev, I. S., 'On the regulation of some phosphorous metabolism enzymes in *E. coli*, in *Reaction Mechanisms and Control Properties of Phosphotransferases, Joint Biochemical symposium UdSSR–DDR, Reinhardsbrunn*, Academie-Verlag, Berlin, 1973, p. 82.

234

541. Nesmeyanova, M. A., Dmitriev, A. D., and Kulaev, I. S., High-molecular polyphosphates and polyphosphate-metabolizing enzymes during the growth of a culture of *Escherichia coli*' (in Russian), *Mikrobiologiya*, **42**, 213 (1973).

542. Nesmeyanova, M. A., Dmitriev, A. D., and Kulaev, I. S., 'The regulation of phosphorus metabolism and the level of polyphosphates in *Escherichia coli* K-12 by exogenous orthophosphate' (in Russian), *Mikrobiologiya*, **43**, 227 (1974).

543. Nesmeyanova, M. A., Gonina, A., Severin, A. I., and Kulaev, I. S., 'The metabolic regulation of phosphohydrolases in *Escherichia coli*' (in Russian), *Mikrobiologiya*, **43**, 410 (1974).

544. Nesmeyanova, M. A., Gonina, S. A., and Kulaev, I. S., 'The biosynthesis of polyphosphatase in *Escherichia coli* under the control of genes in common with alkali phosphatase' (in Russian), *Dokl. Akad. Nauk SSSR*, **224**, 710 (1975).

545. Nesmeyanova, M. A., Maraeva, O. B., Severin, A. I., and Kulaev, I. S., 'The localization of polyphosphatase in cells of *Escherichia coli* with repression and derepression of the biosynthesis of this enzyme' (in Russian), *Dokl. Akad. Nauk SSSR*, **223**, 1266 (1975).

546. Nesmeyanova, M. A., Maraeva, O. B., Severin, A. I., Zaichkin, É. I., and Kulaev, I. S., 'The action of detergents and proteolytic enzymes on membrane-bound polyphosphohydrolases in cells of *Escherichia coli* with repressed and derepressed biosynthesis of these enzymes' (in Russian), *Biokhimiya*, **41**, 1256 (1976).

547. Neuberg, C., and Fischer, H., 'Enzymic splitting of triphosphoric acid. II. Hydrolysis by means of an enzyme found in *Aspergillus niger* and yeast. Analytical reactions of triphosphate' *Enzymologia*, **2**, 241 (1938).

548. Neville, M. M., Suskind, S. R., and Roseman, S., 'A derepressible active transport system for glucose in *Neurospora crassa*', *J. Biol. Chem.*, **246**, 1294 (1971).

549. Nickerson, W. I., Dependance, in yeasts, of phosphate uptake and polymerization upon the occurrence of glucose polymerization', *Experientia* **5**, 202 (1949).

550. Niemeyer, R., 'Cyclic condensed metaphosphates and linear polyphosphates in brown and red algae', *Arch. Microbiol.*, **108**, 243 (1976).

551. Niemeyer, R., and Richter, G., 'Rapidly labelled polyphosphates and metaphosphates in the blue–green alga *Anacystis nidulans*' *Arch. Mikrobiol.*, **69**, 54 (1969).

552. Niemeyer, R., and Richter, G., 'Rapidly labelled polyphosphates in *Acetabularia*', in *Biology and Radiobiology of Anucleate Systems. II. Plant Cells*, Academic Press, New York and London, 1972, p. 225.

553. Niemierko, S., 'Metafosforany *Galleria mellonella*', *Acta Physiol. Pol.*, **1**, 101 (1950).

554. Niemierko, S., 'O meta- i polifosforanach organizmach zywych, *Postepy Biochem.* **1**, 50 (1953).

555. Niemierko, S., 'Pyro- and polyphosphates in insects', *Colloq. Int. Acids Ribonucleiques et Polyphosphates*', **106**, 615 (1962).

556. Niemierko, S., and Niemierko, W., 'Metaphosphate in the excreta of *Galleria mellonella*' *Acta Biol. Exp.*, **15**, 111 (1950).

557. Niemierko, S., and Wojtezak, A., 'Badania nad metafosfataza i pyrofosfatasa u *Galleria mellonella*', *Acta Physiol. Pol.*, **3**, 217 (1952).

558. Nihei, T., 'A phosphorylative process, accompanied by photochemical liberation of oxygen, occurring at the stage of nuclear division in *Chlorella* cells. I', *J. Biochem. (Tokyo)*, **42**, 245 (1955).

559. Nihei, T., 'A phosphorylative process accompanied by photochemical liberation of oxygen, occurring at the stage of nuclear division in *Chlorella* cells. II', *J. Biochem. (Tokyo)*, **44**, 389 (1957).

560. Niklowitz, W., 'Über den Feinbau der Mitochondrien des Schlempilzen *Badhamia utricularis*', *Exp. Cell. Res.*, **13**, 591 (1957).

561. Niklowitz, W., and Drews, G., 'Zur elektronenmikroskopischen Darstellung der

Feinstruktur von *Rhodospirillum rubrum* (Ergebhisse einer neuen, einfachen Dûnnschnittmethode)', *Arch. Microbiol.*, **23**, 123 (1955).

562. Nishi, A., 'Enzymatic studies on the phosphorus metabolism in germinating spores of *Aspergillus niger*', *J. Biochem. (Tokyo)*, **48**, 758 (1960).

563. Nishi, A., 'Role of polyphosphate and phospholipid in germinating spores of *Aspergillus niger*', *J. Bacteriol.*, **81**, 10 (1961).

564. Nordlie, R. G., and Arion, W. J., 'Evidence for the common identity of glucose-6-phosphatase, inorganic pyrophosphatase and pyrophosphate–glucose phosphotranferase', *J. Biol. Chem.*, **239**, 1690 (1965).

565. Northcote, D. H., and Korn, R. W., 'The chemical composition and structure of the yeast cell wall', *Biochem. J.*, **51**, 232 (1952).

566. Ohashi, S., 'Chromatography of phosphorus oxyacids', *Pure Appl. Chem.*, **44**, 415 (1975).

567. Ohashi, S., and Van Wazer, J. R., 'Paper chromatography of very long chain polyphosphates', *Analyt Chem.*, **35**, 1984 (1964).

568. Okorokov, L. A., 'The role of polymeric metal orthophosphates in the regulation of biochemical processes', in *Tezisy Dokl. na Vsesoyuznoi Konferentsii 'Regulyatsia Biokhimicheskikh Protsessov u Mikroorganizmov' (Abstracts of the All-Union Conference on 'The Regulation of Biochemical Processes in Microorganisms')*, Pushchino na Oke, 1972, p. 217.

569. Okorokov, L. A., Kadomtseva, V. M. and Kulaev I. S., 'High-molecular complex phosphates of calcium, magnesium, and iron, which are of biological origin' (in Russian), *Tezisy III Vsesoyuznogo Soveshch. po Fosfatam (Abstracts of the Third All-Union Conference on Phosphates)*, Izd. Znanie. Izv. Akad. Nauk SSSR, Ser. Neorg. Materialy, UP, 1666 (1971).

570. Okorokov, L. A., Kadomtseva, V. M., and Kulaev, I. S., 'The detection of polymeric calcium, magnesium, and manganese orthophosphates in *Penicillium chrysogenum*' (in Russian), *Dokl. Akad. Nauk SSSR*, **209**, 735 (1973).

571. Okorokov, L. A., Kholodenko, V. P., and Kulaev, I. S., 'Polymeric iron orthophosphate, a new compound in certain fungi' (in Russian), *Dokl. Akad. Nauk SSSR*, **199**, 1204 (1970).

572. Okorokov, L. A., Kulaev, I. S., 'A polymeric iron phosphate in mycelia of *Penicillium chrysogenum*' (in Russian), *Biokhimiya*, **33**, 800 (1968).

573. Okorokov, L. A., Lichko, L. P., and Kulaev, I. S., 'A comparative investigation of the functions of polymeric metal orthophosphates and condensed polyphosphates, in *Penicillium chrysogenum*' (in Russian), *Izv. Akad. Nauk SSSR, Ser. Biol.*, No. 1, 86 (1973).

574. Okorokov, L. A., Lichko, L. P., Kadomtseva, V. M., Kholodenko, V. P., and Kulaev, I. S., 'The metabolism and physicochemical state of Mg^{2+} ions in fungi' (in Russian), *Mikrobiologiya*, **43**, 410 (1974).

575. Okorokov, L. A., Perov, N. A., and Kulaev, I. S., 'The localization of polymeric iron orthophosphate in *Penicillium chrysogenum*' (in Russian), *Tsitologiya*, **15**, 481 (1973).

576. Okuntsov, M. M., and Grebennikov, A. S., 'The effect of light on the content of inorganic polyphosphates and certain organophosphorus compounds in *Chlorella pyrenoidosa* Chick.' (in Russian), *Biokhimiya*, **42**, 21 (1977).

577. Onishi, T., and Hagihara, B., 'Preparations of yeast mitochondria by an enzymatic procedure', *J. Biochem.* **56**, 484 (1964).

578. Oparin, A. I., *Vozniknovenie Zhizni na Zemle (The Origin of Life on Earth)*, Izd. Akad. Nauk SSSR, Moscow, 1957.

579. Oparin, A. I., 'The origin of life and the origin of enzymes', *Advan. Enzymol.*, **27**, 347 (1965).

580. Oparin, A. I., 'Modes of origination of metabolic processes, and their artificial modelling in coacervate droplets' (in Russian), in *Proiskhozhdenie Predbiolo-*

gicheskikh Sistem (The Origins of Prebiological Systems) (translated into Russian), Mir, Moscow, 1966, p. 335.

581. Oro, G., *Etapy i Mekhanismy Predbiologicheskogo Organicheskogo Sinteza (Stages and Mechanisms of Prebiological Organic Synthesis)* (translated into Russian), Mir, Moscow, 1966, p. 144.

582. Osterberg, R., and Orgel, L. E., 'Polyphosphate and trimetaphosphate formation under potentially prebiotic conditions', *J. Mol. Evolution*, **1**, 241 (1972).

583. Overbeck, S., 'Der Tageszyklus des Phosphataushalt von *Scenedesmus quadricauda* (Turp.) Breb im Freiland', *Naturwissenschaften*, **48**, 127 (1961).

584. Overbeck, S., 'Untersuchungen sum Phosphataushalt von Grünalgen. III. Das Verhalten der Zellfraktionen von *Scenedesmus quadricauda* (Turp.) Breb in Tagescyclus unter verschiedenen Belichtungsbedingungen und bei verschiedenen Phosphatverbindungen', *Arch. Mikrobiol.*, **41**, 11 (1962).

585. Pakhomova, M. V., Darkanbaeva, G. T., and Zaitseva, G. N., 'The effect of light and darkness on the acid-soluble phosphorus compounds of the green alga *Scenedesmus obliquus* (Kütz)' (in Russian), *Biokhimiya*, **31**, 1237 (1966).

586. Pakhomova, M. V., Surgai, V. V., and Zaitseva, G. N., 'Changes in the composition of acid-soluble phosphorus compounds in the green alga *Asteromonas gracilis* with respect to the nutrient nitrogen source' (in Russian), *Izv. Akad. Nauk SSSR, Ser. Biol.*, No. 4, 624 (1967).

587. Pasynskii, A. G., and Pavlovskaya, 'The formation of biologically important compounds at the prebiological stages of the development of the Earth' (in Russian), *Usp. Khim.*, **33**, 1198 (1964).

588. Penniall, R., and Griffin, J. B., 'Studies of phosphorus metabolism by isolated nuclei. IV. Formation of polyphosphate', *Biochim. Biophys. Acta*, **90**, 429 (1964).

589. Penniall, R., Saunders, J., and Lin-Min, 'Studies of phosphorus metabolism by isolated nuclei. II. Investigation of optimal conditions for its demonstration', *Biochemistry*, **3**, 1454 (1964).

590. Penniall, R., and Saunders, J. P., 'Studies of phosphorus metabolism by isolated nuclei. III. Some fundamental properties of the system', *Biochemistry*, **3**, 459 (1964).

591. Perlmann, G., 'Metaphosphate–protein complexes', *Biochem. J.*, **32**, 931 (1938).

592. Peters, T., and Rieman, W., 'Analyses of mixtures of the condensed phosphates by ion-exchange chromatography. II.′ Mixtures of ortho-, pyro-, tri-, tetra-, trimeta- and tetrametaphosphates and Graham's salt', *Analyt. Chim. acta*, **14**, 131 (1956).

593. Petras, E., 'Nucleinsaure- und Phosphataushalt von *Phycomyces blakesleanus* (Bgtt)', *Arch. Microbiol.*, **30** 433 (1958).

594. Pfrengle, O., 'Die Papierchromatographie der kondensierter Phosphate', *Z. Analyt. Chem.*, **158**, 81 (1957).

595. Phaff, H. J., 'Cell wall of yeasts', *Ann. Rev. Microbiol.*, **17**, 15 (1963).

596. Pierpoint, W. S., 'The phosphatase and metaphosphatase activities of pea extract', *Biochem. J.*, **65**, 67 (1957).

597. Pierpoint, W. S., 'The phosphoesterase of pea plants (*Pisum sativum* L.)', *Biochem. J.*, **67**, 644 (1957).

598. Pierpoint, W. S., 'Polyphosphates excreted by wax-moth larve (*Galleria mellonella* and *Achroea grisella* Fabr.)', *Biochem. J.*, **67**, 624 (1957).

599. Pine, M., 'Metaphosphate accumulation in sulfur starved *Escherichia coli* and its effect on the alcohol soluble protein fraction', *Abstr. VIII Int. Congr. Microbiol., Montreal, Canada*, 1962, p. 36.

600. Pine, M., 'Alcohol-soluble protein of microorganisms', *J. Bacteriol.*, **85**, 301 (1963).

601. Pirson, A., and Kuhl, A., 'Über den Phosphataushalt von Hydrodictyon. I', *Arch. Mikrobiol.*, **30**, 211 (1958).

602. Ponnamperuma, C., 'The non-biological synthesis of some components of nucleic acids' (in Russian), in *Proiskhozhdenie Predbiologicheskikh Sistem (The Origins of Prebiological Systems)* (translated into Russian), Mir, Moscow, 1966, p. 224.

603. Ponnamperuma, C., Sagan, C., and Mariner, R., 'Synthesis of adenosine triphosphate under possible primitive earth conditions', *Nature, Lond.*, **199**, 222 (1963).

604. Prat, S., 'Beitrag zur Kenntniss der Organisation der Cyanophyceae', *Arch. Protistenkd.*, **52**, 142 (1925).

605. Pryanishnikov, D. N., *Azot v Zhizni Rastenii i Zemledelii SSSR (Nitrogen in Plant Life and Agriculture in the USSR)*, Izd. Akad. Nauk SSSR, Moscow, 1945.

606. Prokof'eva-Bel'govskaya, A. A., and Kats, L. A., 'The chemical nature of volutin in Actinomycetes' (in Russian), *Mikrobiologiya*, **29**, 826 (1960).

607. Rabinowitz, J., Chang, S., and Ponnamperuma, C., 'Phosphorylation by way of inorganic phosphate as a potential protobiotic process', *Nature, Lond.*, **218**, 442 (1968).

608. Rabinowitz, J., Flores, J., Krebsbach, R., and Rogers, G., 'Peptide formation in the presence of linear or cyclic polyphosphates', *Nature, Lond.*, **224**, 795 (1969).

609. Rafter, G. W., 'Studies on trimetaphosphate metabolism in yeast', *Arch. Biochem. Biophys.*, **81**, 238 (1959).

610. Rahman, Q., and Krishnan, P. S., 'Phosphorus, nitrogen and carbohydrate composition of the seeds of *Cuscuta*', *Phytochemistry*, **10**, 1751 (1971).

611. Racker, É. *Bioénergeticheskie Mekhanizmy (Bioenergetic Mechanisms)* (translated into Russian), Mir, Moscow, 1967.

612. Ramauah, A., Hathaway, J. A., and Atkinson, D. E., 'Adenylate as a metabolic regulator. Effect on yeast phosphofructokinase kinetics', *J. Biol. Chem.*, **239**, 3619 (1964).

613. Rautanen, N., and Mikkulainen, P., 'On the phosphorus fractions and the uptake of phosphorus by *Torulopsis utilis*', *Acta, Chem. Scand.*, **5**, 89 (1951).

614. Reichenow, E., 'Untersuchungen an *Haematococcus pluvialis* nebst Bemerkungen über andere Flagellaten', *Arb. Kaiserl. Gesundh.*, **33**, Heft 1, 11 (1909).

615. Richter, C., 'Pulse-labelling of nucleic acids and polyphosphates in normal and annucleated cell of *Acetabularia*', *Nature, Lond.*, **212**, 1363 (1966).

616. Riemersma, J. C., 'Ion transport and phosphorylation in dried and rehydrated yeast cells', *Nature, Lond.*, **215**, 1078 (1967).

617. Rotenbach, E. E., and Hinkelmann, S., 'Über Poly- und Metaphosphatasen der Gerste', *Naturwissenschaften*, **41**, 555 (1954).

618. Rose, A., 'Acetate kinase of bacteria (acetokinase)', *Methods Enzymol.*, **1**, 591 (1955).

619. Rosenberg, H., 'The isolation and identification of "volutin" granules from *Tetrahymena*', *Exp. Cell. Res.*, **41**, 397 (1966).

620. Rubtsov, P. M., Efremovich, N. V., and Kulaev, I. S., 'The light-dependent synthesis of pyrophosphates in chloroplasts' (in Russian), *Dokl. Akad. Nauk SSSR*, **230**, 1236 (1976).

621. Rubtsov, P. M., Efremovich, N. V., and Kulaev, I. S., 'The absence of high-molecular polyphosphates from the chloroplasts of *Acetabularia mediterranea*' (in Russian), *Biokhimiya*, **42**, 890 (1977).

622. Rubtsov, P. M., and Kulaev, I. S., 'Some pathways of biosynthesis and breakdown of polyphosphates in the green alga *Acetabularia mediterranea*' (in Russian), *Biokhimiya*, **42**, 1083 (1977).

623. Ruska, H. G., Bringmann, W., Nechel, J., and Schuster, G., 'Über die Entwicklung sowie den morphologischen und cytologischen Aufbau von *Mycobacterium avium* (Chester)', *Z. Wiss. Mikroskop.*, **60**, 425 (1951/52).

624. Salhany, J. M., Yamane, T., Shulman, R. G., and Ogawa, S., 'High resolution

238

^{31}P nuclear magnetic resonance studies of intact yeast cells', *Proc. Nat. Acad. Sci. USA*, **72**, 4966 (1975);

625. Sall, T., Mudd, S., and Davis, J. C., 'Factors conditioning the accumulation and disappearance of metaphosphate in cells of *Corynebacterium diphtheriae*', *Arch. Biochem. Biophys.*, **60**, 130 (1956).
626. Sall, T., Mudd, S., and Takagi, A., 'Phosphate accumulation and utilization as related to synchronized cell division of *Corynebacterium diptheriae*', *J. Bacteriol.* **76**, 640 (1958).
627. Samuelson, O. *Primenenie Ionnogo Obmena v Analiicheskoi Khimii (Ion Exchange in Analytical Chemistry)* (translated into Russian), Izd. IL, Moscow, 1955.
628. Sansoni, B., and Baumgartner, L., 'Trennung von Phosphaten durch Papierelektrophorese. V. Hochspannungspapierelektrophorese Kondensierte Phosphate und Metaphosphate', *Z. Analyt. Chem.*, **158**, 241 (1957).
629. Sauer, H. W., Babcock, K. L., and Rusch, H. P., 'High molecular weight phosphorus compound in nucleic acid extracts of the slime mould *Physarum polycephalum*', *J. Bacteriol.*, **99**, 650 (1969).
630. Schaefer, W. B., and Lewis, C. W., 'Effect of oleic acid on growth and cell structure of *Mycobacteria*', *J. Bacteriol.*, **90**, 1438 (1965).
631. Schlegel, H. G., 'Untersuchungen über den Phosphatstoffwechsel der Wasserstoffoxydierenden Bakterien', *Arch. Mikrobiol.*, **21**, 127 (1951).
632. Schlegel, H. G., 'Die Specherstoffe von *Chromatium okenii*', *Arch. Microbiol.*, **42**, 110 (1962).
633. Schlegel, H. G., and Gottschalk, G., 'Poly-hydroxybuttersäure, ihre Verbreitung, Funktion und Biochemistry', *Angew, Chem.*, **74**, 342 (1962).
634. Schlegel, H. G., and Kaltwasser, H., 'Veränderungen des Polyphosphatgehaltes während des Wachstums von Knalegasbakterien unter Phosphatmangel', *Flora*, **150**, 259 (1961).
635. Schmidt, G., 'The biochemistry of inorganic pyrophosphates and metaphosphates', in *Phosphorus Metabolism*, Eds. McElroy, W. D., and Glass, B., Johns Hopkins Press, Baltimore, 1951, Vol. 1, p. 443.
636. Schmidt, R. R., 'Intracellular control of enzyme synthesis and activity during synchronous growth of *Chlorella*', in *Cell Synchrony Studies in Biosynthetic Regulation*, Eds. Cameron, I. L., and Padilla, G. M., Academic Press, New York, 1966, p. 189.
637. Schmidt, S., 'Untersuchungen über den Phosphorstoffwechsel der Blätter in Beziehung zum entwicklungs physiologischen Verhalten von Apfelbäumen (*Malus domestica*)', *Arch. Gartenbau*, **19**, 157 (1971).
638. Schmidt, S., 'Einfluss der Temperatur auf die Synthese säurenunslöslicher Polyphosphate der Blätter im Verlauf der Entwicklung von Apfelbäumen', *Arch. Gartenbau*, **20**, 419 (1972).
639. Schmidt, S., and Buban, T., 'Beziehungen zwischen dem ^{32}P-Einbau in anorganische Polyphosphate der Blätter und dem Beginn der Blütendifferenzierung bei *Malus domestica*', *Biochem. Physiol. Pflanzen*, **162**, 265 (1971).
640. Schmidt, G., Hecht, L., and Thannhauser, S., 'Enzymic formation and accumulation of large amounts of a metaphosphate in baker's yeast under certain conditions', *J. Biol. Chem.*, **166**, 775 (1946).
641. Schmidt, G., Hecht, L., and Thannhauser, S. J., 'The effect of potassium ions on the absorption of orthophosphate and the formation of metaphosphate by baker's yeast', *J. Biol. Chem.*, **178**, 733 (1949).
642. Schmidt, G., Seraidarian, K., Greenbaum, L. M., Hickey, M. D., and Thannhauser, S. J., 'The effect of certain nutritional conditions on the formation of purines and of ribonucleic acid in baker's yeast', *Biochem. Biophys. Acta*, **20**, 135 (1956).
643. Schmidt, G., and Thannhauser, S., 'A method for the determination of deoxy-

ribonucleic acid, ribonucleic and phosphoproteins in animal tissues', *J. Biol. Chem.*, **161**, 83 (1945).

644. Schoffstall, A. M., 'Prebiotic phosphorylation of nucleosides in formamide', Origins of Life, **7**, 399 (1976).

645. Schramm, G., The importance of a study of viruses for the understanding of the processes involved in biological reproduction' (in Russian), in *Vozniknovenie Zhizni na Zemle (The Origins of Life on Earth)*, Izd. Akad. Nauk SSSR, Moscow, 1957, p. 222.

646. Schramm, G., 'The synthesis of nucleosides and polynucleotides using metaphosphate esters' (in Russian), in *Proiskhozhdenie Predbiologicheskikh Sistem (The Origins of Prebiological Systems)*, Mir, Moscow, 1966, p. 303.

647. Schramm, G., Grotsch, H., and Pollmann, W., 'Non-enzymatic synthesis of polysaccharides, nucleosides and nucleic acids and the origin of self-reproducing systems', *Angew. Chem., Int. Ed. Engl.*, **1**, 1 (1962).

648. Schramm, G., Läzmann, G., and Bechman, F., 'Synthesis of α- and β-Adenosine and of α- and β-2-deoxyadenosine with phenyl polyphosphate', *Biochim. Biophys. Acta*, **145**, 221 (1967).

649. Schülke, U., 'Darstellung von Octametaphosphaten M^1 (P_8O_{24})', *Z. Anorg. Allg. Chem.*, **360**, 231 (1968).

650. Schwartz, A. W., 'Phosphate: solubilization and activation on the primitive earth', in *Chemical Evolution and the Origin of Life*, Ed. Buvet, R., and Ponnamperuma, C., North-Holland, Amsterdam, 1971, p. 207.

651. Schwartz, A. W., 'The sources of phosphorus on the primitive earth–on inquiry', in *Molecular Evolution: Prebiological and Biological*, Ed. Rohifing, L., and Oparin, A. I. 1972, p. 129.

652. Schwartz, A., Bradley, E., and Fox, S., 'The thermal condensation of cytidylic acid in the presence of polyphosphoric acid' (in Russian), in *Proiskhozhdenie Predbiologicheskikh Sistem (The Origins of Prebiological Systems)* (translated into Russian), Mir, Moscow, 1966, p. 321.

653. Schwartz, A., and Ponnamperuma, C., 'Phosphorylation of adenosine with linear polyphosphate salts in aqueous solution', *Nature, Lond.*, **218**, 443 (1968).

654. Sentandreu, R., and Elorza, M. V., 'The biosynthetic pathway of "yeast mannan"', *Symp. on Yeast Protoplasts, Salamanca, Spain, 2–5 Oct., 1972*, 1972, p. 39.

655. Severin, A. I., Lusta, K. A., Kulaev, I. S., and Nesmeyanova, M. A., 'Microsome membrane-bound polyphosphatasc in *Escherichia coli*' (in Russian), *Biokhimiya*, **41**, 357 (1975).

656. Shabalin, Yu. A., Vagabov, V. I., Tsiomenko, A. B., Zemlyanukhina, O. A., and Kulaev, I. S., 'Polyphosphate kinase activity in vacuoles of yeasts' (in Russian), *Biokhimiya*, **42**, 1642 (1977).

657. Shadi, A., Mansurova, S. É., Tsydendambaev, V. D., and Kulaev, I. S., 'The biosynthesis of polyphosphates in chromatophores of *Rhodospirillum rubrum*' (in Russian), *Mikrobiologiya*, **45**, 333 (1976).

658. Shakhov, Yu. A., Mansurova, S. É., and Kulaev, I. S., 'The detection of inorganic pyrophosphate synthesis in association with the respiratory chain in submitochondrial particles from beef heart mitochondria' (in Russian), *Dokl. Akad. Nauk SSSR*, **219**, 1017 (1974).

659. Shaposhnikov, V. N., and Fedorov, V. D., 'An investigation of phosphorus metabolism in green photosynthesising sulphur bacteria in relation to the fixation of carbon dioxide' (in Russian), *Biokhimiya*, **25**, 487 (1960).

660. Shepherd, C. J., 'The enzymes of carbohydrate metabolism in *Neurospora*', *Biochem. J.*, **48**, 483 (1951).

661. Shepherd, C. J., 'Changes occurring in the composition of *Aspergillus nidulans* conidia during germination', *J. Gen. Microbiol.*, **16**, 1 (1957).

662. Shiokawa, K., and Yamana, Y., 'Demonstration of polyphosphate and its

240

possible role in RNA synthesis during early development of *Rana japonica* embryons', *Exp. Cell. Res.*, **38**, 180 (1965).
663. Simonis, W., and Kating, H., 'Untersuchungen zur lichtabhängigen Phosphorilierung von Algen durch Glucosegaben', *Z. Naturforsch.*, **11B**, 115 (1956).
664. Singh, C., 'New staining reagents for the detection of condensed phosphates', *J. Sci. Ind. Res.*, **188**, 249 (1959).
665. Singh, C., 'The interaction between anionic and cationic compounds as typified by inorganic condensed phosphates and antibiotics' (in Russian), *Ref. Sektsionnykh Soobshchenii V Mezhdunar. Biokhim. Kongressa (Abstracts of Section Papers at the Fifth International Biochemical Congress)*, Izd, Akad. Nauk SSSR, Moscow, 1961, Vol. 2, p. 511.
666. Singh, C., 'Interactions between cationic and anionic compounds of biological importance with special reference to inorganic condensed phosphates and polymeric antibiotics', *Indian J. Chem.*, **2**, 67 (1964).
667. Skryabin, K. G., Rubtsov, P. M., Verteletskaya, N. L., and Kulaev, I. S., 'The detection of high-molecular polyphosphates in nuclei of the fungus *Endomyces magnusii*' (in Russian), *Dokl. Vyssh. Shkoly 'Biologicheskie Nauki'*, No. 10, 84 (1973).
668. Skulachev, V. P., *Akkumulatsiya Énergii v Kletke (The Accumulation of Energy in the Cell)*, Izd. Nauka, Moscow, 1969.
669. Skulachev, V. P., *Transformatsiya Énergii v Biomembranakh (The Transformation of Energy in Biological Membranes)*, Izd. Nauka, Moscow, 1972.
670. Smillie, R. M., and Krotkov, G., 'Phosphorus-containing compounds in *Euglena gracilis* grown under different conditions',*Arch. Biochem. Biophys.* **89**, 83 (1960).
671. Smith, I. W., Wilkinson, J. F., and Duguid, J. P., 'Volutin production in *Aerobacter aerogenes* due to nutrient imbalance', *J. Bacteriol.*, **68**, 450 (1954).
672. Sobolev, A. M., 'The distribution, formation, and utilization of phytin in higher plants' (in Russian), *Usp. Biol. Khim.*, **4**, 248 (1962).
673. Sommer, A. L., and Booth, T. E., 'Meta- and pyrophosphate within the algal cell', *Plant Physiol.*, **13**, 199 (1938).
674. Souzu, H., 'Decomposition of polyphosphate in yeast cell by freeze–thawing', *Arch. Biochem. Biophys.*, **120**, 338 (1967).
675. Souzu, H., 'Location of polyphosphate and polyphosphatase in yeast cell and damage to the protoplasmic membrane of the cell by freeze–thawing', *Arch. Biochem. Biophys.*, **120**, 344 (1967).
676. Spiegelman, S., and Kamen, M. D., 'Some basic problems in the relation of nucleic acid turnover to protein synthesis', *Cold Spring Harbor Symp. Quant. Biol.*, **12**, 211 (1947).
677. Spoerl, E., and Looney, D., 'Phosphorus metabolism in yeast accompanying an inhibition of cell division by X-rays', *J. Bacteriol.*, **76**, 63 (1958).
678. Spoerl, E., and Looney D., and Synchronized budding of yeast cells following X-irradiation', *Exp. Cell Res.*, **17**, 320 (1959).
679. Spoerl, E., Looney, D., and Kazmierczak, J. E., 'Polyphosphate metabolism in X-irradiation division-inhibited yeast cells', *Radiat. Res.*, **11**, 793 (1959).
680. Stahl, A. J. C., 'Sur la biosynthese des polyphosphates dans la levure', *Bull. Soc. Chim. Biol.*, **51**, 1211 (1969) .
681. Stahl, A. J. C., Bakes, J., Weil, J. H., and Ebel, J. P., 'Étude du transfert du phosphore des polyphosphates inorganiques dans les acides ribonucleiques chez la levure, II. Distribution du phosphore polyphosphorique dans les diverses fractions ribonucleiques', *Bull. Soc. Chim. Biol.*, **46**, 1017 (1964).
682. Stahl, A. J. C., and Ebel J. P., Étude des complexes entre acides ribonucleiques et polyphosphates inorganiques dans le levure. III. Role des cations divalents dans la formation des complexes', *Bull. Soc. Chim. Biol.*, **45**, 887 (1963).

683. Stahl, A. J. C., and Felter, S., 'Kinetic properties of the yeast polyphosphate–ADP–phosphotransferase', *Abstr. 7th Meeting FEBS, Varna*, 1971, p. 121.

684. Stahl, A. J. C., Felter, S., and Ebel, J. P., 'Recherche de polyphosphates inorganiques dans les mitochondries de foie de rat', *Bull. Soc. Chim. Biol.*, **49**, 1863 (1967).

685. Stahl, A. J. C., Muller-Felter, S., and Ebel, J. P., 'Étude du transfer du phosphore des polyphosphates inorganiques dans les acides ribonucleiques chez la levure. I. Mise en evidence d'un transfert du phosphore polyphosphorique dans differentes fraction phosphorées', *Bull. Soc. Chim. Biol.*, **46**, 1005 (1964).

686. Stanier, A., and van Niel, C. B., 'The concept of bacteria', *Arch. Microbiol.*, **42**, 17, (1962).

687. Starka, J., and Zavada, J., 'Relation between metaphosphate and volutin formation in yeast', *Čse. Mikrobiol.*, **1**, 70 (1956).

688. Steinman, G. D., Lemmon, R. M., and Calvin, M., 'Cyanamide, a possible key compound in chemical evolution', *Proc. Nat. Acad. Sci. USA*, **52**, 27 (1964).

689. Stetten, M. R., 'Metabolism of inorganic pyrophosphate', *J. Biol. Chem.*, **239**, 3576 (1964).

690. Van Steveninck, J., 'The role of polyphosphate in glucose uptake by yeast cells', *Biochem. J.*, **88**, 387.

691. Van Steveninck, J., 'The influence of nickelous ions on carbohydrate transport in yeast cells', *Biochim. Biophys. Acta*, **126**, 154 (1966).

692. Van Steveninck, J., 'The influence of metal ions on the hydrolysis of polyphosphates', *Biochemistry*, **5**, 1998 (1966).

693. Van Steveninck, J., and Booij, H. L., 'The role of polyphosphate in the transport mechanism of glucose in yeast cells', *J. Gen. Physiol.*, **58**, 43 (1964).

694. Van Steveninck, J., and Rothstein, A., 'Sugar transport and metal binding in yeast', *J. Gen. Physiol.*, **49**, 235 (1965).

695. Stich, H., 'Der Nachweis und das Verhalten von Metaphosphaten in normal verdunkelten und Trypaflavin-behandelten Acetabularien', *Z. Naturforsch.*, **8B**, 36 (1953).

696. Stich, H., 'Synthese und Abbau der Polyphosphate von Acetabularia nach autoradiographischen Untersuchungen des ^{32}P-Stoffwechsels', *Z. Naturforsch.*, **10B**, 282 (1955).

697. Stich, H., 'Andergungen von Kern und Polyphosphates in Abgangigkeit von dem Energiegehalt des Cytoplasmas bei Acetabularia', *Chromosoma*, **7**, 693 (1956).

698. Strauss, U. P., and Smith, E. H., 'Polyphosphates as polyelectrolytes. II. Viscosity of aqueous solutions of Graham's salts', *J. Am. Chem. Soc.*, **75**, 6186 (1953).

699. Strauss, U. P., Smith, E. H., and Wineman, P. Y., 'Polyphosphates as polyelectrolytes. I. Light scattering and viscosity of sodium polyphosphates in electrolyte solutions', *J. Am. Chem. Soc.*, **75**, 3935 (1953).

700. Strauss, U. P., and Treitler, T. L., 'Chain branching in glassy polyphosphates: dependence on the Na/P ratio and rate of degradation at 25°, *J. Am. Chem. Soc.*, **77**, 1473 (1955).

701. Strauss, U. P., and Treitler, T. L., 'Degradation of polyphosphates in solution. I. Kinetics and mechanism of the hydrolysis at branching points in polyphosphate chains', *J. Am. Chem. Soc.*, **78**, 3553 (1955).

702. Streshinskaya, G. N., Naumova, I. B., and Gololobov, A. D., 'The polyphosphate content of some species of yeast grown on various carbon sources' (in Russian), *Dokl. Akad. Nauk SSSR*, **190**, 227 (1970).

703. Khun-yui Sun, 'Phosphorus metabolism in the P-176 variant of *Aspergillus niger* T-1, obtained by ultraviolet irradiation. II. Phosphorus compounds in the conidia and mycelia of *Aspergillus niger* T-1' (in Russian), *Mikrobiologiya*, **29**, 475 (1960).

704. Khun-yui Sun, 'Phosphorus metabolism in the P-176 variant of *Aspergillus niger*

242

T-1, obtained by ultraviolet irradiation. III. The metabolic role of phosphatase' (in Russian), *Mikrobiologiya*, **29**, 806 (1960).

705. Sundberg, I., and Nilshammer-Holmvall, M., 'The diurnal variation in phosphate uptake and ATP level in relation to deposition of starch, lipid and polyphosphate in synchronized cells of *Scenedesmus*', *Z. Pflanzenphysiol.*, **76**, 270 (1975).

706. Suzuki, H., Kanenko, T., and Ikeda, Y., 'Properties of polyphosphate kinase prepared from *Mycobacterium smegmatis*', *Biochim. Biophys. Acta*, **268**, 381 (1972).

707. Svihla, G., and Schlenk, F., 'The influence of culture conditions on the formation of spheroplasts from *Candida utilis*', *Symposium on Yeast Protoplasts, Jena, 1965*, Akademie-Verlag, Berlin, 1966, p. 15.

708. Svihla, G., Schlenk, F., and Dainko, J. L., 'Spheroplasts of the yeast *Candida utilis*', *J. Bacteriol.*, **82**, 808 (1961).

709. Szulmajster, J., and Gardiner, R. C., 'Enzymic formation of polyphosphate in an anaerobic bacterium', *Biochim. Biophys. Acta*, **39**, 165 (1960).

710. Szymona, M., 'Utilization of inorganic polyphosphates for phosphorylation of glucose in *Mycobact. phlei*' *Bull. Acad. Pol. Sci., Ser. Sci. Biol.*, **5**, 379 (1957).

711. M. Szymona, 'The inorganic polyphosphate-utilizing hexokinase system of *Mycobacterium phlei*' (in Russian), *Trudy V Mezhdunarod. Biokhim. Kongressa, Ref. Sekts. Soobshch. (Soobshch. 1–12)* [*Proceedings of the Fifth International Biochemical Congress, Abstracts Section (papers 1–12)*], Izd. Akad. Nauk SSSR, 1961, p. 258.

712. Szymona, M., 'Purification and properties of the new hexokinase utilizing inorganic polyphosphate', *Acta Biochim. Pol.*, **9**, 165 (1962).

713. Szymona, M., Communication in discussion, in *Acides Ribonucleiques et Polyphosphates. Structure, Synthèse et Fonctions, Colloq. Int. CNRS, Strasbourg, 1961*, CNRS, Paris, 1962, p. 541.

714. Szymona, M., 'Fosfotransferazy w metabolizmie nie-organicznych polifosfora-now', *Postepy Biochem.*, **10**, 295 (1964).

715. Szymona, M., 'The phosphorylation of glucose and glucosamine in an acetone powder of *Mycobacterium phlei* (in Russian), *Byull. Pol'sk Akad. Nauk, Sekt.* **2, 4**, 123 (1966).

716. Szymona, O., Kowalska, H., and Szymona, M., 'Search for inducible sugar kinases in *Mycobacterium phlei*', *Ann. Univ. Mariae Curie-Sklodowska*, **24**, 1 (1969).

717. Szymona, M., and Ostrowski, W., 'Inorganic polyphosphate glucokinase of *Mycobacterium phlei*', *Biochim. Biophys. Acta*, **85**, 283 (1964).

718. Szymona, O., and Szumilo, T., 'Adenosine triphosphate and inorganic polyphosphate–fructokinases of *Mycobacterium phlei*', *Acta Biochim. Pol.*, **17**, 129 (1966).

719. Szymona, M., and Szymona, O., 'Participation of volutin in the hexokinase reaction of *Corynebacterium diphtheriae*', *Bull. Acad. Pol. Sci. Ser. Sci. Biol.*, **9**, 371 (1961).

720. Szymona, M., Szymona, O., and Kulesza, S., 'On the occurrence of inorganic polyphosphate hexokinase in some microorganisms', *Acta Microbiol. Pol.*, **11**, 287 (1962).

721. Szymona, O., Uryson, S. O., and Kulaev, I. S., 'The detection of polyphosphate glucokinase in some microorganisms' (in Russian), *Biokhimiya*, **32**, 495 (1967).

722. Tamiya, H., Iwamura, T., Shibata, K., Hase, E., and Nihei, T., 'Correlation between photosynthesis and light independent metabolism in the growth of *Chlorella*', *Biochim. Biophys. Acta*, **12**, 23 (1953).

723. Tamiya, K., and Miyachi, S., 'The distribution and conversion of phosphorus compounds in growing cells of *Chlorella*' (in Russian), *Ref. Sektsionnykh Soobshch. V mezhdunar. Biokhim. Kongressa (Abstracts of Section Papers at the Fifth International Biochemical Congress)*, Izd. Akad. Nauk SSSR, Moscow, 1961, Vol. 2, p. 77.

724. Tanzer, J. M., and Krichevsky, M. I., 'Polyphosphate formation by caries-conductive *Streptococcus* sl-1', *Biochim. Biophys. Acta*, **216**, 368 (1970).

725. Tanzer, J. M., Krichevsky, M. I., and Chassy, M., 'Separation of polyphosphates by anion-exchange thin-layer chromatography', *J. Chromatog.*, **38**, 526 (1968).

726. Tardieux-Roche, A., 'Sur la formation de polyphosphates par diverses bacteries du sol', *Ann. Inst. Pasteur.*, **107**, 565 (1964).

727. Täumer, L., 'Morphologie, Cytologie und Fortpflanzung von *Rhopalocystis gleifera* (Schussnig)', *Arch. Protistenkd.*, **104**, 265 (1959).

728. Teichert, W., and Rinnmann, K., 'Sodium metaphosphate', *Acta Chem. Scand.*, **2**, 225 (1948).

729. Terra, de N., and Tatum, E. L., 'A relationship between cell wall structure and colonial growth in *Neurospora crassa*', *Am. J. Bot.*, **50**, 669 (1963).

730. Terry, K. R., and Hooper, A. B., 'Polyphosphate and orthophosphate content of *Nitrosomonas europaea* as a function of growth', *J. Bacteriol.*, **103**, 199 (1970).

731. Tetas, M., and Lowenstein, J., 'The effect of bivalent metal ions on the hydrolysis of adenosine di- and triphosphate', *Biochemistry*, **2**, 350 (1963).

732. Tewari, K. K., and Krishnan, P. S., 'Further studies on the metachromatic reaction of metaphosphate', *Arch. Biochem. Biophys.*, **82**, 99 (1959).

733. Tewari, K. K., and Krishnan, P. S., 'Studies on the role of metaphosphate in molds. III. A report of experiments on the attempted coupling of metaphosphates with hexokinase and glycerokinase in *Aspergillus niger*', *Enzymologia*, **23**, 287 (1961).

734. Tewari, K. K., and Singh, M., 'Acid-soluble and acid-insoluble inorganic polyphosphates in *Cuscuta reflexa*', *Phytochemistry*, **3**, 341 (1964).

735. Thilo, E., 'Fortschritte auf dem Gebiet der Kondensierten Phosphate und Arsenate', *Forsch. Fortschr.*, **26**, 284 (1950).

736. Thilo, E., 'Die kondensierten Phosphate', *Angew. Chem.*, **67**, 141 (1955).

737. Thilo, E., 'Zur Chemie der kondensierten Phosphate', *Chem. Techn.*, **8**, 251 (1956).

738. Thilo, E., 'The structural chemistry of condensed phosphates' (in Russian), *Zh. Prikl. Khim.*, **29**, 1621 (1956).

739. Thilo, E., 'Die Chemie der kondensierten Phosphate', *Acta Chim. Hung.*, **12**, 221 (1957).

740. Thilo, E., 'Die kondensierten Phosphate', *Naturwissenschaften*, **46**, 367 (1959).

741. Thilo, E., 'Polyphosphate und Arsenate', in *Inorganic Polymers, International Symposium, Nottingham*, Durlington House, London, 1961, p. 33.

742. Thilo, E., 'Chemie der Polyphosphate', in *Acides Ribonucleiques et Polyphosphates. Structure, Synthese et Fonction, Colloq. Int. CNRS, Strasbourg, 1961*, CNRS, Paris, 1962, p. 491.

743. Thilo, E., 'Zur Chemie und Konstitution der Kondensierten Phosphate', *Bull. Soc. Chem.*, numero speciale, 1725 (1968).

744. Thilo, E., 'Entwicklung der Chemie der Oligomeren und polymeren Phosphate in ihren Grundzügen', *Z. Chem.* **12**, 169 (1972).

745. Thilo, E., Grünze, E., Hämmerling, J., and Werz, G., 'Über Isolierung und Identifizierung der Polyphosphate aus *Acetabularia mediterranea*', *Z. Naturforsch.*, **116**, 266 (1956).

746. Thilo, E., and Schülke, U., 'Über Metaphosphate die mehr als fier Phosphoratome in Ringanion erhalten', *Z. Anorg. Allg. Chem.*, **341**, 293 (1965).

747. Thilo, E., Schulz, G., and Wichmann, E. M., 'Die Konstitution des Grahamschen und Kurrolschen Salzes ($NaPO_3$)', *Z. Anorg. Allg. Chem.*, **272**, 182 (1953).

748. Thilo, E., and Sonntag, A., 'Zur Chemie der Kondensierten Phosphate und Arsenate. XIX. Über Bildung und Eigenschaften vernetzter Phosphate des Natriums', *Z. Anorg. Allg. Chem.*, **291**, 186 (1957).

749. Thilo, E., and Wieker, W., 'Zur Chemie der kondensierten Phosphate und

244

Arsenate. XVIII. Anionenhydrolyse der kondensierten Phosphate in verdünster wässriger Lösung', *Z. Anorg. Allg. Chem.*, **291**, 164 (1957).
750. Tonino, G. J. M., and Rozijn, T. H., 'On the occurrence of histones in yeast', *Biochim. Biophys. Acta.*, **124**, 427 (1966).
751. Topley, B., 'Condensed phosphates', *Quart. Rev.*, **3**, 345 (1949).
752. Trevelyan, W. E., Mann, P. F. E., and Harrison, J. S., 'The phosphorylase reaction. III. The glucose-I-phosphate and orthophosphate content in baker's yeast in relation to polysacharide synthesis', *Arch. Biochem. Biophys.*, **50**, 81 (1954).
753. Trevithick, J. R., and Metzenberg, R. J., 'The invertase isozyme formed by *Neurospora* protoplasts', *Biochem. Biophys. Res. Commun.*, **16**, 4 (1964).
754. Tsiomenko, A. B., Augustin, I., Vagabov, V. M., and Kulaev, I. S., 'The interrelationship of the metabolism of inorganic polyphosphates and mannan in yeasts' (in Russian), *Dokl. Akad. Nauk SSSR*, **215**, 478 (1974).
755. Tsiomenko, A. B., Shabalin, Yu. A., Dmitriev, V. V., Vagabov, V. M., Fikhte, B. A., and Kulaev, I. S., 'The biosynthesis of components of the cell wall by yeast protoplasts in the presence of cycloheximide and 2-desoxyglucose' (in Russian), *Dokl. Akad. Nauk SSSR*, **225**, 1446 (1975).
756. Tsiomenko, A. B., Vagabov, V. M., and Kulaev, I. S., 'The effect of 2-desoxyglucose on the synthesis and secretion of acid phosphatase and mannan by yeast protoplasts' (in Russian), *Dokl. Akad. Nauk SSSR*, **220**, 250 (1975).
757. Ullrich, W. R., 'Zur Wirkung von Sauerstoff auf die ^{32}P-Markierung von Polyphosphaten und organischen Phosphaten bei *Ankistrodesmus braunii* im Licht', *Planta*, **90**, 272 (1970).
758. Ulrich, E., and Bock, E., 'Auf- und Abbau der Polyphosphat-fraction in Zellen von *Nitrobacter winogradskyi*', *Arch. Mikrobiol.*, **81**, 367 (1972).
759. Ullrich, W., and Simonis, W., 'Die Bildung von Polyphosphaten bei *Ankistrodesmus braunii* durch Monophosphorylierung im Rotlicht von 683 und 712 nm unter Stickstoffatmosphäre', *Planta*, **84**, 358. (1969).
760. Umnov, A. M., Egorov, S. N., Mansurova, S. É., and Kulaev, I. S., 'Some properties of polyphosphatases from *Neurospora crassa*' (in Russian), *Biokhimiya*, **39**, 373 (1974).
761. Umnov, A. M., Umnova, N. S., and Kulaev, I. S., 'The isolation, purification, and properties of *Neurospora crassa* polyphosphate phosphohydrolase' (in Russian), *Molek. Biol.*, **9**, 594 (1975).
762. Umnov, A. M., Umnova, N. S., Steblyak, A. G., Mansurova, S. E., and Kulaev, I. S., 'The possible physiological role of the high-molecular polyphosphate–polyphosphate phosphohydrolase system' (in Russian), *Mikrobiologiya*, **44**, 41 (1974).
763. Urbanek, H., 'Wplyw 8-azaadeniny na biosyntheze niktorych swiazkow fosforowych w grzybni *Neurospora crassa*', *Acta Soc. Bot. Pol.*, **36**, 347 (1967).
764. Uryson, S. O., and Kulaev, I. S., 'The presence of polyphosphate glucokinase in some bacteria' (in Russian), *Dokl. Akad. Nauk SSSR*, **183**, 957 (1968).
765. Uryson, S. O., and Kulaev, I. S., 'The glucokinase activity of some green and blue–green algae' (in Russian), *Biokhimiya*, **35**, 601 (1970).
766. Uryson, S. O., Kulaev, I. S., Agre, N. S., and Egorova, S. N., 'Polyphosphate glucokinase activity as a criterion of the differentiation of the Actinomycetes' (in Russian), *Mikrobiologiya*, **42**, 1067 (1973).
767. Uryson, S. O., Kulaev, I. S., Bogatyreva, T. G., and Aseeva, I. V., 'The detection of polyphosphate glucokinase activity in stained coccoid microorganisms obtained from lithophilic lichens' (in Russian), *Mikrobiologiya*, **43**, 83 (1974).
768. Vagabov, V. M., 'The metabolism of inorganic polyphosphates and other phosphorus compounds in certain green algae' (in Russian), *Candidate's Thesis*, Moscow, 1965.

769. Vagabov, V. M., and Kulaev, I. S., 'Inorganic polyphosphates in the roots of maize' (in Russian), *Dokl. Akad. Nauk SSSR*, **158**, 218 (1964).

770. Vagabov, V. M., and Kulaev, I. S., 'The metabolism and biosynthesis of inorganic polyphosphates in green algae (in Russian), *Tezisy II Otchetnoi Konf. Biofaka (Abstracts of the Second Reporting Conference of the Biological Faculty*, 1965, Izd. MGU, Moscow, 1966.

771. Vagabov, V. M., and Serenkov, G. P., 'An investigation of the polyphosphates present in two species of green algae' (in Russian), *Vestn. Mosk. Gos., Univ. (Biol. Pochvoved)*, No. 4, 38 (1963).

772. Vainio, H., Baltschevskii, M., Baltschevskii, H., and Azzi, A., 'Energy-depending changes in membranes of *Rhodospirillum rubrum* chromatophores as measured by 8-anilino-naphthalene-1-sulfonic acid', *Eur. J. Biochem.*, **30**, 301.

773. Valikhanov, M. N., 'A comparative study of the functions of phytin and high molecular weight polyphosphates in the regulation of orthophosphate levels in some organisms' (in Russian), in *Tezisy Dokl. na Vsesoyuznoi Konferentsii Regulyatsiya Biokhimicheskikh Protsessov u Mikroorganismov*' (Abstracts of the All-Union Conference on the Regulation of Biochemical Processes in *Microorganisms*'), Pushchino-na-Oke, 1972, p. 34.

774. Valikhanov, M. N., and Kulaev, I. S., 'Phosphorus metabolism during the germination of seeds of cotton' (in Russian), *Tezisy II Otchetnoiy Konf. Biofaka (Abstracts of the Second Reporting Conference of the Biological Faculty*), 1965, Izd. MGU, Moscow, 1966.

775. Valikhanov, M. N., Kulaev, I. S., and Belozersky, A. N., 'A study of the metabolism of phosphorus during the germination of seeds of cotton' (in Russian), in *Biologiya Avtotrofnykh Organizmov (The Biology of Autotrophic Organisms), Trudy MOIP (Proceedings of the Moscow Society of Naturalists)*, **24**, 81 (1966).

776. Van Gool, A. P., Tobback, P. P., and Fischer, I., 'Autotrophic growth and synthesis of reserve polymers in *Nitrobacter winogradskyi*', *Arch. Mikrobiol.*, **76**, 252 (1971).

777. Verbina, N. M., 'The physiological significance of polyphosphates in microorganisms' (in Russian), *Usp. Mikrobiol.*, **1**, 75 (1964).

778. Villanueva, J. N., 'Trends in yeast research', *Symposium on Yeast Protoplasts, Jena, 1965*, Akademie-Verlag, Berlin, 1966, p. 1.

779. Vishniac, W., 'The antagonism of sodium tripolyphosphate and adenosine tripolyphosphate in yeast', *Arch. Biochem.*, **26**, 167 (1950).

780. Voelz, H., Voelz, V., and Ortihoza, R. P., 'The "Polyphosphate overplus" phenomenon in *Myxococcus xanthus* and its influence on the architecture of the cell', *Arch. Microbiol.*, **53**, 371 (1966).

781. Vorobyeva, G. I., Grigoryeva, S. P., Kulaev, I. S., and Maksimova, G. N., 'Studies on the regularities of phosphate metabolism in yeasts cultivated on hydrocarbons', *Proceedings of the 3rd Int. Speci. Symp. on Yeasts, Otaniemi/Helsinki*, 1973, p. 51.

782. De Waart, C., 'Some aspects of phosphate metabolism·of *Claviceps purpurea* II. Relationship between nucleic acids synthesis and ergot alkaloid production', *Can. J. Microbiol.*, **7**, 883 (1961).

783. De Waart, C., and Taber, W. A., 'Some aspects of phosphate metabolism of *Claviceps purpurea*' *Can. J. Microbiol.*, **6**, 675 (1960).

784. Wade, H. E., and Morgan, D. M., 'Fractionation of phosphates by paper ionophoresis and chromatography', *Biochem. J.*, **60**, 264 (1955).

785. Wang, D., and Mancini, D., 'Studies on ribonucleic acid–polyphosphate in plants', *Biochim. Biophys. Acta*, **129**, 231 (1966).

786. Wassink, E. C., 'Phosphate in the photosynthetic cycle in *Chlorella*', in *Research in Photosynthesis*, Ed. Gaffron M. H. International Publishers, New York, 1957, p. 333.

246

787. Wassink, E. C., Wintermans, I. F. G. M., and Tjia, I. E., 'Phosphate exchanges in *Chlorella* in relation to conditions for photosynthesis', *Proc. Kon. Ned. Akad. Wet.*, **54**, 41 (1951).
788. Van Wazer, J., 'Structure and properties of condensed phosphates. V. Molecular weight of the polyphosphates from viscosity data', *J. Am. Chem. Soc.*, **72**, 906 (1950).
789. Van Wazer, J. R., 'A new perspective for phosphorus chemistry and its possible significance to biochemistry', in *Acides Ribonucleiques et Polyphosphates. Structure, Synthèse et Fonction, Colloq. Int. CNRS, Strasbourg, 1961*, CNRS, Paris, 1962, p. 513.
790. Van Wazer, J. R., Phosphorus and its Compounds' (translated into Russian), Izd. IL, Moscow, 1962.
791. Van Wazer, J. R., and Griffith, E. Z., 'Structure and properties of the condensed phosphates. X. General structural theory', *J. Am. Chem. Soc.*, **77**, 6140 (1955).
792. Van Wazer, J., and Holst K., 'Structure and properties of the condensed phosphates. I. Some general considerations about phosphoric acids', *J. Am. Chem. Soc.*, **72**, 639 (1950).
793. Van Wazer, J. R., and Karl-Kroupa, E., 'Existence of ring phosphates higher than the tetrametaphosphate', *J. Am. Chem. Soc.*, **78**, 1772 (1956).
794. Weber, H., 'Über das Vorkommen von kondensierten Phosphaten in Purpurbakterien', *Z. Allg. Mikrobiol.*, **5**, 315 (1965).
795. Weil, J. H., Bakes, J., Before, N., and Ebel, J. P., 'Contribution a l'étude du RNA messenger de la levure', *Bull. Soc. Chim. Biol.*, **46**, 1073 (1964).
796. Weimberg, R., 'Effect of potassium chloride on the uptake and storage of phosphate by *Saccharomyces mellis*', *J. Bacteriol.*, **103**, 37 (1970).
797. Weimberg, R., and Orton, W. L., 'Evidence for an exocellular site for the acid phosphatase of *Saccharomyces mellis*', *J. Bacteriol.*, **88**, 1743 (1964).
798. Weimberg, R., and Orton, W. L., 'Synthesis and breakdown of the polyphosphate fraction and acid phosphomonoesterase of *Saccharomyces mellis* and their location in the cell', *J. Bacteriol.*, **89**, 740 (1965).
799. Well, J. H., Hebding, N., and Ebel, J. P., 'Séparation des acides ribonucleiques et des polyphosphates. III. Essais de separation des acides ribonucleiques et des polyphosphates per difference de solubilité de leurs combinaisons avec les sels d'ammonium quaternaire', *Bull. Soc. Chim. Biol.*, **45**, 595 (1963).
800. Westman, A., Scott, A., and Redley, I., 'Filter paper chromatography of the condensed phosphates', *Chem. Can.*, **10**, 35 (1952).
801. Wiame, J. M., 'Sur l'existance d'un nouveau composé phosphore dans les levures', *C.R. Soc. Biol.*, **139**, 194, 5 (1945).
802. Wiame, J., 'Basophillie et métabolisme du phosphore chez la levuré, *Bull. Soc. Chim. Biol.*, **28**, 552 (1946).
803. Wiame, J. M., 'Remarque sur la metachromasie des cellules de levure', *C.R. Soc. Biol.*, **140**, 897 (1946).
804. Wiame, J. M., 'Étude d'une substance polyphosphorée, basophillie et metachromatique chez les levures', *Biochim. Biophys. Acta*, **1**, 234 (1947).
805. Wiame, J. M., 'Metaphosphate of yeast', *Fed. Proc.*, **6**, 302 (1947).
806. Wiame, J. M., 'The metachromatic reaction of hexametaphosphate', *J. Am. Chem. Soc.*, **69**, 3146 (1947).
807. Wiame, J. M., 'Metaphosphate et corpuscules metachromatiques chez la levure', *Rev. Ferment. Ind. Aliment*, **3**, 83 (1948).
808. Wiame, J. M., 'The occurrence and physiological behaviour of two metaphosphate fractions in yeast', *J. Biol. Chem.*, **178**, 919 (1949).
809. Wiame, J. M., 'Accumulation de l'acide phosphorique (phytine, polyphosphates)', in *Handbuch der Pflanzenphysiologie*, Springer-Verlag, Berlin, Gottingen, and Heidelberg, 1958, Vol. 9, p. 136.

810. Wiame, J. M., and Lefebvre, P. H., 'Conditions de formation dans la levure d'un compose nucleique polyphosphore', *C.R. Soc. Biol.*, **140**, 921 (1946).

811. Wiemken A. Eigenschaften der Hefevakuole PhD thesis No. 4340 Swiss Fed. Inst. Technol.

812. Widra, A., 'Metachromatic granules of microorganisms', *J. Bacteriol.*, **78**, 664 (1959).

813. Wieker, W., 'Zur Chemie der kondensierten Phosphate und Arsenate. XXXII. Darstellung und Eigenschaften von ^{32}P-markierten Penta-. Hexa-. Hepta- und Oktaphosphaten', *Z. Elektrochem. Ber. Bunsenges. Phys. Chem.*, **64**, 1047 (1960).

814. Williamson, D. H., and Wilkinson, J. P., 'The isolation and estimation of the poly-β-hydroxybutyrate inclusions of *Bacillus* species', *J. Gen. Microbiol.*, **19**, 198. (1958).

815. Wilkinson, J. P., and Duguid, J. P., 'The influences of cultural conditions on bacterial cytology', *Int. Rev. Cytol.*, **9**, 1 (1960).

816. Wilsenach, R., and Kessel, M., 'The role of lomazomes in wall formation in *Penicillium verticulatum*', *J. Gen. Microbiol.*, **40**, 401 (1965).

817. Winder, F., and Denneny, J., 'Metaphosphate in mycobacterial metabolism', *Nature, Lond.*, **174**, 353 (1954).

818. Winder, F., and Denneny, J. M., 'Studies on a "metaphosphokinase" from *Mycobacterium smegmatis*', *Res. 3eme Congr. Int. Biochim., Bruxelles*, 1955, p. 31.

819. Winder, F. G., and Denneny, J. M., 'Utilization of metaphosphate for phosphorylation by cell-free extracts of *Mycobacterium smegmatis*', *Nature, Lond.*, **175**, 636 (1955).

820. Winder, F. G., and Denneny, J. M., 'Phosphorus metabolism of mycobacteria Determination of phosphorus compounds in some *Mycobacteria*' *J. Gen. Microbiol.*, **15**, 1 (1956).

821. Winder, F. G., and Denneny, J. M., 'The metabolism of inorganic polyphosphate in *Mycobacteria*', *J. Gen. Microbiol.*, **17**, 573 (1957).

822. Winder, F., and Denneny, J. M., 'Effect of iron and zinc on nucleic acid and protein synthesis in *Mycobacterium smegmatis*', *Nature, Lond.*, **184**, 742 (1959).

823. Winder, F. G., and Roche, H. G., 'The accumulation of inorganic polyphosphate brought about by tetrahydrofurfuryl alcohol in *Mycobacterium smegmatis*', *Biochem. J.*, **103**, 57 (1967).

824. Windisch, F., Hinkelmann, St., and Stierand, D., 'Plasmochromatische Zellreaktionen bei regenerierender. Phosphat-Applikation nach P-Mangelzüchtung', *Protoplasma (Wien)*, **48**, 177 (1957).

825. Wintermans, J. F. G., 'On the formation of polyphosphates in *Chlorella* in relation to conditions for photosynthesis', *Proc. Kon. Ned. Acad. Wet.*, **57**, 574 (1954).

826. Wintermans, J. F. G., 'Polyphosphate formation in *Chlorella* in relation to photosynthesis', *Meded. Landbouwhogesch. Wageningen*, **55**, 69 (1955).

827. Wintermans, J. F. G., and Tjia, J. E., 'Some observations on the properties of phosphate compounds in *Chlorella* in relation to conditions for photosynthesis', *Proc. Kon. Ned. Akad. Wet.*, **55**, 34 (1952).

828. Wojtezak, A., 'Chromatograficzne rozdzielanie orto-, pyro- i metafosforanow z wydalin gusienie mola weskowego', *Acta Physiol. Pol.*, **5**, 590 (1954).

829. Wool, S. H., and Held, A. A., 'Polyphosphate granules in encysted zoospores of *Rozella allomycis*', *Arch. Microbiol.*, **108**, 145 (1976).

830. Ykawa, M., 'Bacterial phosphatides and natural relationships', *Bacteriol. Rev.*, **31**, 54 (1967).

831. Yoshida, A., 'Biochemical studies on metaphosphate', *Sci. Pap. Coll. Gen. Educ. Univ. Tokyo*, **3**, 151 (1953).

248

832. Yoshida, A., 'Metaphosphate. II. Heat of hydrolysis of metaphosphate extracted from yeast cells', *J. Biochem. (Tokyo)*, **42**, 163 (1955).
833. Yoshida, A., 'On the metaphosphate of yeast. III. Molecular structure and molecular weight of yeast metaphosphate', *J. Biochem. (Tokyo)*, **42**, 381 (1955).
834. Yoshida, A., 'Polyphosphate in microorganisms, their structure and their role in nucleic acid synthesis', in *Acides ribonucleiques et Polyphosphates. Structure, Synthèse et Fonctions, Colloq. Int. CNRS, Strasbourg, 1961*, CNRS, Paris, 1962, p. 575.
835. Zaitseva, G. N., 'Oxidative phosphorylation and polyphosphate synthesis in *Azotobacter vinelandii*' (in Russian), *Ref. Sekts. Soobshch. V Mezhdunarod. Biokhim. Kongressa (Abstracts Section of the Communications to the Fifth International Biochemical Congress)*, Izd. Akad. Nauk SSSR, Moscow, 1961, Vol. 2, p. 391.
835a. Zaitseva, G. N., 'Nitrogen and phosphorus metabolism in developing Azotobacter' (in Russian), *Doctoral Thesis*, Moscow, 1963.
835b. Zaitseva, G. N., and Belozersky, A. N., 'The chemistry of Azotobacter. Phosphorus compounds in *Azotobacter agile* at different growth stages and with different nitrogen sources' (in Russian), *Mikrobiologiya*, **27**, 308 (1958).
836. Zaitseva, G. N., and Belozersky, A. N., 'The production and utilization of polyphosphates by an enzyme isolated from *Azotobacter vinelandii*' (in Russian), *Dokl. Akad. Nauk SSSR*, **132**, 820 (1960).
837. Zaitseva, G. N., Belozersky, A. N., and Novozhilova, L. P., 'Phosphorus compounds in *Azotobacter vinelandii* during the development of cultures' (in Russian), *Biokhimiya*, **24**, 1054 (1959).
838. Zaitseva, G. N., Belozersky, A. N., and Novozhilova, L. P., 'A study of phosphorus compounds using ^{32}P during the development of *Azotobacter vinelandii*' (in Russian), *Biokhimiya*, **25**, 198 (1960a).
839. Zaitseva, G. N., Belozersky, A. N., and Novozhilova, L. P., 'The effect of the calcium ion on nitrogen and phosphorus metabolism in *Azotobacter vinelandii*' (in Russian), *Mikrobiologiya*, **29**, 343 (1960b).
840. Zaitseva, G. N., Belozersky, A. N., and Frolova, L. Yu., 'Oxidative phosphorylation and polypshosphate synthesis in cells of *Azotobacter vinelandii*' (in Russian), *Dokl. Akad. Nauk SSSR*, **132**, 470 (1960).
841. Zaitseva, G. N., and Frolova, L. Yu., 'The effect of chloramphenicol on phosphorus and nucleic acid metabolism in *Azotobacter vinelandii*' (in Russian), *Biokhimiya*, **26**, 200 (1961).
842. Zaitseva, G. N., and Li Tszyun-in, 'The effect of X-rays on phosphorus and nitrogen metabolism in *Azotobacter agile*' (in Russian), *Mikrobiologiya*, **30**, 197 (1961).
843. Zaitseva, G. N., Khmel', I. A., and Belozersky, A. N., 'Biochemical changes in a synchronous culture of *Azotobacter vinelandii*' (in Russian), *Dokl. Akad. Nauk SSSR*, **141**, 740 (1961).
844. Zalokar, M., 'Growth and differentiation of *Neurospora hyphae*', *Am. J. Bot.*, **46**, 602 (1959).
845. Zalokar, M., 'Integration of cellular metabolism', in *The Fungi, an Advanced Treatise*, Eds. Ainsworth, G. C., and Sussman, A. S., Academic Press, New York, 1965, Vol. 1, p. 377.
846. Zhdan-Pushkina, S. M., and Pinevich, A. V., 'The relationship between the resistance of bacteria to histones and the accumulation of polyphosphates' (in Russian), *Mikrobiologiya*, **47**, 782 (1972).
847. Zuelzer, M., 'Über *Spirocheta plicatilis* (Ehrba) und deren Verwandschaftsbeziehungen', *Arch. Protistenkd.*, **24**, 1 (1912).
848. Zuleski, F. R., and McGrinnes, E. T., 'Use of phosphorus-32 labeled polyphosphoric acid for synthesis of labeled purine and pyrimidine 5'-ribonucleotides', *Analyt. Biochem.*, **47**, 315 (1972).

Index of
Generic Names

250

Subject Index